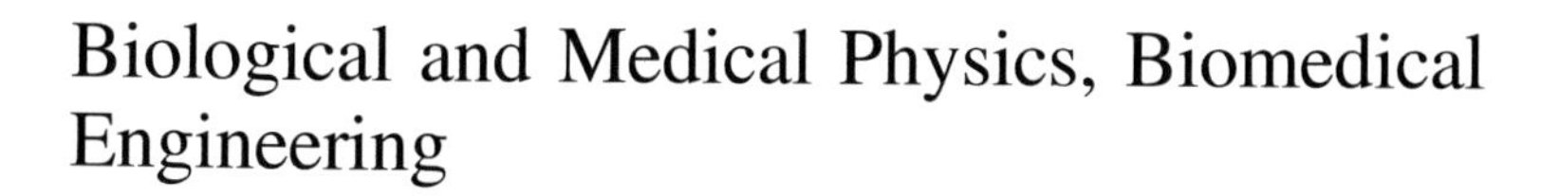

More information about this series at http://www.springer.com/series/3740

BIOLOGICAL AND MEDICAL PHYSICS, BIOMEDICAL ENGINEERING

The fields of biological and medical physics and biomedical engineering are broad, multidisciplinary and dynamic. They lie at the crossroads of frontier research in physics, biology, chemistry, and medicine. The Biological and Medical Physics, Biomedical Engineering Series is intended to be comprehensive, covering a broad range of topics important to the study of the physical, chemical and biological sciences. Its goal is to provide scientists and engineers with textbooks, monographs, and reference works to address the growing need for information.

Books in the series emphasize established and emergent areas of science including molecular, membrane, and mathematical biophysics; photosynthetic energy harvesting and conversion; information processing; physical principles of genetics; sensory communications; automata networks, neural networks, and cellular automata. Equally important will be coverage of applied aspects of biological and medical physics and biomedical engineering such as molecular electronic components and devices, biosensors, medicine, imaging, physical principles of renewable energy production, advanced prostheses, and environmental control and engineering.

Tatiana Koshlan · Kirill Kulikov

Mathematical Modeling of Protein Complexes

Tatiana Koshlan
Saint Petersburg State University
Saint Petersburg, Russia

Kirill Kulikov
Peter the Great St. Petersburg
Polytechnic University
Saint Petersburg, Russia

ISSN 1618-7210 ISSN 2197-5647 (electronic)
Biological and Medical Physics, Biomedical Engineering
ISBN 978-3-030-07481-4 ISBN 978-3-319-98304-2 (eBook)
https://doi.org/10.1007/978-3-319-98304-2

This Springer imprint is published by the registered company Springer Nature Switzerland AG
The registered company address is: Gewerbestrasse 11, 6330 Cham, Switzerland

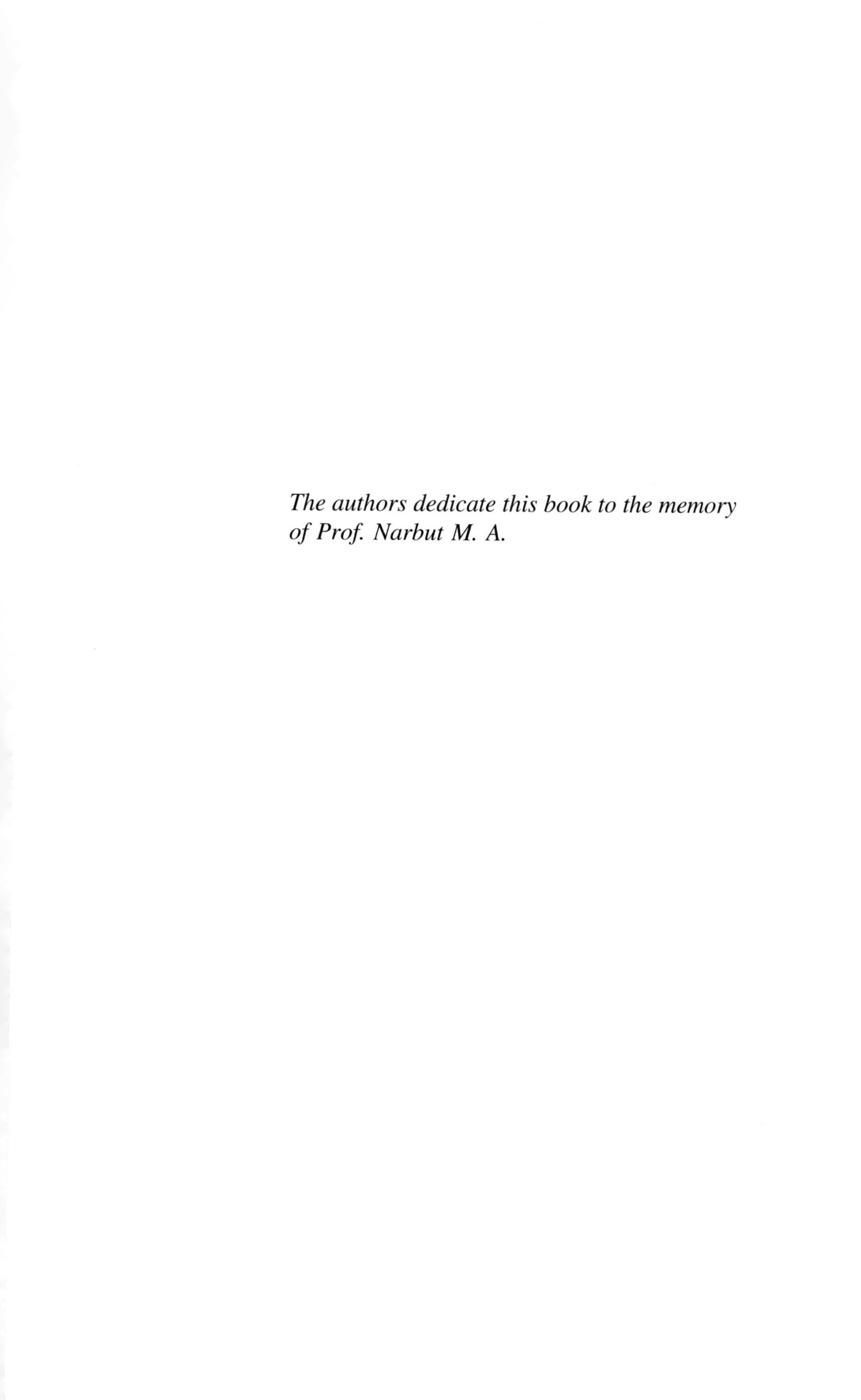

The authors dedicate this book to the memory of Prof. Narbut M. A.

Preface

Protein molecules are one of the most important types of macromolecules of the living world. They perform many functions: enzymatic (catalytic), structural, contractile, etc. Each protein molecule is characterized by its native conformation, which is determined by the amino acid sequence, and staying in it allows it to fulfill its biological functions. The interaction between such protein molecules is carried out in certain areas, the structure of which is determined by their conformation. The determination of such active sites responsible for the binding of protein complexes at the present time is a difficult task since even the exact conformation of proteins is not always known. For this reason, it is very difficult for the experimenter to determine the affinity of the various regions of proteins, their reactivity and the stability of the formed biological complex, as well as the probability of participation of different domains in the formation of a biological complex. It is also worth noting that even for native proteins, there is always the possibility of partial disturbance of the native structure, especially under stress conditions (e.g., heat, oxidation, or osmotic). When the native structure is violated, the proteins stop performing their biological functions, become less stable, and may tend to aggregate, which can lead to a wide range of pathological states of the cell and the whole organism.

Note that for modern proteomics, research and prediction of protein interactions are very important tasks, since they determine the function of proteins at levels from the cell to the whole organism. For proteins whose structure is known, the search for intermolecular interactions according to known data on the conformation of their tertiary structure reduces to the problem of searching for geometric complementarity of the sections of two interacting molecular surfaces and modeling their contacts, the so-called molecular docking. The task of molecular docking is the task of a conformational search algorithm, which reduces to a search for the conformational space of the formed biological complex due to the variation of the torsion angles of protein molecules.

Modern conformational search algorithms in most cases find conformations that are generally close to the experimentally found structures in a relatively short time. However, there are factors that also have a significant impact on the success of the docking, which are often not taken into account in standard algorithms. One such factor is the conformational mobility of the target protein. The mobility range can be different—beginning with a small ≪adjustment≫ of the side chains and ending with scale domain movements. These movements play an important role. At first glance, the most logical solution to this problem is to take into account the mobility of the protein in a docking program. Unfortunately, modern computational tools do not allow such modeling to be performed in an acceptable time frame since a protein molecule is very large, and allowing for mobility over all degrees of freedom can lead to a so-called combinatorial explosion (an astronomical increase in the number of possible variants). Only in some programs is there a limited mobility of protein-binding sites (usually at the level of a small adaptation of conformations of the side chains of the active center residues). Another approach to this problem consists in docking the same protein in several different conformations and then selecting the best solutions from each docking run. The third approach is to find a universal structure of the target protein in which docking would produce fairly good results for different classes of ligands. In this case, the number of ≪missed≫ (but correct) solutions decreases, but the number of incorrect options also increases significantly. It should also be noted that most programs for the theoretical docking of proteins work according to the following principle: One protein is fixed in space, and the second is rotated around it in a variety of ways.

At the same time, for each rotation configuration, estimates are made for the evaluation function. The evaluation function is based on surface complementarity (the mutual correspondence of complementary structures (macromolecules, radicals), determined by their chemical properties), electrostatic interactions, van der Waals repulsion, and so on. The problem with this approach is that calculations throughout the configuration space require a lot of time, rarely leading to a single solution, which in turn does not allow us to speak of the uniqueness of the target protein and ligand interaction variant.

So while modeling by the methods of molecular dynamics, from 200 to 10,000 possible combinations of the formation of a protein complex with a ligand were found (see Chap. 5).

Such a large number of modifications, along with the lack of a criterion for selecting the most probable variants of the bound structures of biological complexes (which would allow a radical reduction in their number), make it very difficult to interpret the theoretical results obtained for practical use, namely the finding of catalytic centers and a qualitative assessment of the dissociation constant of interacting substances.

The mathematical approach developed and physically substantiated in the monograph, in addition to the works on molecular dynamics, will allow theoretically predicting the passage of the biochemical reaction in the chosen direction with the given amino acid sequences, studying the behavior of dimers in vitro in solutions with different concentration of the monovalent salt in the formation of the nucleosome nucleus, the temperature on the stability of protein dimers, determining the regions of protein molecules responsible for the aggregate the effect of phosphorylation of the amino acid residues of the polypeptide chain on the formation of biological complexes, determining the presence of active regions of proteins and detecting the stability of various regions of protein complexes by analyzing the matrix of the potential energy of electrostatic interaction between different sites of the biological complex, and also investigating the influence of point mutations in BH3 peptides on the stability of the biological one complex with pro-apoptotic proteins Bcl-2 family and accurately determine the dissociation constants in the binding of different BH3-peptides to proteins Bcl-2 and Bcl-xl.

In the future, this will help to solve fundamental and applied problems of medicine.

The monograph is addressed to researchers and specialists in biomedical physics, molecular biology interested in the development and application of mathematical methods in medical research.

The monograph consists of eight chapters.

In Chap. 1, we consider various experimental approaches for determining the structure of molecules, the mutual arrangement of domains in space, the identification of active protein centers, and their advantages and limitations in the study of various physical properties of biological complexes.

In Chap. 2, we construct a physical model of the interactions between protein molecules and study of their propensity to form biological complexes.

In Chap. 3, we discuss mathematical model of the temperature effect on the character of linking between monomeric proteins in aqueous solutions.

In Chap. 4, we construct mathematical model that will allow us to describe the behavior of biological complexes in vitro on the example of the formation of two histone dimers H2A–H2B and H3–H4 from the corresponding monomeric proteins H2A, H2B, H3, and H4 in solutions with different concentrations of monovalent salt.

In Chap. 5, two algorithms are developed: algorithm 1 and algorithm 2. Algorithm 1 was developed in order to search for the interaction site of a polypeptide chain of a full-length protein with short active region. Algorithm 2 was developed to determine the most active sites of interaction between full-length proteins when dimers are formed in the direction from the N-terminus to C-terminus

In Chap. 6, we construct a physical model of phosphorylation the amino acid residues of the polypeptide chain of a protein on the formation of biological complexes with other proteins.

In Chap. 7, we discuss a new method that allows one to qualitatively determine the effect of point mutations in peptides on the stability of the formed complex with full-length proteins. On the basis of the developed approach, a qualitative correlation of the obtained results with the dissociation constant was revealed using the example of the formation of the BH3 peptide biological complex of Bmf, Puma, Bad, Hrk, Bax, Bik, Noxa, Bid, Bim, and Bak proteins with the Bcl-xl protein and the BH3 peptides protein Bax with the Bcl-2 protein, taking into account the replacement of amino acid residues.

In Chap. 8, two algorithms have been developed, first was developed specifically for the selection of amino acid residues in peptides to improve their affinity in the interaction of peptides with whole proteins, and second was developed to search for scattered active regions of the protein when bound to the peptide.

Before closing, we want to acknowledge my sincere thanks to my colleague Prof. A. P. Golovitsky for a critical reading of the manuscript. Our thanks are to Springer Nature, in particular Dr. Habil. Claus E. Ascheron.

Note that Sect. 2.5 (Physical Interpretation of Condition Number) is written in cooperation with Prof. A. P. Golovitskii.

The results of the work were obtained using computational resources of Peter the Great Sainte-Petersburg Polytechnic University Supercomputing Center (www.spbstu.ru).

Note that when using different versions of MATLAB, the numerical estimates obtained may differ, however, their analysis will be of qualitative agreement.

The calculations were performed in MATLAB computing environment 2017a, operating system CentOS Linux 7.

Saint Petersburg, Russia

Kirill Kulikov
Tatiana Koshlan

Contents

Chapter 1
Physical Methods for Studying Proteins

Abstract This chapter discusses various experimental approaches for determining the structure of molecules, the mutual arrangement of domains in space, the identification of active protein centers, and their advantages and limitations in the study of various physical properties of biological complexes.

1.1 Electrophoresis Methods

The basis of the electrophoresis method is the motion of charged particles under the action of an electric field [1–7]. A large number of such important biological molecules as proteins, amino acids, nucleic and ribonucleic acids have different ionizable groups and therefore there are various electric charged particles in a solution, either as anions or as cations. These particles migrate either to the anode or to the cathode under the action of an applied electric field. Note that many complex biological molecules can have several different charged groups and will move in a medium with the applied electric field depending on the resultant charge.

Consider in more detail how the charged biological particles are separated in the process of electrophoresis.

To separate the selected biological objects in the medium, a potential difference (voltage) must be applied to the electrodes, which creates an electric field E, which corresponds to the applied voltage V and the distance d between the electrodes. Thus, when the electric field E acts on a molecule that has a charge q(C), the force Eq(H) arises. This force leads to the motion of the charged molecule to the electrode. During the movement, a friction force arises that delays the movement of the molecule. The friction force is hydrodynamic in nature and depends on the shape and size of the molecule and on the pore size of the medium in which the electrophoresis is carried out as well as the viscosity of the buffer. In practice, the most common concept of electrophoretic mobility of an ion is that of the ratio of the ion's velocity to the field strength. Molecules are divided according to their size even if they have the same total charge since they are affected by different frictional forces. Thus, molecules begin to move in the buffer at different rates and separate depending on their electrophoretic mobility when applying a potential difference to the electrodes.

T. Koshlan and K. Kulikov, *Mathematical Modeling of Protein Complexes*, Biological and Medical Physics, Biomedical Engineering,
https://doi.org/10.1007/978-3-319-98304-2_1

When the electric field is turned off before all the molecules reach the electrodes, the components of the biological sample mixture are separated depending on their electrophoretic mobility. The determination and localization of the separated components of the biological mixture are determined by autoradiography or by staining with a suitable dye. Electrophoregrams of protein-enzymes make it possible to study changes in the activity and isoenzyme spectrum of proteins under the influence of external and internal factors in both humans and other organisms.

Electrophoresis is a convenient method that allows one to separate a mixture of proteins after various experiments and draw conclusions about the stability of biological complexes: homodimers, heterodimers, tetramers, oligopolymers, etc. During the electrophoresis, most of the generated power is dissipated as heat. Heating of the medium in which electrophoresis is performed has negative consequences, namely: -the decrease in the viscosity of the buffer with a decrease in the resistance of the medium, -the formation of convection currents, leading to the mixing of separated samples, -an increase in the diffusion rate of sample and buffer ions, leading to a broadening of the zones of the separated samples. -the thermal destruction of samples, which are quite often sensitive to heat. This can cause protein denaturation.

The electrophoresis method can only confirm or infirm the formation of a molecular complex or its dissociation, but does not tell us anything about the nature of the interaction or the causes of dissociation, nor about the structure of the complex formed [7]. It does not reveal the structure of the catalytic active centers of the molecule, nor does it make it possible to calculate the thermodynamic constants of molecules entering into interaction or to draw any conclusions about the stability of biological complexes.

1.2 Chromatographic Methods

The chromatographic method is a method of separation of substances [8–17] developed by the Russian botanist Mikhail Tsvetom.

The chromatographic method is based on the distribution of the analyzed substances between two immiscible phases, which is described by the distribution coefficient K_d. Two different phases A and B correspond to a coefficient of distribution at a given temperature, which is the value of the constant [18]

$$K_d = \frac{N_A}{N_B},$$

where N_A is the concentration in phase A, and N_B is the concentration in phase B.

In any chromatographic system, there is a fixed (stationary) phase and a mobile phase. The fixed phase can be a solid carrier, gel, liquid, or a mixture of solid and liquid. The mobile phase can be liquid or gaseous and penetrates through the stationary phase after applying the mixture to be separated. In the process of chromatography, substances come into contact with two phases and differences in the coefficient of distribution of substances lead to separation of the mixture under study. At the moment, there are two basic options for carrying out chromatographic separations: thin-layered and columnar. Let's consider in more detail column chromatography. In this method, the stationary phase is in a metal or glass column. A mixture of substances to be analyzed is applied to the column and we then begin to pass a mobile phase through it, which is called an eluent. This method of column chromatography is most often used for analytical purposes in biochemistry. During the passage of the eluent through the column, the analyzed solutions are separated depending on their distribution coefficients and leave the column as a solution of the eluate. In a thin layer chromatography, a stationary phase on a suitable matrix is covered with a thin layer of glass, metal foil-plastic. The mixture to be separated is applied to the edge of the plate in the form of spots or stripes. Further, in the plate, located in a horizontal or vertical position, the mobile phase begins to act under the action of capillary forces, causing the substances to be analyzed to migrate to the opposite edge of the plate with determined velocity for each substance. The advantage of this method over column chromatography is the ability to analyze several samples at once. In order to improve the separation of the substances to be analyzed, the composition of the mobile phase can be changed, for example by modifying the salt concentration, the pH value, or the polarity. A successful chromatography yields all components of the mixture in their pure form. Let us consider in more detail the methods of analysis in chromatography, depending on the purpose of the experiment. Chromatographic separation of the samples under study is carried out at a qualitative or quantitative level. Qualitative analysis is performed to confirm the presence in the sample of a certain component. Quantitative analysis is carried out for the purpose of finding and confirming the presence in the sample of a certain substance and measuring its quantity. The quantitative analysis is based on determining and measuring the peak area and determining the substance using an appropriate calibration curve. The chromatographic method is suitable for determining the substances and their quantity in the mixture, but this method does not allow us to make any assumptions about the structural organization of the present molecules, about the presence in them of certain domains responsible for binding the protein complex or about the nature of their interaction. It does not reveal the structure of the catalytic active centers of the molecule. It also does not make it possible to calculate the thermodynamic constants of the molecules entering into interactions to draw any conclusions about the stability of biological complexes.

1.3 Mass Spectrometry

Mass spectrometry is a direct method that allows one to directly determine the following physical parameters of the studied substances: the molecular weight, the elemental composition of molecules, their fragments, the relationship between each other and their relative location, and to study the mechanisms of fragmentation [19]. Based on the data obtained during mass spectrometry, correlation dependences are obtained between the structural characteristics of molecules and ions formed as a result of the decay of molecules upon ionization [20–23]. A mass-spectrometric experiment also studies the processes of energy transfer during the interaction of molecules with electrons ions, the processes of atom rearrangement in the formed ions and the influence of certain functional and structural groups on ionization and fragmentation processes.

1.3.1 Electron Impact

Electron impact is the main and most frequently used ionization method in mass spectrometry. It consists in the following: molecules in a gas phase are bombarded by a beam of electrons emitted by a cathode heated to a high temperature and accelerated to a given energy by a potential difference between the cathode and the anode. The cathode is usually made of tungsten or rhenium wire. The ionization process takes place in a vacuum. For most of the organic molecules studied, the ionization energy is 7–12 eV. The efficiency of ionization increases with increasing energy of ionizing electrons, reaching a maximum at 30–40 eV, then slowly decreases. During the ionization process, under the influence of ionizing electrons, the molecules begin to acquire excess energy, which causes the destruction of the molecular ion $[M]^+$ with the formation of fragmented and rearranged ions characterizing the structure of the substance under study. E-impact mass spectrometry is a compound research method that produces the most reproducible mass spectra of individual compounds. The main contribution to the average relative measurement error is low-intensity peaks of ions, which, as a rule, are not used for structural relationships. The average relative error in the values of the ion peaks recorded on different mass spectrometers under the same mode is approximately 15%. Mass spectrometry electron impact nevertheless has a significant drawback. Thermolabile compounds (mostly natural compounds), high molecular weight substances, organic salts, metal complex compounds and even some classes of organic compounds (hydrocarbons, aliphatic alcohols and amines, etc.) that do not give a peak in the mass spectrometry of electronic shock fall out of the scope of its application as a structural method.

1.3.2 Chemical Ionization

Another method of ionization is chemical ionization with the formation of positive and negative ions (positive and negative chemical ionization) and is based on the flow of ion-molecular reactions in a gas phase. The first mass spectra with chemical ionization of simple substances were obtained by Field and Manson in 1965. At present, this variant of mass spectrometry with chemical ionization has found wide application in the practice of scientific research, and also in analytical chemistry. The method of chemical ionization with the formation of positive ions can increase the intensity of the peak of the molecular ion as in the case of labile molecules and more clearly trace the main directions of fragmentation. In order to obtain optimal mass spectra from the ratio of the peaks of molecular and fragment ions, it is necessary to select a reagent gas and ionization conditions. The sensitivity of this method with the use of chemical ionization depends on the nature of the substance and the ionization conditions but does not exceed the sensitivity of the method with ionization by electron impact. The advantage of the method of mass spectrometry with negative chemical ionization is very high sensitivity and selectivity to compounds that have a high affinity for the electron, for example, halogen-containing substances. However, in this method, just as in the method of positive chemical ionization, it is necessary to select a reagent gas and ionization conditions. The sensitivity of the method depends on the ionization conditions and on the structure of the molecules. It should be noted that it is not always possible to evaporate many organic substances without decomposition, and therefore, they cannot be ionized by electron impact and chemical ionization. The possibilities of electron and chemical ionization decrease dramatically in the transition from amino acids to peptides due to even lower volatility and increased thermolability of these compounds [24]. Such biological objects include most biological molecules (proteins, DNA, etc.), physiologically active substances, polymers, as well as many highly polar substances. Also, significant disadvantages can be attributed to the relatively large size and high cost of modern mass analyzers.

1.4 X-ray Analysis of Protein Crystals

X-ray analysis is one of the most important experimental methods that allow us to determine with atomic precision the spatial coordinates of all the atoms in the object [25]. After determining the position of each atom, the following parameters of the molecule structure can be calculated: interatomic distances, valence angles, rotation angles around bonds, and surface charge distribution. The data obtained by X-ray diffraction analysis are valuable information for chemists, biochemists, biophysicists and biologists who study the different relationships between the structural characteristics of the biological molecules and their functional properties, as well as for specialists studying the electronic structure of molecules and molecular interactions. Today, the structures of about 15 thousand proteins and their complexes with biologically important molecules are known. The method of X-ray structural analysis is based on the diffraction of X-rays on a crystal lattice, so it can be applied only to substances that are in a crystalline state. If the sample consists of a large number of randomly oriented identical molecules in solution, then the scattering pattern will be determined from the averaged directions, which will to a considerable extent prevent obtaining detailed information on the atomic structure. We note that this method is based on the phenomenon of X-ray diffraction on a three-dimensional crystal lattice. For a successful X-ray examination of the structure of a biological object, it is necessary to use monochromatic X-ray radiation. For this purpose, various filters and monochromators are used. To obtain a diffraction pattern, the object under study is placed in an X-ray beam and the intensity of the radiation scattered in different directions is measured. The easiest way is to place the beam of the film on the way and judge the intensity of scattering in this direction by the degree of darkening of the spot after the exposition. At the output, a set of intensities of the rays scattered in different directions, or diffraction pattern, is obtained. The next stage is the analysis of the obtained diffraction pattern and obtaining information about the atomic structure of the object under study. One of the drawbacks of the method of X-ray structural analysis in the study of protein structures is that the proteins are in vitro and in vivo in solution, and, when tested, they are subjected to crystallization processes [18]. A logical question arises: is there any fundamental distortion of the structure of protein molecules during crystallization. It is considered that strong distortions don't occur after all. However, as evidenced by a number of experimental studies [26–28] in the field of investigation of various regions of protein binding of the histone chaperone Nap1, the results of the X-ray diffraction analysis are not fully consistent with the results obtained in the study of the thermodynamic properties of the same protein in solution. In particular, the results of the work contain conflicting data on the involvement of the flexible ends of the histone chaperone molecule when it binds to dimers of histone proteins H2A–H2B and H3–H4. The flexible ends of the histone-chaperone molecule, according to X-ray diffraction analysis data, do not interact with dimers of histone proteins, but experiments using targeted site-specific mutagenesis in solution showed that the ends of the histone chaperone can make a synergistic contribution to the interaction with histone dimers. Thus, despite the high accuracy of the X-ray structural method in the study of proteins, it must be supplemented with other physical methods for studying biological objects.

1.5 Methods of Spectral Analysis

Let us analyze the spectral nature of the interaction of electromagnetic radiation with matter. This interaction is of a quantum nature and depends on the radiation property, as well as on the material. The main physical characteristics of electromagnetic radiation are frequency, wavelength, and intensity. Let us consider the interaction of electromagnetic radiation with the energy levels of the substances under study. Electrons in atoms and molecules of the substances under study can be at different energy levels, but in principle, they tend to occupy the level with the lowest energy-the basic level. In order to effect the transition of an electron from a lower to a higher energy level (into an excited state), the system must transmit a certain amount of energy, and if the source of this energy is electromagnetic radiation, then an absorption spectrum appears. The molecule of the irradiated substance absorbs a strictly defined amount of energy, which corresponds to the difference in the energy levels that the electron occupies. In the transition of an atom or a molecule from a higher to a lower energy level, one quantum of energy is emitted, which in turn is accompanied by the appearance of a radiation spectrum. Thus, transitions in atoms and molecules, which are usually observed in the form of absorption spectrum, emission spectrum or fluorescence spectrum in the ultraviolet or visible parts of the spectrum, are the cause of the appearance of electronic spectra.

1.5.1 *Spectroscopy in the Ultraviolet and Visible Range*

Consider the general law of light absorption by matter, the Bouguer–Lambert–Baer law. This law is based on two relationships: one of the relationships relates the absorption of light to the concentration of the absorbing substance, and the other relates the length of the light path or the thickness of the layer to the absorbing substances. The ratio of the intensity of the incident and transmitted light is called transmission T: $T = I/I_0$, where I_0 is the intensity of incident light, I is intensity of transmitted light. Note that intensity is the product of the photon energy by the number of photons colliding on a surface unit per time unit. At $T = 100\%$, the substance is absolutely transparent and does not absorb radiation, since the intensity of the incident light is equal to the intensity of the light passing through the substance, and at $T = 0$ the substance is completely opaque and completely absorbs the incident radiation. Substances with intermediate transmission values are characterized by the extinction of E [18]:

$$E = \lg\left[\frac{1}{T}\right] = \lg\left[\frac{I_0}{I}\right].$$

When performing spectrophotometric analysis, it is customary to prepare a series of standard solutions, with the help of which a calibration curve of the dependence of absorption on concentration is constructed [29, 30]. After that, the absorption of the analyzed solution is measured, and its concentration is found from the calibration curve by interpolation. The main advantage of the spectrophotometer is the ability to scan the full wavelength range of the ultraviolet and visible light regions and obtain absorption spectra [18]. These spectra obtained for each substance reflect the dependence of the absorption on the wavelength. Note that the wavelength of the absorbed light is determined by the corresponding electron and therefore specific absorption peaks can be attributed to known molecular fragments. Qualitative analysis in the ultraviolet and visible wavelength ranges allows identification of certain classes of compounds in pure form, as well as in mixtures, for example, proteins, nucleic acids, cytochromes, and chlorophylls. The method serves to identify changes in the chemical structure of compounds. The quantitative analysis is based on the fact that some chromophore groups, such as aromatic amino acids in proteins or heterocyclic bases of the nucleic acids, absorb light at a certain wavelength. Proteins can be analyzed at a wavelength of 280 nm, and nucleic acids-at 260 nm, although in any case, it is necessary to take into account the possibility of the influence of impurities from other substances if they are also present in the solution. To account for this effect, it is customary to measure the absorption of additional impurities at two wavelengths: at the wavelength at which the analyze is absorbed, and at the wavelengths at which it is not absorbed. This method allows one to make an assumption about changing the structure of biological molecules in their binding and to draw conclusions about the presence or absence of interaction between the selected objects; if one knows in advance about the presence of the active center of the molecule, the method allows one to observe the presence of changes in the structure of this center. The method of spectroscopy in the ultraviolet and visible ranges can find good application in the study of already known biological molecules with a given structure and known catalytic centers. Usually, this method is used in conjunction with other physical methods of biological objects research to confirm or supplement the changes in the previously known absence or presence of interaction between biological objects in the solution, to make assumptions about their nature of binding, which must be supplemented and confirmed by additional physical methods of research [18].

1.6 Spectrofluorimetry

Fluorescence is the phenomenon of light emission by matter, i.e. when a molecule moves from a state with a higher energy level to a state with a lower energy level. This transition is recorded by the intensity of the emitted radiation. In order to carry out such a possibility of transitions from a higher to a lower energy level, it is necessary to excite the molecule by exposure to electromagnetic radiation. In this case, the wavelength of the emitted radiation is larger, and the energy, therefore, is less than the wavelength of the absorbed light. The difference between these two wavelengths is called the Stokes shift, and usually, the best results are obtained with those compounds in which the Stokes shift is greater. The fluorescence spectra provide information on events that occur within a time interval of less than 10^{-8} s. The quantum yield Q is determined by the expression [18]:

$$Q = \frac{Q_1}{Q_2}$$

where $Q_1$is the number of emitted quanta, Q_2 is the number of absorbed quanta. At low substance concentrations, the fluorescence intensity I_f depends on the incident light intensity I_0 [18]:

$$I_f = 2.3 I_0 \varepsilon_\lambda c d Q,$$

where c is the molar concentration of the fluorescent solution, d is length of light path in solution (cm), ε_λ is molar extinction coefficient of material at wavelength λ(dm/(mole·cm). Fluorescence analysis is widely used, despite the fact that only a few molecules are able to fluoresce. If it is necessary to identify the substance, the absorption and fluorescence spectra are compared. To clarify the known structure, an analysis of the effect of pH and solvent composition is performed, as well as fluorescence polarization. If the substance is non-fluorescent, fluorescent labels are attached to its molecules and the radiation of the attached label is monitored. Then, as with the natural fluorescence of proteins, the fluorescence of aromatic groups in the side chains of amino acid residues is monitored. The label is used for both qualitative and quantitative analysis. The main application of fluorimetry is the quantitative determination of substances whose concentration is too low. Similarly, the fluorometry method is suitable for studying the kinetics of enzymatic processes. Since our work is devoted to the study of the interaction of proteins, we will consider in more detail what allows us to investigate the fluorometry method in this direction. The presence of tryptophan or flavin adenine dinucleotide in the protein amino acid molecule, which takes an active part in many redox biochemical processes as a cofactor, makes it possible to obtain the natural fluorescence of the biological objects under study. Binding and release of cofactors, inhibitors, and substrates near the fluorescent group lead to a change in the fluorescence spectrum, which allows obtaining information about the relative conformation changes, denaturation, and aggregation of molecules. The change in the conformation of the analyzed protein, which occurs when the ligand is bound, is reflected in the nature of the fluorescence. The method of fluorescence due to resonance energy transfer allows determining the localization of metals in metalloproteins, various conformational changes in enzymes and receptors when binding substrates to ligands, and the distance between different pairs of proteins on the ribosome. Thus, the method of spectrofluorimetry makes it possible to obtain accurate results when analyzing samples with very low concentrations. This method is also characterized by high selectivity since the Stokes shift allows the use of two monochromators-one for exciting light, the other for selective light. The disadvantages of the method are its high sensitivity to changes in temperature, pH, and polarity of the solvent; it is also impossible to predict the ability of a particular substance to fluoresce. The main disadvantage is the suppression of fluorescence [18]. The reason for this phenomenon is that the energy that could be released in the form of fluorescence is transmitted to other molecules during a collision.

1.7 Circular Dichroism

Circular dichroism is the effect of optical anisotropy which manifests itself in the difference between the absorption coefficients of light polarized in the right and left circles [31]. This method is based on measuring the angle of rotation of the polarization plane after the polarized light passes through the solution containing the chiral substance. We note that chirality is a property of a molecule that does not coincide in space with its mirror image. A plane polarized light can be represented as a sum of two circularly polarized rays with right and left polarization. Asymmetrical chiral molecules in their nature interact differently with these components, so the rays propagate in the sample at a different speed (they have different refractive indices). As a result, the plane of polarization of the light beam passing through the sample rotates relative to the plane of polarization of the initial beam. In the spectral regions where the substance absorbs light, circular dichroism is observed. The right and left rays are not only refracted in different ways by the chiral substance but are also absorbed to varying degrees. As a result, an initially plane polarized light becomes elliptically polarized. Since it is difficult to measure the ellipticity, it is therefore measured separately for the absorption of the left and right rays. The CD spectrum represents the dependence of the ellipticity on the wavelength. This spectrum makes it possible to obtain information on the three-dimensional structure of molecules, the relative content of elements of the secondary structure of the protein (alpha helices, beta layers and irregular sequences) in solution, which becomes particularly relevant when studying the effect of denaturing agents or temperature on the three-dimensional structure of selected proteins. This method has proven itself in studies of protein denaturation in which there is a change in alpha and beta-structures in irregular sequences during protein denaturation. However, the application of CD to the analysis of spatial three-dimensional structures of biological molecules is limited due to a lack of understanding of the influence of individual parts of molecules on the formation of this level of structure. In due time, the CD spectra of poly-L-amino acids were obtained, which were used as standards in determining the content of each form of the secondary structure in proteins. For known proteins, the method of approximation of curves was used. An important advantage of the CD method is that it can be used to study the conformation changes of various substances when interacting with other proteins, DNA, and ligands. The drawbacks of the CD method are that it does not allow one to calculate the distance between different paramagnetic proteins or to reveal the structure of the catalytic active sites of the molecule and does not make it possible to calculate the thermodynamic constants of molecules entering into interaction or to draw any conclusions about the stability of biological complexes [18].

1.8 Conclusion

The main objective of this chapter, despite the fact that it is of an overview nature and contains a number of known facts, is to consider various experimental approaches associated with the study of the structure of molecules, the mutual arrangement of domains in space, and the detection of active protein centers: electrophoresis methods, chromatographic analysis methods, mass spectrometry, X-ray crystal analysis of proteins, spectral analysis methods, and circular dichroism. Their advantages and disadvantages are listed. It should be noted that most of the presented experimental approaches have significant limitations in the study of various physical properties of biological complexes. To study the whole variety of physical characteristics of protein complexes, it is necessary to combine a large number of different experimental approaches, each of which makes it possible to determine a narrow list of desired physical parameters. We also note that most of the biochemical reactions with the given chemical elements must be verified experimentally, which is a fairly labor-consuming and expensive method, requiring a large amount of time for its conduct. The above limitations, which involve the study of the physical characteristics of biological structures, require the development of a new mathematical approach that would allow:

- to theoretically predict the passage of a biochemical reaction in the chosen direction with given amino acid (a.a.) sequences;
- to study the behavior of dimers in vitro in solutions with different concentrations of monovalent salt for the production of nucleosome histone nucleus, to investigate the influence of temperature on the stability of protein dimers H2A–H2B H3–H4, some sections of amino acid sequences of the input solution, taking into account the contribution of hydrophobic amino acid residues in the formation of the structure of dimers;
- to determine the regions of protein molecules responsible for the aggregation of proteins in aqueous solution under different temperature regimes from 20–40 °C;
- to study the effect of phosphorylation of amino acid residues of polypeptide chain on the formation of biological complexes on the example of phosphorylated flexible n-end protein p53 for two amino acid residues;
- to determine the location of active protein sites and to detects the stability of different protein sites by analyzing the potential energy-electrostatic interaction matrix between different sites of the microbiological complex, such as histone chaperone Nap1-sNap1, protein heterodimer p53-Mdm2 and Mdm2-Mdm2 homodimer, which are responsible for the entry of an entire protein molecule into biochemical reactions;
- to study the effect of point mutations in BH3 peptides on the stability of the formed or biological complex with pro-apoptotic proteins of the Bcl-2 family, as well as the qualitative determination of the dissociation constant for binding different BH3 peptides to Bcl-2 and Bcl-xl proteins.

References

1. E. Gaal, G. Medbesi, L. Veretski, *Electrophoresis in the Division Biological Macromolecules* (Moscow, 1982), 448 pp
2. V.F. Gavrilenko, M.E. Ladygina, L.M. Khandobina, *A Large Practical Workshop on Plant Physiology* (Higher School, Moscow, 1975), 327 pp
3. V.M. Golitsyn, Non-denaturing electrophoresis. Fractionation photosynthetic pigment-protein complexes and plasma proteins. Biochemistry **64**, 64–70 (1999)
4. V.S. Kamyshnikov, *Reference Book on Clinical and Biochemical Laboratory Diagnostics (in 2 Volumes)* (Minsk, 2000), 463 pp
5. L.A. Osterman, *Methods for Studying Proteins and Nucleic Acids: Electrophoresis and Ultracentrifugation* (Nauka, Moscow, 1981), 288 pp
6. V.I. Safonov, M.P. Safonova, Research of proteins and enzymes of plants by electrophoresis in a polyacrylamide gel. *Biochemical Methods in Plant Physiology* (Nauka, Moscow, 1971), pp. 113–137
7. V.N. Titov, V.A. Amelyushkina, *Electrophoresis of Proteins Blood Serum* (Optium Press, 1994), 62 pp
8. E. Heftman, T. Caster, A. Niderweiser, N. Katsimpulas, A. Kuksis, R. Krotyau, R. Ronald, in *Chromatography. Practical Application of the Method. Part 1*, ed. by E. Heftmann (Moscow, 1986), 336 pp
9. H. Schluter, Reversed-phase chromatography. J. Chromatogr. Libr. **61**, 147–234 (2000)
10. A. Jungbauer, C. Machold, Chromatography of proteins. J. Chromatogr. Libr. **69. Part B**, 669–737 (2004)

11. M. Caude, A. Jardy, Normal-phase liquid chromatography. Chromatogr. Sci. Ser. **78**, 325–363 (1998)
12. M. Rogner, Size exclusion chromatography. J. Chromatogr. Libr. **61**, 89–145 (2000)
13. F. Anspach, Affinity chromatography. J. Chromatogr. Libr. **69. Part**, 139–169 (2004)
14. P. Cuatrecasas, M. Wilchek, Affinity chromatography. Encycl. Biol. Chem. **1**, 51–56 (2004)
15. D. Hage, M. Nelson, Chromatographic immunoassays. Anal. Chem. **73**, 198A–205A (2001)
16. A. Henschen, K. Xyp, F. Lotshpayh, V. Walter, *Highly Effective Liquid Chromatography in Biochemistry* (Moscow, 1988), 688 pp
17. L. Osterman, *Chromatography of Proteins and Nucleic Acids* (Nauka, Moscow, 1985), 536 pp
18. K. Wilson, J. Walker, *Principles and Methods of Biochemistry and Molecular Biology* (Binom, Moscow, 2013)
19. N.A. Klyuev, E.S. Brodsky, Modern methods mass spectrometric analysis of organic compounds. J. Russ. Chem. Soc. named after D.I. Mendeleyev's **XLVI**(4) (2002)
20. R. Aebersold, M. Mann, Mass spectrometry-based proteomics. Nature. **422**(6928), 198–207 (2003)
21. M. Mann, R. Hendrickson, A. Pandey, Analysis of proteins and proteomes by mass spectrometry. Annu. Rev. Biochem. **70**, 437–473 (2001)
22. J. Fenn, M. Mann, C. Meng, S. Wong, C. Whitehouse, Electrospray ionization for mass spectrometry of large biomolecules. Science **246**(4926), 264–271 (1989)
23. J. Fenn, M. Mann, C. Meng, S. Wong, C. Whitehouse, Electrospray ionization-principles and practice. Mass Spectrom. Rev. **9**(1), 37–70 (1990)
24. A.T. Lebedev, K.A. Artemenko, TYu. Samghina, *Basics Mass Spectrometry of Proteins and Peptides* (Technosphere VMSO, Moscow, 2012)
25. T. Blandel, L. Johnson, *Crystallography of Protein* (Moscow, 1979)
26. S. D'Arcy, K.W. Martin, T. Panchenko, X. Chen, S. Bergeron, L.A. Stargell, B.E. Black, K. Luger, Chaperone Nap1 shields histone surfaces used in a nucleosome and can put H2A-H2B in an unconventional tetrameric form. Mol. Cell **51**(5), 662–677 (2013)
27. A.J. Andrews, G. Downing, K. Brown, Y.J. Park, K. Luger, A thermodynamic model for Nap1-histone interactions. J. Biol. Chem. **283**(47), 32412–32418 (2008)
28. C. Aguilar-Gurrieri, Structural evidence for Nap1-dependent H2A-H2B deposition and nucleosome assembly. EMBO J. **35**(13), 1465–1482 (2016)
29. B. Williams, K. Wilson, *Methods of Practical Biochemistry* (Moscow, 1978)
30. Workshop on Biochemistry, ed. by S.E. Severin, G.A. Solovyovoy (Moscow State University, Moscow, 1989)
31. G.D. Fasman, *Circular Dichroism and the Conformational Analysis of Biomolecules* (Springer, Berlin, 1996)

Chapter 2
Mathematical Simulation of Complex Formation of Protein Molecules Allowing for Their Domain Structure

Abstract This chapter we construct a physical model of the interactions between protein molecules and study of their propensity to form biological complexes. The reactivities of proteins have been studied using electrostatics methods based on the example of histones H2A, H2B, H3, H4. The capability of proteins to form stable biological complexes that allow for different segments of amino acid sequences has been analyzed. The ability of protein molecules to form compounds has been considered by calculating matrices of electrostatic potential energy of amino acid residues constituting the polypeptide chain. The method of matrices has been used in the analysis of the ability of protein molecules to form complex biological compounds.

2.1 Introduction

The present chapter is dedicated to developing a mathematical model that will be able to theoretically predict a biochemical reaction that occurs in a chosen direction that involves given proteins with known amino acid sequences. There are a number of papers that should be noted in which different amino acid sequences of the chosen proteins have been analyzed.

In [1] hydrogen–deuterium exchange in combination with mass-spectrometry was used to reveal binding sites of the H2A–H2B dimer with Nap1. Authors identify the interaction surface between H2A–H2B and Nap1, and confirm its relevance both in vitro and in vivo.

In [2] investigated contacts between histones H2A, H2B, H3 and H4 in the nucleosome. The authors used the experimental structures chromosome containing histones and calculated the number of contacts between different histones.

In this work the number of contacts between histone intrudes in the nucleosome was found that the H2A–H2B and H3–H4 heterodimers have the greatest number of contacts between pairs of heterodimers.

T. Koshlan and K. Kulikov, *Mathematical Modeling of Protein Complexes*,
Biological and Medical Physics, Biomedical Engineering,
https://doi.org/10.1007/978-3-319-98304-2_2

A quantitative method for studying the affinity of Nap1 to histones was developed in [3] The binding affinity between Nap1 and H2A–H2B was found to be on the nanomolar level. It was noted that each Nap1 dimer binds two H2A–H2B dimers; the termini of the Nap1 molecule were shown to give a synergetic contribution to binding with histones.

No clear criteria were given in the mentioned papers, which should allow one to determine the reactivities of different protein domains that are responsible for the participation of whole molecules in various biochemical reactions. Thus, the present work aims at developing a mathematical model that should facilitate the processing of existing experimental data, predict theoretically a biochemical reaction passing in a chosen direction involving given amino acid sequences, and reveal protein sites responsible for interactions between different protein molecules, which actually determines the importance of the task. The work consists of several parts. In the first section, basic principles of the nucleosome formation are given, the principles of the formation of histone heterodimers H2A–H2B and H3–H4. The second part is dedicated to developing a physical model of interactions between proteins with the formation of biological complexes based on electrostatic interactions of protein molecules allowing for their amino acid sequences. Biological systems are considered in detail in the third part, including the formation of H2A–H2B and H3–H4 heterodimers; the results of interaction simulations in the chosen biological systems are presented. All calculations were conducted in order to allow different amino acid sequences to participate in the studied protein reactions.

The developed algorithms allow to determine the interaction of different domains of proteins or whole proteins.

2.2 General Principles of the Formation of Biological Complexes

Consider general principles of biological complex formation in regard to our research. We will analyze protein interactions H2A, H2B, H3, H4 which actively participate in the nucleosome formation process. Chromatin assembly is a stepwise process that starts with the association of a tetramer of histone $(H3–H4)_2$ with the DNA, followed by the incorporation of two H2A–H2B dimers to form the nucleosome [4, 5].

The nucleosome assembly process is facilitated by several partially redundant pathways and is aided by histone chaperone proteins, such as nucleoplasmin, antisilencing factor 1 (Asf1), histone regulator (HIR), chromatin assembly factor 1 (CAF-1), and nucleosome assembly protein 1 (Nap1) [6, 7].

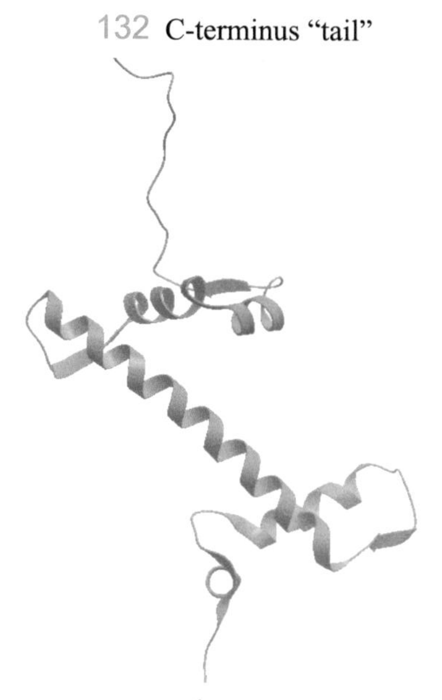

Fig. 2.1 Three-dimensional structure of histone H2A with indication of the N–terminus and C–terminus of the amino acid sequence [8]

2.2.1 Formation of Heterodimers (H3–H4) and (H2A–H2B)

Histones H2A, H2B, H3 and H4 are the basic proteins [7], which form dimers, the structure of which is called ≪handshake≫ in this structure, the formation of dimers H2A–H2B or H3–H4 occurs in opposite directions of each histone [5].

In Fig. 2.1 scheme of the H2A protein showing the N-tail and the C-tail of the protein is presented. In this chapter, we analyze different protein sequences and allow for their domain structure to determine possible compound formations (H2A–H2B), (H3–H4). Note that domains arise due to combination, alternation, and α-helices and β-sheets, between which less dense structure appear [9].

Amino acid sequences of histone proteins H2A, H2B, H3, H4, as well as their secondary structures were taken from [10]; entry numbers for proteins were P04911, A6ZKU6, P61830, P02309 respectively. In this chapter we have investigated yeast proteins.

2.3 Description of the Physical Model

Earlier experiments [11] revealed that the interaction of protein molecules is determined by the potential energy of the electrostatic interaction. For this reason, this study is devoted to a theoretical analysis of the electrostatic interactions between protein molecules.

Let us describe a physical model of the electrostatic interaction between the amino acid sequences of different proteins.

Each amino acid is represented as a uniformly charged sphere with its own radius value. The protein is represented as a free-articulated polyamino acid sequence [12].

When studying the interaction of charged protein molecules, a number of approximations were used:

1. the energy of protein interaction is determined only by the forces of electrostatic interaction;
2. a protein molecule is modeled as interconnected amino acid residues;
3. each amino acid residue of the protein is represented as a uniformly charged sphere.

The sphere radius size of each amino acid residue was taken from the work [13]:

$$R_A = 0.6\,\text{nm},\ R_R = 0.809\,\text{nm},\ R_N = 0.682\,\text{nm},\ R_D = 0.665\,\text{nm},\ R_C = 0.629\,\text{nm},$$
$$R_Q = 0.725\,\text{nm},\ R_E = 0.714\,\text{nm},\ R_G = 0.537\,\text{nm},\ R_H = 0.732\,\text{nm},\ R_I = 0.735\,\text{nm},$$
$$R_L = 0.734\,\text{nm},\ R_K = 0.737\,\text{nm},\ R_M = 0.741\,\text{nm},\ R_F = 0.781\,\text{nm},\ R_P = 0.672\,\text{nm},$$
$$R_S = 0.615\,\text{nm},\ R_T = 0.659\,\text{nm},\ R_W = 0.826\,\text{nm},\ R_Y = 0.781\,\text{nm},\ R_V = 0.654\,\text{nm}.$$

Note that the ability of molecules to interact with different amino acid residues strongly depends on their environment, which is due to their polar and non-polar parts.

Since in this chapter we are considering the problem of electrostatic interaction between histones, we make the following assumption: we divide interactions between 20 amino acid residues into 10 classes from $0.05e$ to $1e$ in accordance with [14].

Thus, we assign to the amino acid residues a charge that is less than or equal to the charge of the electron. Note that the square of the modulus of the wave function, which describes the state of an electron in a multi-center model, determines the probability density of an electron or the density of an electron cloud, which characterizes the unequal probability of finding an electron in a selected part of the volume of an electron cloud in a polyatomic amino acid residue. It follows that the redistribution of electron charges in polyatomic amino acid residues may well lead to the fact that the probability of finding a valence electron in the neighborhood of the residue is less than one, and the average value of the charge of the residue is less than the charge of the electron.

The distances between the two interacting amino acid residues of neighboring proteins were determined from the following assumptions [15, 16]:

1. the distance between the oppositely charged amino acid residues was 0.15 nm;
2. between like-charged ones, the distance was 0.4 nm;
3. The distance between the amino acid residues that form, presumably, one hydrogen bond was 0.35 nm;

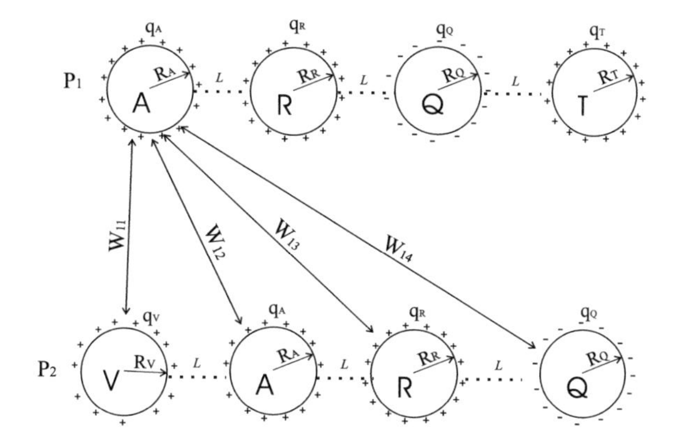

Fig. 2.2 Schema of the interaction of amino acid residues of two interacting proteins P_1 and P_2. Each amino acid is represented as a uniformly charged sphere of a given diameter

4. The distance between amino acid residues, which, presumably, can form more than one hydrogen bond, was set to 0.25 nm.

Below we have considered the problem of modelling the processes of electrostatic interaction of the formation of dimers from the complete amino acid sequences of selected proteins within the framework of the classical electrostatic theory (Fig. 2.2).

2.4 The Problem of the Electrostatic Interaction of Two Conducting Spheres

Let there be two conducting spheres with radii R_1, R_2 and charges Q_1 and Q_2, located at a distance L between the centers. Then, following the works [17–20] we can write the expression for the potential energy of the electrostatic interaction of the spheres between their centers:

$$W_q = \left[\frac{1+\gamma}{2\alpha}\right] \frac{\alpha^2 c_{11} - 2\alpha c_{12} + c_{22}}{c_{11}c_{22} - c_{12}^2}, \tag{2.1}$$

where $Q_1\, Q_2$ are the charges of the first and second spheres, $\gamma = R_2/R_1$,$\alpha = Q_2/Q_1$, c_{11}, c_{12}, c_{22} are the capacitive coefficients [20]:

$$c_{11} = 4\pi\varepsilon_0\varepsilon R_1\gamma \sinh(\beta) \sum_{n=1}^{\infty} \left[\gamma \sinh(n\beta) + \sinh[(n-1)\beta)]\right]^{-1}, \tag{2.2}$$

$$c_{22} = 4\pi\varepsilon_0\varepsilon R_1\gamma \sinh(\beta) \sum_{n=1}^{\infty} \left[\sinh(n\beta) + \gamma \sinh[(n-1)\beta)]\right]^{-1}, \tag{2.3}$$

$$c_{12} = -4\pi\varepsilon_0\varepsilon R_1\gamma\frac{\sinh(\beta)}{(1+\gamma)h}\sum_{n=1}^{\infty}[\sinh(n\beta)]^{-1}, \tag{2.4}$$

where ε is the dielectric constant of the medium, ε_0 is the dielectric constant. In this case, $(R_1 + R_2)$ is chosen as a unit length, so that h is the dimensionless distance between the centers of the spheres, which is defined as $h = L/(R_1 + R_2)$, and β is associated with the distance between the centers of the spheres as follows [17–20]:

$$h = l/(R_1 + R_2)$$

The parameter β associated with the distance between the centers of the balls as follows [17–19]:

$$\cosh(\beta) = \frac{h^2(1+\gamma)^2 - (1+\gamma^2)}{2\gamma} \tag{2.5}$$

Note that the capacitive coefficients (2.2)–(2.4) are defined in units of R_1/k, $k = 1/4\pi\varepsilon_0$. Then, as the unit of energy measurement, we choose the value $kQ_1Q_2/(R_1 + R_2)$, in this case for the dimensionless energy of the electrostatic interaction of the spheres and, respectively, for the capacitive coefficients c_{11}, c_{12}, c_{22}, we obtain the following expressions:

$$\widetilde{W}_q = \left[\frac{1+\gamma}{2\alpha}\right]\frac{\alpha^2 c_{11} - 2\alpha c_{12} + c_{22}}{c_{11}c_{22} - c_{12}^2}, \tag{2.6}$$

$$c_{11} = \varepsilon\gamma\sinh(\beta)\sum_{n=1}^{\infty}\left[\gamma\sinh(n\beta) + \sinh[(n-1)\beta)]\right]^{-1}, \tag{2.7}$$

$$c_{22} = \varepsilon\gamma\sinh(\beta)\sum_{n=1}^{\infty}\left[\sinh(n\beta) + \gamma\sinh[(n-1)\beta)]\right]^{-1}, \tag{2.8}$$

$$c_{12} = -\varepsilon\gamma\frac{\sinh(\beta)}{(1+\gamma)h}\sum_{n=1}^{\infty}[\sinh(n\beta)]^{-1}. \tag{2.9}$$

We perform the following transformations for expressions (2.7)–(2.9). We introduce the variable $z = \exp(-\beta)$, then we obtain the following expressions:

$$c_{11} = 2\varepsilon\gamma\sqrt{\cosh(\beta)^2 - 1}\times$$

$$\times\sum_{n=1}^{\infty}\frac{z^n}{(1-z^{2n})[(\gamma+\cosh(\beta)) - \sqrt{\cosh(\beta)^2 - 1}(1+z^{2n})/(1-z^{2n})]},$$

$$c_{22} = 2\varepsilon\gamma\sqrt{\cosh(\beta)^2 - 1}\times$$

Fig. 2.3 Representation of the potential energy matrix of the electrostatic interaction $W_{i,j}, i = \overline{1,4}, j = \overline{1,4}$ of two proteins P_1 P_2

	A	R	Q	T
V	W_{11}	W_{12}	W_{13}	W_{14}
A	W_{21}	W_{22}	W_{23}	W_{24}
R	W_{31}	W_{32}	W_{33}	W_{34}
Q	W_{41}	W_{42}	W_{43}	W_{44}

$$\times \sum_{n=1}^{\infty} \frac{z^n}{(1-z^{2n})[(1-\gamma \cdot \cosh(\beta)) - \gamma \cdot \sqrt{\cosh(\beta)^2 - 1}(1+z^{2n})/(1-z^{2n})]},$$

$$c_{11} = -\varepsilon \frac{2\gamma\sqrt{\cosh(\beta)^2 - 1}}{h(1+\gamma)} \sum_{n=1}^{\infty} [(z^{2n})/(1-z^{2n})],$$

where $\cosh(\beta)$ is defined by the expression (2.5).

The resulting values of the potential energy of electrostatic interaction between the corresponding amino acid residues, which we represent as charged spheres, are written into the matrix (see Fig. 2.3).

In our model, we will assume that each amino acid residue of one protein molecule can interact with any other amino acid residue of another protein.

To analyze the biochemical processes we use the notion of condition number $\mathrm{cond}(W_0)$ of the matrix W_0: In order to analyze the relation between biochemical processes with histones, we use the condition number ($\mathrm{cond}(W_0)$) concept, which measures the degree of biochemical structures stability given the physical conditions.

$$\mathrm{cond}(W_0) = ||W_0|| \cdot ||W_0^{-1}||.$$

where $||W_0||$ is the norm of the matrix of the potential energy of the pair electrostatic interaction between peptides. For the calculation of the condition number we use the singular value decomposition (SVD) [21]. Then we have the following expression:

$$\mathrm{cond}(W_0) = \frac{\sigma_{\max}(W_0)}{\sigma_{\min}(W_0)}, \tag{2.10}$$

where $\sigma_{\max}(W_0)$, $\sigma_{\min}(W_0)$ are largest and smallest singular values of the potential energy matrix of pairwise electrostatic interaction between amino acid protein sequences.

In this physical formulation of the problem, it will characterize the degree of stability of the configuration of the biological complex. In order to choose a more stable biochemical compound between proteins, we select the matrix of potential energy of electrostatic interaction with the **smallest** value of the condition number.

Fig. 2.4 A system modeling the formation of a biocomplex

2.5 Physical Interpretation of Condition Number

Consider the biological complex (see Fig. 2.4).

It can be represented as a polyatomic molecule. In this case, we assume that the chains themselves (black and red) under the external conditions under consideration are stable, having elastic, but strong links between the elements of each individual chain with high discontinuity energies. The newly formed links between individual elements of different chains will be considered less stable (green), i.e. their bond-dissociation energy is lower. Then we can consider the resulting formation (bio-complex) as a kind of quasi-molecule, consisting of bound undeformed amino acid residues (hereinafter called ≪fragments≫). Under external influence, i.e. upon impact with its ≪neighbors≫, a molecule receives energy of the order of kT(k is Boltzmann constant). In this case, vibrations develop in the molecule and a change in its structure (bond dissociation) can occur. But it is not a fact that the weakest bond will break: in a molecule, it can be connected with a strong one.

The wave function describing the entire quasi-molecule in the general case depends on the $3N$ variables of the geometric coordinates of each fragment. Since 6 of them describe the translational and rotational motion of a quasi-molecule as a whole, the wave function describing the oscillations inside the quasi-molecule $\Psi^{vib}(q)$ itself will depend on the $3N-6$ coordinates: $q=(q_1,q_2,\ldots,q_{3N-6}),\, q_i\equiv(x_i,y_i,z_i)$.

The Schrodinger equation for $\Psi^{vib}(q)$

$$\left[-\frac{\hbar}{2}\sum_{i,j}^{3N-6}a_{ij}\frac{\partial^2}{\partial q_i\partial q_j}+W(q)\right]\Psi^{vib}(q)=E^{vib}\Psi^{vib}(q),$$

where $W(q)$ is the coordinate dependence of the interaction potential between the fragments. The coefficients a_{ij}, generally speaking, depend on the coordinates, but for small oscillation amplitudes (quasi-harmonic oscillations) they can be assumed to be constant. We set the oscillation amplitudes to small and expand $W(q)$ in a neighborhood of the equilibrium coordinate q_0 in the Taylor series:

$$W(q) = W(q_0) + \frac{1}{2}\sum_{i,j}^{3N-6} \left.\frac{\partial^2 W(q)}{\partial q_i \partial q_j}\right|_{q_0} (q_i - q_i^{(0)})(q_j - q_j^{(0)}) + \cdots$$

First, let's move the coordinate to the point $q_0 = q_1^{(0)}, q_2^{(0)}, \ldots$, and the energy will be evaluated from its minimum value, i.e. from $W(q_0) = W(q_1^{(0)}, q_2^{(0)}, \ldots,)$.

$$W(q) = \frac{1}{2}\sum_{i,j}^{3N-6} \left.\frac{\partial^2 W(q)}{\partial q_i \partial q_j}\right|_{q_0} q_i q_j + \cdots \tag{2.11}$$

For small oscillations (2.11) is represented by the quadratic form

$$W(q) = \frac{1}{2}\sum_{i,j}^{3N-6} k_{ij} q_i q_j,$$

where

$$k_{ij} \equiv \left.\frac{\partial^2 W(q)}{\partial q_i \partial q_j}\right|_{q_0}$$

are power constants of bonds. They form a symmetric square matrix $K = (k_{ij})$. If this matrix is nondegenerate, then it is always possible to perform such an orthogonal coordinate transformation that only the diagonal elements in the new coordinates Q_i will be non-zero. In such coordinates, called normal, we get (for small oscillations).

$$W(Q) = \frac{1}{2}\sum_{i=1}^{3N-6} \widetilde{k_i} Q_i^2, \tag{2.12}$$

where

$$\widetilde{k_i} = \left.\frac{\partial^2 W(Q)}{\partial Q_i^2}\right|_{Q_0}$$

are power constants (hereinafter - PC) of the ith mode of the normal oscillation form the diagonal matrix $\widetilde{K}$. The observed lines of vibrational transitions in molecules correspond precisely to mode (and not interconnected) oscillations. We emphasize that the $\widetilde{k_i}$ do not coincide with the PC of the links between the fragments (forming the complete matrix K), but are elements of the diagonal matrix of the PC oscillation modes $\widetilde{K}$. Mathematically, the values of $\widetilde{k_i}$ are the roots of the characteristic polynomial, i.e. the eigenvalues of the matrix K. Since all $\widetilde{k_i}$ are positive, the matrix K is positive-definite and, moreover, square and symmetric. Hence it can be represented as $K = A^T A$, where A is a nondegenerate matrix [22, 23]. For matrices of this form, the eigenvalues coincide with its singular numbers [22, 23]; the latter can be found by performing a singular expansion of the matrix K.

After the transition to normal coordinates, the Schrodinger equation for small oscillations looks like as

$$\left[\sum_{i=1}^{3N-6}\left[-\frac{\hbar}{2\mu_i}\frac{\partial^2}{\partial Q_i^2}+\frac{1}{2}\widetilde{k}_i Q_i^2\right]\right]\Psi^{vib}(Q)=E^{vib}\Psi^{vib}(Q), \tag{2.13}$$

where μ_i are the values called reduced masses and are inverse to the diagonal elements of the matrix $\widetilde{A}$, into which the matrix a_{ij} passed after the coordinate transformation. Note that (2.13) makes it possible to represent the wave function $\Psi^{vib}(Q)$ as the product of individual wave functions for each ith mode

$$\Psi^{vib}(Q)=\Psi^{vib}(Q_1)\cdot\Psi^{vib}(Q_2)\cdot\Psi^{vib}(Q_3)\cdot\ldots\cdot\Psi^{vib}(Q_{3N-6}),$$

which depends on one variable Q_i. Then

$$\left[-\frac{\hbar}{2\mu_i}\frac{\partial^2}{\partial Q_i^2}+\frac{1}{2}\widetilde{k}_i Q_i^2\right]\Psi^{vib}(Q_i)=E_i^{vib}\Psi^{vib}(Q_i), i=1,2,\ldots,3N-6, \tag{2.14}$$

$$E_i^{vib}=\hbar\sqrt{\frac{\widetilde{k}_i}{\mu_i}}(n_i+1/2), n_i=1,2,\ldots \tag{2.15}$$

where E_i^{vib} are the values of the energies of (small) oscillations of the ith mode.

$$E_i^{vib}=\hbar\sum_{i=1}^{3N-6}\left[\sqrt{\frac{\widetilde{k}_i}{\mu_i}}(n_i+1/2)\right], n_i=1,2,\ldots \tag{2.16}$$

The expression (2.16) is the total energy of oscillations of the quasimolecule.

We associate the PC of the $\widetilde{k}_i$ with the parameters of the quasimolecular potential. We note that the Schredinger equation (2.14) for each of the modes is similar to that for a diatomic molecule. The Morse potential is a very successful and frequently used approximation for the interaction potential of atoms of diatomic molecules, which provides a good accuracy of the potential curves of real molecules. Therefore, we assume that the interaction potential for any ith mode of oscillations is described by the Morse potential (see Fig. 2.5):

$$W(Q_i)=W_i^0\left[1-e^{-\alpha_i(Q_i-Q_0)}\right]^2,$$

where W_i^0 is the stationary potential energy of the ith mode, which is equal to the dissociation energy of the ith mode, Q_0 is the equilibrium of the coordinate. For small values of Q_i-Q_0 we expand the expression $e^{-\alpha_i(Q_i-Q_0)}$ in a Taylor series near $Q_i=Q_0$ we get

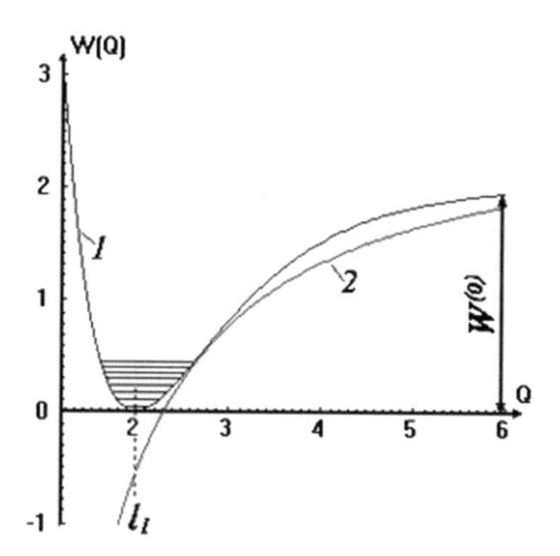

Fig. 2.5 Curves of the potential binding energy between the fragments biological complex: 1 is Morse potential, 2 is Coulomb potential

$$W(Q_i) \approx W_i^0[1 - [1 - \alpha_i(Q_i - Q_0)]]^2 = W_i^0\alpha_i^2[Q_i - Q_0]^2. \tag{2.17}$$

It follows that the expression (2.17) is a quadratic function of $(Q_i - Q_0)$, which coincides with the expression for the potential energy of a harmonic oscillator

$$W(Q_i) = \frac{1}{2}\widetilde{k}_i(Q_i - Q_0)^2.$$

Then

$$\frac{1}{2}\widetilde{k}_i(Q_i - Q_0)^2 = W_i^0\alpha_i^2(Q_i - Q_0)^2. \tag{2.18}$$

$$\widetilde{k}_i = 2W_i^0\alpha_i^2, \tag{2.19}$$

where $\widetilde{k}_i$ are PC of the ith mode.

Substitute (2.19) into the expression (2.16):

$$E_i^{vib} = \hbar\sum_{i=1}^{3N-6}\left[\sqrt{\alpha_i\frac{2W_i^0}{\mu_i}}(n_i + 1/2)\right], n_i = 1, 2, \ldots \tag{2.20}$$

To estimate the coefficients α_i, we represent the relationship between the constituent elements of an effective (in the sense of (2.14)) linear harmonic oscillator of an individual mode in the form of an attractive force F working against the repulsive force. The latter can be represented as a spring that is compressed by the force F and has a force constant of k (see Fig. 2.6). Due to its electrostatic nature, the force F will be determined as

$$F = \frac{|q_1q_2|}{\kappa l_1^2}$$

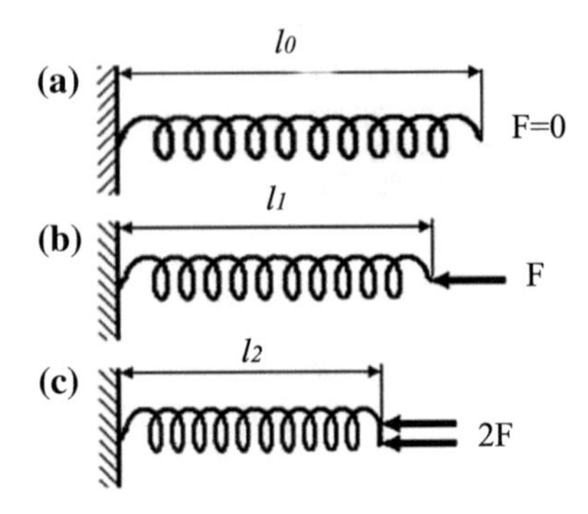

Fig. 2.6 Model of the forces and potential energies near the minimum of the potential curve

in absolute value, where $\kappa = 4\pi\varepsilon\varepsilon_0$, the l_0 is length of the uncompressed spring without affecting the compressive force (see Fig. 2.6).

In the equilibrium position, the values of the attractive and repulsive forces must be equal, i.e. $F = k(l_0 - l_1)$, and $2F = k(l_0 - l_2)$. Let $\Delta l = l_2 - l_1$, then we get

$$F = k\Delta l, \tag{2.21}$$

and also that

$$\Delta l = l_0 - l_1, \tag{2.22}$$

In equilibrium (see Fig. 2.6b) under the action of the force F, the expression (2.21) taking into account (2.22) will be rewritten as

$$\frac{|q_1 q_2|}{\kappa l_1^2} = k(l_0 - l_1) = k\Delta l, \tag{2.23}$$

taking into account (2.22) we get

$$W^0 = \frac{|q_1 q_2|}{\kappa l_1} - \frac{k\Delta l^2}{2},$$

Then

$$\frac{|q_1 q_2|}{\kappa l_1^2} = \frac{W^0}{l_1} + \frac{k\Delta l^2}{2l_1}.$$

Taking into account the expression (2.23), we get

$$k\Delta l = \frac{W^0}{l_1} + \frac{k\Delta l^2}{2l_1} \Rightarrow W^0 = k\Delta l\left[l_1 - \frac{\Delta l}{2}\right].$$

Taking into account the expression (2.19), we get

$$\alpha^2 = \frac{1}{\Delta l(2l_1 - \Delta l)}. \tag{2.24}$$

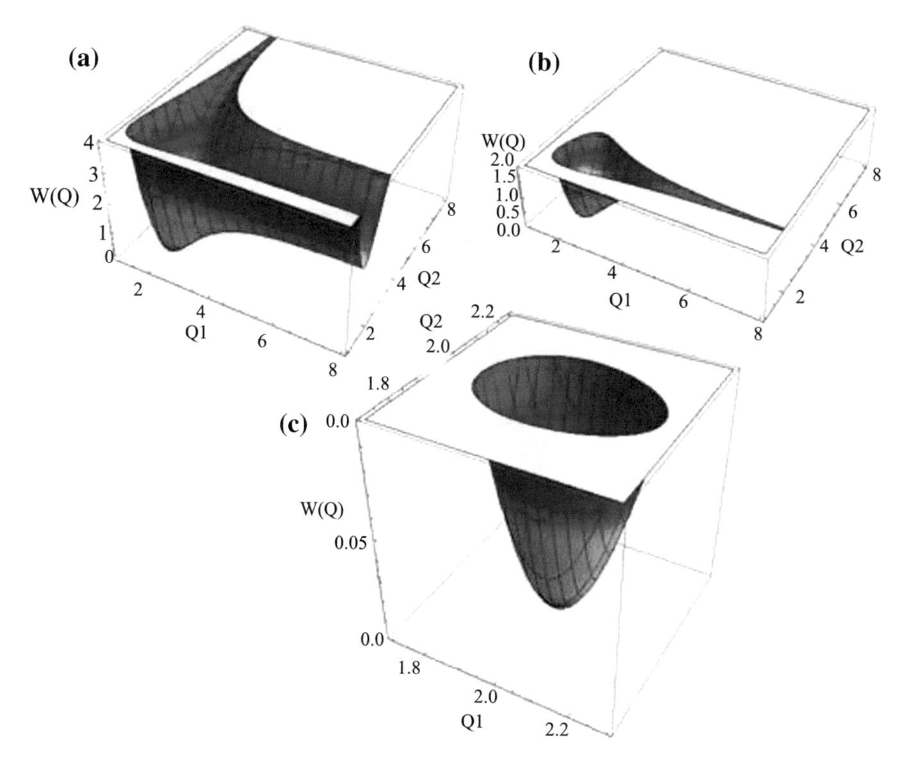

Fig. 2.7 Surface of the potential energy of the oscillator modes described by the two-dimensional Morse potential at ($W^0_{max}/W^0_{min} \approx 2$) and clipped at the top: **a**—at the level of greater dissociation energy, **b**—at the level of lower dissociation energy, with—at the level of applicability of the linear harmonic oscillator approximation

As is known from the literature, the bond lengths (both single l_1 and double l_2) between fragments in different biomolecules depend only slightly on the composition of the fragments [24]. In the first approximation, they can be made constant. Then, in accordance with (2.24), one can assume the practical constancy of the coefficients α.

For the thermodynamic equilibrium state in question, the most probable value of the quasimolecule vibration energy is kT (k is Boltzmann constant). Figure 2.7 shows the potential energy surface (PES) $W(Q)$ for the two-dimensional Morse potential. Consider the sections of the PES with the plane $w_t = kT$. Note that a real multidimensional case should be understood as ≪hyperplane≫.

If W_T exceeds or equals W_i^0 (the stationary potential energy of the ith mode) corresponding to the lowest horizontal asymptote of the potential energy surface, then the mode bond is severed, and the conformation of the quasimolecule breaks (see the Fig. 2.7b).

If $W_T \ll W_i^0$, then the PPE section of plane $W_T = kT$ will be an ellipse (hyper-ellipsoid) in the coordinates Q (see Fig. 2.7c). This hyperellipsoid will be the most extended along the direction Q_i corresponding to the ith oscillation mode, which has the lowest binding energy W_i^0.

The equation of this hyperellipsoid in accordance with (2.17) appears in an implicit form as

$$\sum_{i=1}^{3N-6} W_i^0 \alpha_i^2 (Q_i - Q_0)^2 - W_T = 0. \tag{2.25}$$

Thus, the strong elongation of the hyperellipsoid in some direction Q_i very likely means a greater tendency of the bio-complex to break the conformation in the sense of breaking the connection for the ith mode of oscillations (the weakest coupling is torn, and the weakest mode is torn). Therefore, the ratio of the maximum and minimum lengths of the axes of this hyperellipsoid

$$\left.\frac{\max(Q_i - Q_0)}{\min(Q_i - Q_0)}\right|_{W(Q_i - Q_0) = W_T}$$

equals

$$\sqrt{\frac{\max(\widetilde{k_i})}{\min(\widetilde{k_i})}} = \sqrt{\frac{\max(W_i^0 \alpha_i^2)}{\min(W_i^0 \alpha_i^2)}}$$

can serve as a qualitative indicator of the stability of the bio-complex (the greater this value, the worse the stability). Similarly, the stability indicator can be the ratio

$$\frac{\max(\widetilde{k_i})}{\min(\widetilde{k_i})}.$$

For the matrix $K = (k_{ij})$, which has the above properties, the condition number cond(K) coincides with the ratio of its maximal and minimal eigenvalues, i.e. the ratio of the maximal and minimal elements of the matrix $\widetilde{K}$:

$$\mathrm{cond(K)} = \frac{\max(\widetilde{k_i})}{\min(\widetilde{k_i})}$$

The PC of the ith mode is proportional to W_i^0 (the stationary potential binding energy of the mode). Assuming that the coefficients α_i, as shown above, depend weakly on i, we get

$$\frac{\max(\widetilde{k_i})}{\min(\widetilde{k_i})} = \frac{\max(W_i^0)}{\min(W_i^0)}$$

The last relation coincides with $\mathrm{cond(W_0)}$, where $(\mathrm{W_0}) = (\mathrm{W_{ij}})$ is the matrix of stationary potential binding energies between the fragments of the bio-complex (it is also symmetric and positive-definite).

So, the criterion for the stability of a biocomplex can be the value $\mathrm{cond}(W_0)$, where $\mathrm{W_0} = (\mathrm{W_{ij}})$– is the matrix of stationary potential binding energies between fragments of the biocomplex.

2.6 Numerical Simulation of Interaction of Biological Systems. Conclusion

We simulated interactions of various sites of amino acid sequences that comprise secondary structures of different protein domains and analyzed the possibility of domain formation by different sites of the amino acid sequences as follows:

$$(1)\mathrm{H2A} + \mathrm{H2B} \rightarrow (\mathrm{H2A} - \mathrm{H2B}),$$

$$(2)\mathrm{H3} + \mathrm{H4} \rightarrow (\mathrm{H3} - \mathrm{H4}).$$

2.6.1 *Heterodimer Formation* H2A–H2B

We stimulated an interaction of histone proteins as they binded into the (H2A–H2B) dimer.

We used different sites of the proteins H2A and H2B and analyzed their ability to form stable biological complexes. Note that the (H2A–H2B) dimer is formed by ≪head-to-tail≫ joining of two histones into a ≪handshake≫ motif Fig. (2.8a) [5, 25].

Thus, interaction between different domains of the H2A and H2B proteins was calculated by analyzing the electrostatic interaction potential energy with allowance for the fact that, in the dimer, they are bound in the ≪head-to-tail≫ orientation.

To solve this problem, various segments of the amino acid sequences were taken and the analysis of conditioning numbers for the matrix of electrostatic interaction was performed. Each in the matrix is the matrix of potential energy of electrostatic interaction between two corresponding proteins: $\mathrm{H2A}_{(19-56)}$–$\mathrm{H2B}_{(91-124)}$, $\mathrm{H2A}_{(89-114)}$–$\mathrm{H2B}_{(31-57)}$, $\mathrm{H2A}_{(19-56)}$–$\mathrm{H2B}_{(31-57)}$ and $\mathrm{H2A}_{(89-114)}$–$\mathrm{H2B}_{(91-124)}$.

We considered central parts of the chains.

Tables 2.1 and 2.2 presents the calculation results for different domains of H2A and H2B proteins that allow for two analyzed binding patterns of the proteins to the dimer: ≪head-to-tail≫ and ≪head-to-head≫.

Figure 2.9a, b show the schemes of histone proteins H2A and H2B, sites areas of the amino acid chain, between which the interaction for two cases of formation

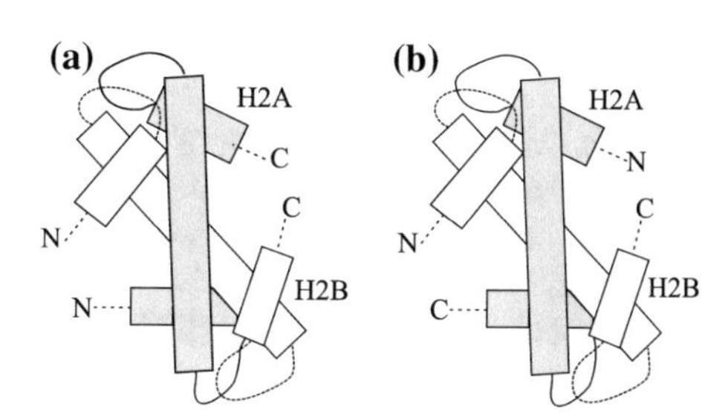

Fig. 2.8 Schematic representation of (H2A–H2B) dimer formation in the case of **a** the ≪tail-to-head≫ and **b** ≪head-to-head≫ structures. The polypeptide chain numbering is from the N-terminus to the C-terminus

Table 2.1 Common logarithm of condition numbers for the electrostatic interaction energy matrices for proteins (H2A–H2B) in the ≪head-to-tail≫ configuration

№	Name of protein	Amino acid sequence	lg(cond(W))
1	$H2A_{(19-56)}$	RSAKAGLTFPVGRVHRLLRRGNYAQRIGSGAPVYLTAV	
2	$H2B_{(91-124)}$	STISAREIQTAVRLILPGELAKHAVSEGTRAVTK	18.416
3	$H2A_{(89-114)}$	IRNDDELNKLLGNVTIAQGGVLPNIH	
4	$H2B_{(31-57)}$	KKRSKARKETYSSYIYKVLKQTHPDTGI	18.611

lg(cond(W)) is common logarithm of condition number.

Table 2.2 Common logarithm of condition numbers for the electrostatic interaction energy matrices for proteins (H2A–H2B) in the ≪head-to-head≫ configuration

№	Name of protein	Amino acid sequence	lg(cond(W))
1	$H2A_{(19-56)}$	RSAKAGLTFPVGRVHRLLRRGNYAQRIGSGAPVYLTAV	
2	$H2B_{(31-57)}$	KKRSKARKETYSSYIYKVLKQTHPDTGI	31.685
3	$H2A_{(89-114)}$	IRNDDELNKLLGNVTIAQGGVLPNIH	
4	$H2B_{(91-124)}$	STISAREIQTAVRLILPGELAKHAVSEGTRAVTK	17.880

lg(cond(W)) is common logarithm of condition number.

of heterodimers: ≪head-to-tail≫ (a) and ≪head-to-head≫ (b). The numbers in the rectangular boxes indicate the common logarithm of the conditioning number for the selected pair of amino acid sequences. The values of the lg(cond(W)) for model of histone binding in the dimer ≪head-to-tail≫ for amino acid sections $H2A_{(19-56)}$–$H2B_{(91-124)}$ and $H2A_{(89-114)}$–$H2B_{(31-57)}$ are 18.416 and 18.611, respectively. The values of the conditioning numbers for the second model of formation of the heterodimer ≪head-to-head≫ with the participation of sections $H2A_{(19-56)}$–$H2B_{(31-57)}$ and $H2A_{(89-114)}$–$H2B_{(91-124)}$ are 31.685 and 17.880 respectively.

A significant increase in the value of lg(cond(W)) for the site $H2A_{(18-56)}$–$H2B_{(31-57)}$ indicates a very small degree of stability of the configuration of the biological complex, which is formed by two histone proteins $H2A_{(18-56)}$–$H2B_{(31-57)}$.

Thus, the model of the formation of the histone dimer ≪head -to-head≫ with the participation of amino acid sequences $H2A_{(18-56)}$–$H2B_{(31-57)}$ and $H2A_{(89-114)}$–$H2B_{(91-124)}$ has a smaller number of stable configurations of interacting amino acid residues in comparison with the ≪head-to-tail≫ model. One stable biological segment in the ≪head-to-head≫ model can not provide the formation of a ≪head-to-head≫ complex.

Results of the performed calculations for amino acid sequences of the histones $H2A_{(18-56)}$, $H2B_{(91-124)}$, $H2A_{(89-114)}$, $H2B_{(31-57)}$ indicate, that histones are more inclined to form heterodimers in the direction of ≪head-to-tail≫, than ≪head-to-head≫.

Thus, with the participation of selected amino acid sequences of different sections of histone proteins H2A and H2B, we can conclude that the formation of a heterodimer in the direction of ≪head-to-tail≫ is more preferable in comparison with the direction of ≪head-to-head≫, since the formation of a heterodimer according to the first model (≪head-to-tail≫) is performed by two interacting proteins regions $H2A_{(19-56)}$–$H2B_{(91-124)}$, $H2A_{(89-114)}$–$H2B_{(31-57)}$.

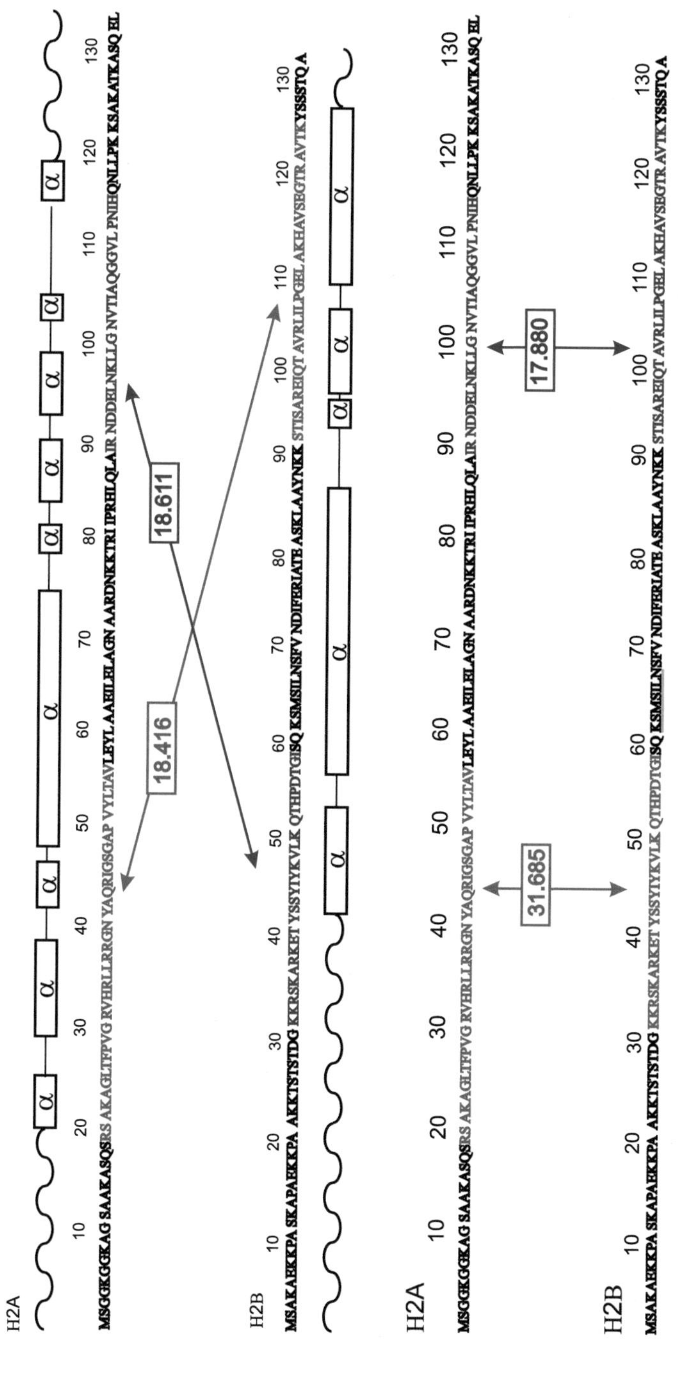

Fig. 2.9 Schemes of histone proteins H2A and H2B

Table 2.3 Common logarithm of condition numbers for the electrostatic interaction energy matrices for proteins (H3–H4) in the ≪head-to-tail≫ configuration

№	Name of protein	Amino acid sequence	lg(cond(W))
1	$H3_{(84-127)}$	RFQSSAIGALQESVEAYLVSLFEDTNLAAIHAKRVTIQKKDIKL	
2	$H4_{(25-50)}$	DNIQGITKPAIRRLARRGGVKRISGL	17.500
3	$H3_{(50-75)}$	REIRRFQKSTELLIRKLPFQRLVREI	
4	$H4_{(76-103)}$	HAKRKTVTSLDVVYALKRQGRTLYGFGG	18.435

lg(cond(W)) is common logarithm of condition number.

In case ≪head-to-head≫, the stability of the heterodimer is realized by the interaction of only one region of histone proteins: $H2A_{(89-114)}$–$H2B_{(91-124)}$.

2.6.2 *Heterodimer Formation* H3–H4

Interaction of histone proteins upon binding into the (H3–H4) dimer allowing for their secondary structure was simulated.

Note that the (H3–H4) dimer is formed by ≪head-to-tail≫ joining of two histones into a ≪ *handshake* ≫ motif as well as the formation of the H2A–H2B heterodimer (see Fig. 2.8a) [5, 25].

Thus, interaction between different domains of the H3 and H4B proteins was calculated by analyzing the electrostatic interaction potential energy with allowance for the fact that, in the dimer, they are bound in the ≪head-to-tail≫ orientation. The possibility of the formation of a ≪head-to-head≫ dimer by the considered histones was also analyzed.

To solve this problem, various segments of the amino acid sequences histones H3, H4 were taken and the analysis of conditioning numbers for the matrix of electrostatic interaction was performed: $H3_{(50-75)}$, $H4_{(76-103)}$, $H3_{(84-127)}$ and $H4_{(25-50)}$.

Figure 2.10a shows the secondary structures of histone proteins H3 and H4 and scheme of interaction of different segments of amino acid sequences for model the dimer formation: ≪head-to-head≫.

Figure 2.10b shows the scheme of interaction between different sections of amino acid sequences for histone proteins H3 and H4, which models the dimer formation: ≪head-to-tail≫.

The numbers in the rectangular boxes indicate the common logarithm of the conditioning number (lg(cond(W))) for the selected pair of amino acid sequences.

Tables 2.3 and 2.4 present the calculation results for different domains of H3 and H4 proteins that allow for two analyzed binding patterns of the proteins to the dimer: ≪head-to-tail≫ and ≪head-to-head≫.

The results of the numerical simulation performed for the model ≪head-to-tail≫ when forming a histone heterodimer, H3–H4 demonstrates the set of values lg(cond(W)) smaller in value than the set of values for the formation model of the heterodimer ≪head-to-head≫. The values for the first model ≪head-to-tail≫ are 18.435 and 17.500, and for the second model, we got the values: 19.394 and 17.845.

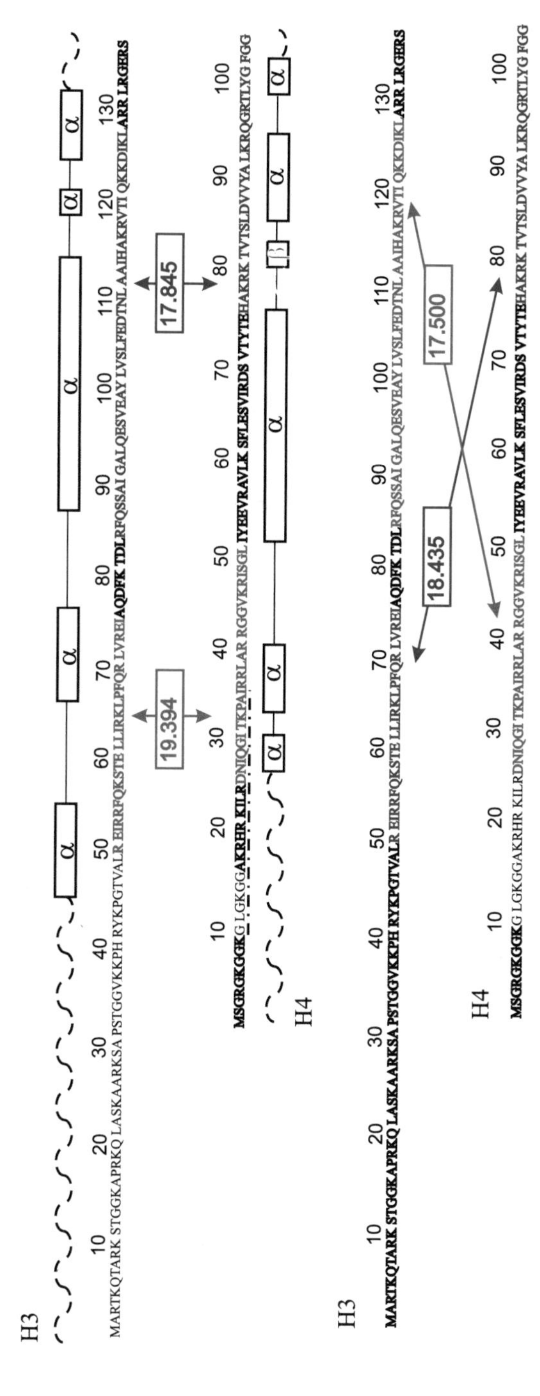

Fig. 2.10 Schemes of histone proteins H3 and H4

Table 2.4 Condition numbers of the for electrostatic interaction energy matrices for proteins (H3–H4) in the ≪head-to-head≫ configuration

№	Name of protein	Amino acid sequence	lg(cond(W))
1	$H3_{(50-75)}$	REIRRFQKSTELLIRKLPFQRLVREI	
2	$H4_{(25-50)}$	DNIQGITKPAIRRLARRGGVKRISGL	19.394
3	$H3_{(84-127)}$	RFQSSAIGALQESVEAYLVSLFEDTNLAAIHAKRVTIQKKDIKL	
4	$H4_{(76-103)}$	HAKRKTVTSLDVVYALKRQGRTLYGFGG	17.845

lg(cond(W)) is common logarithm of condition number.

Thus, the numerical results model of histone heterodimer formation the ≪head-to-tail≫ with the participation between chosen amino acid sequences demonstrates a more stable interaction between amino acid sites of histones H3 and H4, than the model of formation histone heterodimer by the ≪head-to-head≫. These results are in good agreement with earlier experiments [5].

The results of numerical modeling made it possible to establish the most stable interactions between different domains.

Thus, the present study allows one to draw conclusions based on the fact that common logarithm condition number lg(cond(W)) which contains the interaction of different amino acid chains reveals which biological objects form the most stable compounds.

The model showed a remarkable sensitivity to the amino acid composition of the studied proteins.

It allows one to theoretically predict amino acid sequences with the given physical properties, facilitate experimental studies, and reduce their price by decreasing the number of measurements.

It is possible that some other sites of binding exist in the studied biological complexes; a separate simulation should be done if it is necessary to carry out an experiment with a given amino acid sequence.

2.7 MATLAB Script for Mathematical Simulation of Complex Formation of Protein Molecules Allowing for Their Domain Structure

Input parameters:

1. S_1, S_2 are amino acid sequences of biological complexes ($S_1 \geq S_2$). 2. epsilon is the dielectric constant of the medium.

Output parameters:

lg(cond(W)) is the common logarithm of the condition number of the matrix W, where its elements are composed of the electrostatic potential energy which is created based on the interaction between pair of amino acid residues of biological complexes.

Calculation:

lg(cond(W) is the common logarithm of the condition number of the matrix W, which will allow a prediction the reactivity of the studied biological complexes.

```
1  clear all
2  clc
3  format long e
4  epsilon=80.103;
5  %H2A 86-115
6  S_20=['I'  'R' 'N'  'D'  'D'  'E'  'L'  'N'  'K'  'L' ...
7  'L'  'G' 'N'  'V'  'T'  'I'  'A'  'Q'  'G'  'G'  'V' ...
8  'L' 'P'  'N'  'I'  'H' ];
9  %H2B  30-60
10 S_1=[ 'K'  'K'  'R'  'S'  'K'  'A'  'R'  'K'  'E'  'T' ...
11 'Y'  'S'    'S'  'Y'  'I'  'Y'  'K'  'V'  'L'  'K'  'Q' ...
12 'T'  'H'  'P'  'D'  'T'  'G'  'I'];
13 len_S1=length(S_1);
14 len_S20=length(S_20);
15 N1=100*len_S20;
16 [S_1,S_20,Q1,Q2,R1,R2,h,M,N]=potential_giston(S_1,S_20);
17 [A]=electrostatic(Q1,Q2, R1,R2,h,M,N,N1,epsilon);
18 [R1]=condmy(A)
19 %----------------------------------------------------------------
20 %H2A 18-72
21 S_1=['R'  'S' 'A'  'K'  'A'  'G'  'L'  'T'  'F'  'P'  'V' ...
22 'G' 'R'  'V'  'H'  'R'  'L'  'L'  'R'  'R'  'G'  'N' 'Y' ...
23 'A'  'Q'  'R'  'I'  'G'  'S'  'G'  'A'  'P' 'V'  'Y'  'L' ...
24    'T'  'A'  'V'  ];
25 % H2B 91- 124
26 S_20=['S'  'T'  'I'  'S'  'A'  'R'  'E'  'I'  'Q'  'T'  'A'...
27 'V'  'R'  'L'  'I'  'L'  'P'  'G'  'E'  'L' 'A'  'K'  'H'...
28 'A'  'V'  'S'  'E'  'G'  'T'  'R'  'A'  'V'  'T'  'K'];
29 len_S1=length(S_1);
30 len_S20=length(S_20);
31 [S_1,S_20,Q1,Q2,R1,R2,h,M,N]=potential_giston(S_1,S_20);
32 [A]=electrostatic(Q1,Q2, R1,R2,h,M,N,N1,epsilon);
33 [R2]=condmy(A)
34 %----------------------------------------------------------------
35 %H2A 18-72
36 S_1=[   'R'  'S' 'A'  'K'  'A'  'G'  'L'  'T'  'F'  'P'  'V'..
37 'G' 'R'  'V'  'H'  'R'  'L'  'L'  'R'  'R'  'G'  'N' 'Y' ...
38 'A'  'Q'  'R'  'I'  'G'  'S'  'G'  'A'  'P' 'V'  'Y'  'L'...
39 'T'  'A'  'V' ];
40 %H2B  30-60
41 S_20=[ 'K'  'K'  'R'  'S'  'K'  'A'  'R'  'K'  'E'  'T'  'Y'..
42 'S'  'S'  'Y'  'I'  'Y'  'K'  'V'  'L'  'K'  'Q'  'T'  'H'  ...
43 'P'  'D'  'T'  'G'  'I'  ];
44 len_S1=length(S_1);
45 len_S20=length(S_20);
46 [S_1,S_20,Q1,Q2,R1,R2,h,M,N]=potential_giston(S_1,S_20);
47 [A]=electrostatic(Q1,Q2, R1,R2,h,M,N,N1,epsilon);
48 [R3]=condmy(A)
49 %----------------------------------------------------------------
50 %H2A 86-115
51 S_20=[ 'I'  'R' 'N'  'D'  'D'  'E'  'L'  'N'  'K'  'L'  'L'...
52 'G' 'N'  'V'  'T'  'I'  'A'  'Q'  'G'  'G'  'V'  'L' 'P'  ...
53 'N'  'I'  'H' ];
```

```
54  % H2B 91- 124
55  S_1=['S'  'T'  'I' 'S'  'A'  'R'  'E'  'I'  'Q'  'T'  'A'  'V'...
56  'R'  'L'  'I'  'L'  'P'  'G'  'E'  'L' 'A'  'K'  'H'  'A' ...
57  'V'  'S'  'E'  'G'  'T'  'R'  'A'  'V'  'T'  'K'];
58  len_S1=length(S_1);
59  len_S20=length(S_20);
60  [S_1,S_20,Q1,Q2,R1,R2,h,M,N]=potential_giston(S_1,S_20);
61  [A]=electrostatic(Q1,Q2, R1,R2,h,M,N,N1,epsilon);
62  [R4]=condmy(A)
63  %------------------------------------------------------------
64  %H3 52-74
65  S_20=[ 'R' 'E'  'I'  'R'  'R'  'F'  'Q'  'K'  'S'  'T'  'E'...
66  'L'  'L'  'I'  'R'  'K'  'L'  'P'  'F'  'Q'  'R' 'L'  'V' ...
67  'R'  'E'  'I'];
68  %H4  75-103
69  S_1=[ 'H'  'A'  'K'  'R'  'K' 'T'  'V'  'T'  'S'  'L'  'D'...
70  'V'  'V'  'Y'  'A' 'L'  'K'  'R'  'Q'  'G'  'R'  'T'  'L' ...
71  'Y'  'G' 'F'  'G'  'G'];
72  len_S1=length(S_1);
73  len_S20=length(S_20);
74  [S_1,S_20,Q1,Q2,R1,R2,h,M,N]=potential_giston(S_1,S_20);
75  [A]=electrostatic(Q1,Q2, R1,R2,h,M,N,N1,epsilon);
76  [R5]=condmy(A)
77  %----------------------------------------------------------------
78  %H3 84-126
79  S_1=['R'  'F'  'Q'  'S'  'S'  'A'  'I' 'G'  'A'  'L'  'Q'  'E' ...
80  'S'  'V'  'E'  'A'  'Y' 'L'  'V'  'S'  'L'  'F'  'E'  'D'  'T' ...
81  'N'  'L' 'A'  'A'  'I'  'H'  'A'  'K'  'R'  'V'  'T'  'I' 'Q' ...
82  'K'  'K'  'D'  'I'  'K'  'L'  ];
83  %H4 25 -50
84  S_20=[  'D'  'N'  'I'  'Q'  'G'  'I' 'T'  'K'  'P'  'A'  'I' ...
85  'R'  'R'  'L'  'A'  'R' 'R'  'G'  'G'  'V'  'K'  'R'  'I' ...
86  'S'  'G'  'L'];
87  len_S20=length(S_20);
88  [S_1,S_20,Q1,Q2,R1,R2,h,M,N]=potential_giston(S_1,S_20);
89  [A]=electrostatic(Q1,Q2, R1,R2,h,M,N,N1,epsilon);
90  [R6]=condmy(A)
91  %----------------------------------------------------------------
92  %H3 52-74
93  S_1=  [ 'R' 'E'  'I'  'R'  'R'  'F'  'Q'  'K'  'S'  'T'  'E' ...
94  'L'  'L'  'I'  'R'  'K'  'L'  'P'  'F'  'Q'  'R' 'L'  'V'  ...
95  'R'  'E'  'I'];
96  %H4 25 -50
97  S_20=[  'D'  'N'  'I'  'Q'  'G'  'I' 'T'  'K'  'P'  'A'  'I' ...
98  'R'  'R'  'L'  'A'  'R' 'R'  'G'  'G'  'V'  'K'  'R'  'I'  ...
99  'S'  'G'  'L'];
100 len_S1=length(S_1);
101 len_S20=length(S_20);
102 [S_1,S_20,Q1,Q2,R1,R2,h,M,N]=potential_giston(S_1,S_20);
103 [A]=electrostatic(Q1,Q2, R1,R2,h,M,N,N1,epsilon);
104 [R7]=condmy(A)
105 %----------------------------------------------------------------
106 %H3 84-126
107 S_1=['R'  'F'  'Q'  'S'  'S'  'A'  'I' 'G'  'A'  'L'  'Q' ...
```

```
108 'E'  'S'  'V'  'E'  'A'  'Y' 'L'  'V'  'S'  'L'  'F'  'E' ...
109 'D'  'T'  'N'  'L' 'A'  'A'  'I'  'H'  'A'  'K'  'R'  'V' ...
110 'T'  'I' 'Q'  'K'  'K'  'D'  'I'  'K'  'L'  ];
111 %H4  75-103
112 S_20=[ 'H'  'A'  'K'  'R'  'K' 'T'  'V'  'T'  'S'  'L'  'D' ...
113 'V'  'V'  'Y'  'A' 'L'  'K'  'R'  'Q'  'G'  'R'  'T'  'L' ...
114 'Y'  'G' 'F'  'G'  'G'];
115 len_S1=length(S_1);
116 len_S20=length(S_20);
117 [S_1,S_20,Q1,Q2,R1,R2,h,M,N]=potential_giston(S_1,S_20);
118 [A]=electrostatic(Q1,Q2, R1,R2,h,M,N,N1,epsilon);
119 [R8]=condmy(A)
120 %-------------------------------------------------------------
121 function[cond2]=condmy(A)
122 [U,S,V]=SVD_2(A);
123 lambda_max=max(diag(S));
124 lambda_min=min(diag(S));
125 cond_1=(((lambda_max)/(lambda_min)));
126 cond2=(log(cond_1))/(log(10));
127 return
128
129 function [Uout,Sout,Vout] = SVD_2(A)
130           m = size(A,1);
131       n = size(A,2);
132       U = eye(m);
133       V = eye(n);
134       e = eps*fro(A);
135       while (sum(abs(A(~eye(m,n)))) > e)
136           for i = 1:n
137               for j = i+1:n
138                   [J1,J2] = jacobi(A,m,n,i,j);
139                   A = mtimes(J1,mtimes(A,J2));
140                   U = mtimes(U,J1');
141                   V = mtimes(J2',V);
142               end
143               for j = n+1:m
144                   J1 = jacobi2(A,m,n,i,j);
145                   A = mtimes(J1,A);
146                   U = mtimes(U,J1');
147               end
148           end
149       end
150       S = A;
151
152       if (nargout < 3)
153           Uout = diag(S);
154       else
155           Uout = U; Sout = times(S,eye(m,n)); Vout = V;
156       end
157     end
158
159     function [J1,J2] = jacobi(A,m,n,i,j)
```

```
160        B = [A(i,i), A(i,j); A(j,i), A(j,j)];
161        [U,S,V] = tinySVD(B); %
162
163        J1 = eye(m);
164        J1(i,i) = U(1,1);
165        J1(j,j) = U(2,2);
166        J1(i,j) = U(2,1);
167        J1(j,i) = U(1,2);
168
169        J2 = eye(n);
170        J2(i,i) = V(1,1);
171        J2(j,j) = V(2,2);
172        J2(i,j) = V(2,1);
173        J2(j,i) = V(1,2);
174     end
175
176     function J1 = jacobi2(A,m,n,i,j)
177        B = [A(i,i), 0; A(j,i), 0];
178        [U,S,V] = tinySVD(B);
179
180        J1 = eye(m);
181        J1(i,i) = U(1,1);
182        J1(j,j) = U(2,2);
183        J1(i,j) = U(2,1);
184        J1(j,i) = U(1,2);
185     end
186
187     function [Uout,Sout,Vout] = tinySVD(A)
188       t=rdivide((minus(A(1,2),A(2,1))),(plus(A(1,1),A(2,2))));
189       c=rdivide(1,sqrt(1+t^2));
190       s = times(t,c);
191       R = [c,-s;s,c];
192       M = mtimes(R,A);
193       [U,S,V] = tinySymmetricSVD(M);
194       U = mtimes(R',U);
195
196       if (nargout < 3)
197           Uout = diag(S);
198       else
199           Uout = U; Sout = S; Vout = V;
200       end
201     end
202
203 function [Uout,Sout,Vout]=tinySymmetricSVD(A)
204       if (A(2,1) == 0)
205           S = A;
206           U = eye(2);
207           V = U;
208       else
209
210           w = A(1,1);
211           y = A(2,1);
```

```
212            z = A(2,2);
213            ro = rdivide(minus(z,w),times(2,y));
214     t2=rdivide(sign(ro),plus(abs(ro),sqrt(plus(times(ro,ro),1))));
215            t = t2;
216            c = rdivide(1,sqrt(plus(1,times(t,t))));
217            s = times(t,c);
218            U = [c, -s; s, c];
219            V = [c,  s;-s, c];
220            S = mtimes(U,mtimes(A,V));
221            U = U';
222            V = V';
223         end
224 
225         [U,S,V] = fixSVD(U,S,V);
226               if (nargout < 3)
227            Uout = diag(S);
228         else
229           Uout = U; Sout = S; Vout = V;
230         end
231       end
232 
233       function [U,S,V] = fixSVD(U,S,V)
234         Z = [sign(S(1,1)),0; 0,sign(S(2,2))];
235         U = mtimes(U,Z);
236         S = mtimes(Z,S);
237         if (S(1,1) < S(2,2))
238            P = [0,1;1,0];
239            U = mtimes(U,P);
240            S = mtimes(P,mtimes(S,P));
241            V = mtimes(P,V);
242         end
243       end
244 
245       function f = fro(M)
246         f = sqrt(sum(sum(times(M,M))));
247       end
248 
249       function s = sign(x)
250          if (x > 0)
251             s = 1;
252          else
253             s = -1;
254          end
255       end
256 
257 function[S_1,S_2,Q1,Q2,R1,R2,h,M,N]=potential_giston(S_1,S_20);
258 N=length(S_1);
259 M=length(S_20);
260 S_2=S_20;
261 Q1=[];
262 Q2=[];
263 R1=[];
```

```
264 R2=[];
265 for i=1:length(S_1);
266 for j=1:length(S_2);
267 if (S_1(i)=='D' & S_2(j)=='E')| (S_1(i)=='E' & S_2(j)=='D');
268 Q1(i,j)= 0.16e-19;
269 Q2(i,j)= 0.16e-19;
270 else
271 if (S_1(i)=='D' & S_2(j)=='D');
272 Q1(i,j)= 0.07e-19;
273 Q2(i,j)= 0.07e-19;
274 else
275 if (S_1(i)=='D' & S_2(j) =='C')|(S_1(i)=='C' & S_2(j) =='D');
276 Q1(i,j)= 0.05e-19;
277 Q2(i,j)= 0.05e-19;
278 else
279 if (S_1(i)=='D' &S_2(j)=='N')|(S_1(i)=='N' &S_2(j)=='D')|...
280 (S_1(i)=='D' &S_2(j)=='F')|...
281 (S_1(i)=='D' &S_2(j)=='Y')|(S_1(i)=='D' &S_2(j)=='Q')|...
282 (S_1(i)=='D' &S_2(j)=='S')|...
283 (S_1(i)=='F' &S_2(j)=='D')|(S_1(i)=='Y' &S_2(j)=='D')|...
284 (S_1(i)=='Q' &S_2(j)=='D')|...
285 (S_1(i)=='S' &S_2(j)=='D');
286 Q1(i,j)= 0.57e-19;
287 Q2(i,j)= 0.57e-19;
288 else
289 if ((S_1(i)=='D' & S_2(j)=='M')|(S_1(i)=='D' & S_2(j)=='T')|...
290 (S_1(i)=='D' & S_2(j)=='I')|(S_1(i)=='D' & S_2(j)=='G')|...
291 (S_1(i)=='D' & S_2(j)=='V')|...
292 (S_1(i)=='D' & S_2(j)=='W')|(S_1(i)=='D' & S_2(j)=='L')|...
293 (S_1(i)=='D' & S_2(j)=='A')|...
294 (S_1(i)=='M' & S_2(j)=='D')|(S_1(i)=='T' & S_2(j)=='D')|...
295 (S_1(i)=='I' & S_2(j)=='D')|...
296 (S_1(i)=='G' & S_2(j)=='D')|(S_1(i)=='V' & S_2(j)=='D')|...
297 (S_1(i)=='W' & S_2(j)=='D')|...
298 (S_1(i)=='L' & S_2(j)=='D')|(S_1(i)=='A' & S_2(j)=='D'));
299 Q1(i,j)= 0.64e-19;
300 Q2(i,j)= 0.64e-19;
301 else
302 if ((S_1(i)=='D'& S_2(j)=='P')|(S_1(i)=='P'& S_2(j)=='D'));
303 Q1(i,j)= 0.78e-19;
304 Q2(i,j)= 0.78e-19;
305 else
306 if ((S_1(i)=='D' & S_2(j)=='H')|(S_1(i)=='H'& S_2(j)=='D'));
307 Q1(i,j)= 0.99e-19;
308 Q2(i,j)= 0.99e-19;
309 else
310 if ((S_1(i)=='D'& S_2(j)=='K')|(S_1(i)=='K'& S_2(j)=='D'));
311 Q1(i,j)= 1.4e-19;
312 Q2(i,j)= 1.4e-19;
313 else
314 if ((S_1(i)=='D' & S_2(j)=='R')|(S_1(i)=='R'& S_2(j)=='D'));
315 Q1(i,j)= 1.59e-19;
```

```
316 Q2(i,j)= 1.59e-19;
317 else
318 if ((S_1(i)=='E'&S_2(j)=='E'));
319 Q1(i,j)= 0.16e-19;
320 Q2(i,j)= 0.16e-19;
321 else
322 if ((S_1(i)=='E'  & S_2(j)=='C')|(S_1(i)=='E' & S_2(j)=='F')|...
323 (S_1(i)=='E' & S_2(j)=='N')|...
324 (S_1(i)=='C'  & S_2(j)=='E')|(S_1(i)=='F' & S_2(j)=='E')|...
325 (S_1(i)=='N' & S_2(j)=='E'));
326 Q1(i,j)= 0.55e-19;
327 Q2(i,j)= 0.55e-19;
328 else
329 if  ((S_1(i)=='E' & S_2(j)=='Q')|(S_1(i)=='E' & S_2(j)=='Y')|...
330 (S_1(i)=='E' & S_2(j)=='S')|...
331 (S_1(i)=='E' & S_2(j)=='M')|(S_1(i)=='E' & S_2(j)=='T')|...
332 (S_1(i)=='E' & S_2(j)=='I')|...
333 (S_1(i)=='E' & S_2(j)=='G')|(S_1(i)=='E' & S_2(j)=='V')|...
334 (S_1(i)=='E' & S_2(j)=='W')|...
335 (S_1(i)=='E' & S_2(j)=='L')|(S_1(i)=='E' & S_2(j)=='A')|...
336 (S_1(i)=='Q' & S_2(j)=='E')|...
337 (S_1(i)=='Y' & S_2(j)=='E')| (S_1(i)=='S' & S_2(j)=='E')|...
338 (S_1(i)=='M' & S_2(j)=='E')|...
339 (S_1(i)=='T' & S_2(j)=='E')|(S_1(i)=='I' & S_2(j)=='E')|...
340 (S_1(i)=='G' & S_2(j)=='E')|...
341 (S_1(i)=='V' & S_2(j)=='E')|(S_1(i)=='W' & S_2(j)=='E')|...
342 (S_1(i)=='L' & S_2(j)=='E')|...
343 (S_1(i)=='A' & S_2(j)=='E'));
344 Q1(i,j)= 0.64e-19;
345 Q2(i,j)= 0.64e-19;
346 else
347 if ((S_1(i)=='E' & S_2(j)=='P' )|(S_1(i)=='P' & S_2(j)=='E'));
348 Q1(i,j)= 0.78e-19;
349 Q2(i,j)= 0.78e-19;
350 else
351 if ((S_1(i)=='E' & S_2(j)=='H')|(S_1(i)=='H' &S_2(j)=='E'));
352 Q1(i,j)= 0.99e-19;
353 Q2(i,j)= 0.99e-19;
354 else
355 if (S_1(i)=='E'& S_2(j)=='K')| (S_1(i)=='K'& S_2(j)=='E');
356 Q1(i,j)= 1.34e-19;
357 Q2(i,j)= 1.34e-19;
358 else
359 if (S_1(i)=='E' & S_2(j)=='R')|(S_1(i)=='R' & S_2(j)=='E');
360 Q1(i,j)= 1.58e-19;
361 Q2(i,j)= 1.58e-19;
362 else
363 if (S_1(i)=='C' & S_2(j)=='C')|(S_1(i)=='C' & S_2(j)=='F')|...
364 (S_1(i)=='C' & S_2(j)=='Q')|...
365 (S_1(i)=='C'& S_2(j)=='Y')|(S_1(i)=='C' & S_2(j)=='S')|...
366 (S_1(i)=='C' & S_2(j)=='M')|...
367 (S_1(i)=='C' & S_2(j)=='T')|(S_1(i)=='C' & S_2(j)=='I')|...
368 (S_1(i)=='C' & S_2(j)=='G')|...
```

```
369 (S_1(i)=='C' & S_2(j)=='V')|(S_1(i)=='C' & S_2(j)=='W')|...
370 (S_1(i)=='C' & S_2(j)=='L')|...
371 (S_1(i)=='C' & S_2(j)=='L')|(S_1(i)=='C' & S_2(j)=='A')|...
372 (S_1(i)=='F' & S_2(j)=='C')|...
373 (S_1(i)=='Q' & S_2(j)=='C')|(S_1(i)=='Y'& S_2(j)=='C')|...
374 (S_1(i)=='S' & S_2(j)=='C')|...
375 (S_1(i)=='M' & S_2(j)=='C')|(S_1(i)=='T' & S_2(j)=='C')|...
376 (S_1(i)=='I' & S_2(j)=='C')|...
377 (S_1(i)=='G' & S_2(j)=='C')|(S_1(i)=='V' & S_2(j)=='C')|...
378 (S_1(i)=='W' & S_2(j)=='C')|...
379 (S_1(i)=='L' & S_2(j)=='C')|( S_1(i)=='A' & S_2(j)=='C');
380 Q1(i,j)=0.74e-19;
381 Q2(i,j)=0.74e-19;
382 else
383 if (S_1(i)=='C' & S_2(j)=='H')| (S_1(i)=='H' & S_2(j)=='C');
384 Q1(i,j)= 0.99e-19;
385 Q2(i,j)= 0.99e-19;
386 else
387 if (S_1(i)=='C' & S_2(j)=='K')|(S_1(i)=='K' & S_2(j)=='C');
388 Q1(i,j)= 1.34e-19;
389 Q2(i,j)= 1.34e-19;
390 else
391 if (S_1(i)=='C' & S_2(j)=='R')|(S_1(i)=='R' & S_2(j)=='C');
392 Q1(i,j)= 1.59e-19;
393 Q2(i,j)= 1.59e-19;
394 else
395 if (S_1(i)=='N' & S_2(j)=='N')|(S_1(i)=='N' & S_2(j)=='F')...
396 |(S_1(i)=='N' & S_2(j)=='Q')|...
397 (S_1(i)=='N' & S_2(j)=='Y')|(S_1(i)=='N' & S_2(j)=='S')|...
398 (S_1(i)=='N'& S_2(j)=='M')|...
399 (S_1(i)=='F' & S_2(j)=='N')|(S_1(i)=='Q' & S_2(j)=='N')|...
400 (S_1(i)=='Y' & S_2(j)=='N')|...
401 (S_1(i)=='S' & S_2(j)=='N')|(S_1(i)=='M'& S_2(j)=='N');
402 Q1(i,j)=0.74e-19;
403 Q2(i,j)=0.74e-19;
404 else
405 if (S_1(i)=='N' & S_2(j)=='H')|(S_1(i)=='H' & S_2(j)=='N')
406 Q1(i,j)= 0.99e-19;
407 Q2(i,j)= 0.99e-19;
408 else
409 if(S_1(i)=='N' & S_2(j)=='K')|(S_1(i)=='K' & S_2(j)=='N');
410 Q1(i,j)= 1.05e-19;
411 Q2(i,j)= 1.05e-19;
412 else
413 if (S_1(i)=='N' & S_2(j)=='R')|(S_1(i)=='R' & S_2(j)=='N');
414 Q1(i,j)= 1.1e-19;
415 Q2(i,j)= 1.1e-19;
416 else
417 if ((S_1(i)=='F' & S_2(j)=='F')|(S_1(i)=='F' & S_2(j)=='Q'));
418 Q1(i,j)=0.74e-19;
419 Q2(i,j)=0.74e-19;
420 else
```

```
421 if ((S_1(i)=='F' & S_2(j)=='Y')|(S_1(i)=='F' & S_2(j)=='S')|...
422 (S_1(i)=='F' & S_2(j)=='M')|...
423 (S_1(i)=='Q' & S_2(j)=='F')|(S_1(i)=='Y' & S_2(j)=='F'));
424 Q1(i,j)=0.74e-19;
425 Q2(i,j)=0.74e-19;
426 else
427 if (S_1(i)=='S' & S_2(j)=='F')|(S_1(i)=='M' & S_2(j)=='F');
428 Q1(i,j)=0.74e-19;
429 Q2(i,j)=0.74e-19;
430 else
431 if (S_1(i)=='F' & S_2(j)=='H')|(S_1(i)=='H' & S_2(j)=='F');
432 Q1(i,j)= 0.99e-19;
433 Q2(i,j)= 0.99e-19;
434 else
435 if (S_1(i)=='F' & S_2(j)=='K')|(S_1(i)=='K' & S_2(j)=='F');
436 Q1(i,j)= 1.05e-19;
437 Q2(i,j)= 1.05e-19;
438 else
439 if (S_1(i)=='F' & S_2(j)=='R')|(S_1(i)=='R' & S_2(j)=='F');
440 Q1(i,j)= 1.1e-19;
441 Q2(i,j)= 1.1e-19;
442 else
443 % Q
444 if (S_1(i)=='Q' & S_2(j)=='H')|(S_1(i)=='H' & S_2(j)=='Q');
445 Q1(i,j)= 0.99e-19;
446 Q2(i,j)= 0.99e-19;
447 else
448 if (S_1(i)=='Q' & S_2(j)=='K')|(S_1(i)=='K' & S_2(j)=='Q');
449 Q1(i,j)= 1.05e-19;
450 Q2(i,j)= 1.05e-19;
451 else
452 if (S_1(i)=='Q' & S_2(j)=='R')|(S_1(i)=='R' & S_2(j)=='Q');
453 Q1(i,j)= 1.1e-19;
454 Q2(i,j)= 1.1e-19;
455 else
456 if (S_1(i)=='Q' & S_2(j)=='H')|(S_1(i)=='H' & S_2(j)=='Q');
457 Q1(i,j)= 0.99e-19;
458 Q2(i,j)= 0.99e-19;
459 else
460 if (S_1(i)=='Y' & S_2(j)=='K')|(S_1(i)=='K' & S_2(j)=='Y')
461 Q1(i,j)= 1.05e-19;
462 Q2(i,j)= 1.05e-19;
463 else
464 if (S_1(i)=='Y' & S_2(j)=='R')|(S_1(i)=='R' & S_2(j)=='Y');
465 Q1(i,j)= 1.1e-19;
466 Q2(i,j)= 1.1e-19;
467 else
468 if (S_1(i)=='S' & S_2(j)=='H')|(S_1(i)=='H' & S_2(j)=='S');
469 Q1(i,j)= 0.99e-19;
470 Q2(i,j)= 0.99e-19;
471 else
472 if (S_1(i)=='S' & S_2(j)=='K')|(S_1(i)=='K' & S_2(j)=='S');
```

```
473 Q1(i,j)= 1e-19;
474 Q2(i,j)= 1e-19;
475 else
476 if (S_1(i)=='S' & S_2(j)=='R')|(S_1(i)=='R' & S_2(j)=='S');
477 Q1(i,j)= 1.1e-19;
478 Q2(i,j)= 1.1e-19;
479 else
480 if (S_1(i)=='M' & S_2(j)=='H')|(S_1(i)=='H' & S_2(j)=='M');
481 Q1(i,j)= 0.99e-19;
482 Q2(i,j)= 0.99e-19;
483 else
484 if (S_1(i)=='M' & S_2(j)=='K')|(S_1(i)=='K' & S_2(j)=='M');
485 Q1(i,j)= 1e-19;
486 Q2(i,j)= 1e-19;
487 else
488 if (S_1(i)=='M' & S_2(j)=='R')|(S_1(i)=='R' & S_2(j)=='M');
489 Q1(i,j)= 1.1e-19;
490 Q2(i,j)= 1.1e-19;
491 else
492 if (S_1(i)=='T' & S_2(j)=='H')|(S_1(i)=='H' & S_2(j)=='T');
493 Q1(i,j)= 0.99e-19;
494 Q2(i,j)= 0.99e-19;
495 else
496 if (S_1(i)=='T' & S_2(j)=='K')|(S_1(i)=='K' & S_2(j)=='T');
497 Q1(i,j)= 1e-19;
498 Q2(i,j)= 1e-19;
499 else
500 if (S_1(i)=='T' & S_2(j)=='R')|(S_1(i)=='R' & S_2(j)=='T');
501 Q1(i,j)= 1.05e-19;
502 Q2(i,j)= 1.05e-19;
503 else
504 if (S_1(i)=='I' & S_2(j)=='H')|(S_1(i)=='H' & S_2(j)=='I');
505 Q1(i,j)= 0.99e-19;
506 Q2(i,j)= 0.99e-19;
507 else
508 if (S_1(i)=='I' & S_2(j)=='K')|(S_1(i)=='K' & S_2(j)=='I');
509 Q1(i,j)= 1e-19;
510 Q2(i,j)= 1e-19;
511 else
512 if (S_1(i)=='I' & S_2(j)=='R')|(S_1(i)=='R' & S_2(j)=='I');
513 Q1(i,j)= 1.05e-19;
514 Q2(i,j)= 1.05e-19;
515 else
516 if (S_1(i)=='G' & S_2(j)=='H')|(S_1(i)=='H' & S_2(j)=='G');
517 Q1(i,j)= 0.99e-19;
518 Q2(i,j)= 0.99e-19;
519 else
520 if (S_1(i)=='G' & S_2(j)=='K')|(S_1(i)=='K' & S_2(j)=='G');
521 Q1(i,j)= 1e-19;
522 Q2(i,j)= 1e-19;
523 else
524 if (S_1(i)=='G' & S_2(j)=='R')|(S_1(i)=='R' & S_2(j)=='G');
```

```
525 Q1(i,j)= 1.05e-19;
526 Q2(i,j)= 1.05e-19;
527 else
528 if (S_1(i)=='V' & S_2(j)=='H')|(S_1(i)=='H' & S_2(j)=='V');
529 Q1(i,j)= 0.99e-19;
530 Q2(i,j)= 0.99e-19;
531 else
532 if (S_1(i)=='V' & S_2(j)=='K')|(S_1(i)=='K' & S_2(j)=='V');
533 Q1(i,j)= 1e-19;
534 Q2(i,j)= 1e-19;
535 else
536 if (S_1(i)=='V' & S_2(j)=='R')|(S_1(i)=='R' & S_2(j)=='V');
537 Q1(i,j)= 1.05e-19;
538 Q2(i,j)= 1.05e-19;
539 else
540 if (S_1(i)=='W' & S_2(j)=='H')|(S_1(i)=='H' & S_2(j)=='W');
541 Q1(i,j)= 0.99e-19;
542 Q2(i,j)= 0.99e-19;
543 else
544 if (S_1(i)=='W' & S_2(j)=='K')|(S_1(i)=='K' & S_2(j)=='W');
545 Q1(i,j)= 1e-19;
546 Q2(i,j)= 1e-19;
547 else
548 if (S_1(i)=='W' & S_2(j)=='R')|(S_1(i)=='R' & S_2(j)=='W');
549 Q1(i,j)= 1.05e-19;
550 Q2(i,j)= 1.05e-19;
551 else
552 if (S_1(i)=='L' & S_2(j)=='H')|(S_1(i)=='H' & S_2(j)=='L');
553 Q1(i,j)= 0.99e-19;
554 Q2(i,j)= 0.99e-19;
555 else
556 if (S_1(i)=='L' & S_2(j)=='K')|(S_1(i)=='K' & S_2(j)=='L');
557 Q1(i,j)= 1e-19;
558 Q2(i,j)= 1e-19;
559 else
560 if (S_1(i)=='L' & S_2(j)=='R')|(S_1(i)=='R' & S_2(j)=='L');
561 Q1(i,j)= 1.05e-19;
562 Q2(i,j)= 1.05e-19;
563 else
564 if (S_1(i)=='A' & S_2(j)=='H')|(S_1(i)=='H' & S_2(j)=='A');
565 Q1(i,j)= 0.99e-19;
566 Q2(i,j)= 0.99e-19;
567 else
568 if (S_1(i)=='A' & S_2(j)=='K')|(S_1(i)=='K' & S_2(j)=='A');
569 Q1(i,j)= 1e-19;
570 Q2(i,j)= 1e-19;
571 else
572 if (S_1(i)=='A' & S_2(j)=='R')|(S_1(i)=='R' & S_2(j)=='A');
573 Q1(i,j)= 1.05e-19;
574 Q2(i,j)= 1.05e-19;
575 else
576 if (S_1(i)=='P' & S_2(j)=='H')|(S_1(i)=='H' & S_2(j)=='P');
```

```
577 Q1(i,j)= 0.99e-19;
578 Q2(i,j)= 0.99e-19;
579 else
580 if (S_1(i)=='P' & S_2(j)=='K')|(S_1(i)=='K' & S_2(j)=='P');
581 Q1(i,j)= 0.82e-19;
582 Q2(i,j)= 0.82e-19;
583 else
584 if (S_1(i)=='P' & S_2(j)=='R')|(S_1(i)=='R' & S_2(j)=='P');
585 Q1(i,j)= 0.96e-19;
586 Q2(i,j)= 0.96e-19;
587 else
588 if (S_1(i)=='H' & S_2(j)=='H');
589 Q1(i,j)= 0.82e-19;
590 Q2(i,j)= 0.82e-19;
591 else
592 if (S_1(i)=='H' & S_2(j)=='K')|(S_1(i)=='K' & S_2(j)=='H');
593 Q1(i,j)= 0.82e-19;
594 Q2(i,j)= 0.82e-19;
595 else
596 if (S_1(i)=='H' & S_2(j)=='R')|(S_1(i)=='R' & S_2(j)=='H');
597 Q1(i,j)= 0.74e-19;
598 Q2(i,j)= 0.74e-19;
599 else
600 if (S_1(i)=='K' & S_2(j)=='K');
601 Q1(i,j)= 0.54e-19;
602 Q2(i,j)= 0.54e-19;
603 else
604 if (S_1(i)=='K' & S_2(j)=='R')|(S_1(i)=='R' & S_2(j)=='K');
605 Q1(i,j)= 0.41e-19;
606 Q2(i,j)= 0.41e-19;
607 else
608 if (S_1(i)=='R' & S_2(j)=='R');
609 Q1(i,j)= 0.16e-19;
610 Q2(i,j)= 0.16e-19;
611 else
612 Q1(i,j)= 0.824e-19;
613 Q2(i,j)= 0.824e-19;
614 end
615 end
616 end
617 end
618 end
619 end
620 end
621 end
622 end
623 end
624 end
625 end
626 end
627 end
628 end
```

```
629 end
630 end
631 end
632 end
633 end
634 end
635 end
636 end
637 end
638 end
639 end
640 end
641 end
642 end
643 end
644 end
645 end
646 end
647 end
648 end
649 end
650 end
651 end
652 end
653 end
654 end
655 end
656 end
657 end
658 end
659 end
660 end
661 end
662 end
663 end
664 end
665 end
666 end
667 end
668 end
669 end
670 end
671 end
672 end
673 end
674 end
675 end
676 end
677 end
678 end
679 end
680 end
```

```
681 end
682 end
683 end
684 end
685 end
686 end
687 end
688  Q3=[];
689  Q4=[];
690  R1=[];
691  R2=[];
692 for i=1:length(S_1);
693 if (S_1(i)=='A');
694 R1(i)=0.6e-9;
695 else
696 if (S_1(i)=='R');
697 R1(i)=0.809e-9;
698 else
699 if (S_1(i)=='N');
700 R1(i)=0.682e-9;
701 else
702 if (S_1(i)=='D');
703 R1(i)=0.665e-9;
704 else
705 if (S_1(i)=='C');
706 R1(i)=0.629e-9;
707 else
708 if (S_1(i)=='Q');
709 R1(i)=0.725e-9;
710 else
711 if (S_1(i)=='E');
712 R1(i)=0.714e-9;
713 else
714 if (S_1(i)=='G');
715 R1(i)=0.537e-9;
716 else
717 if (S_1(i)=='H');
718 R1(i)=0.732e-9;
719 else
720 if (S_1(i)=='I');
721 R1(i)=0.735e-9;
722 else
723 if (S_1(i)=='L');
724 R1(i)=0.734e-9;
725 else
726 if (S_1(i)=='K');
727 R1(i)=0.737e-9;
728 else
729 if (S_1(i)=='M');
730 R1(i)=0.741e-9;
731 else
732 if (S_1(i)=='F');
```

```
733 R1(i)=0.781e-9;
734 else
735 if (S_1(i)=='P');
736 R1(i)=0.672e-9;
737 else
738 if (S_1(i)=='S');
739 R1(i)=0.615e-9;
740 else
741 if (S_1(i)=='T');
742 R1(i)=0.659e-9;
743 else
744 if (S_1(i)=='W');
745 R1(i)=0.826e-9;
746 else
747 if (S_1(i)=='Y');
748 R1(i)=0.781e-9;
749 else
750 if (S_1(i)=='V');
751 R1(i)=0.694e-9;
752 end
753 end
754 end
755 end
756 end
757 end
758 end
759 end
760 end
761 end
762 end
763 end
764 end
765 end
766 end
767 end
768 end
769 end
770 end
771 end
772 for j=1:length(S_2);
773 if (S_2(j)=='A');
774 R2(j)=0.6e-9;
775 else
776 if (S_2(j)=='R');
777 R2(j)= 0.809e-9;
778 else
779 if (S_2(j)=='N');
780 R2(j)=0.682e-9;
781 else
782 if (S_2(j)=='D');
783 R2(j)=0.665e-9;
784 else
```

```
785 if (S_2(j)=='C');
786 R2(j)=0.629e-9;
787 else
788 if (S_2(j)=='Q');
789 R2(j)=0.725e-9;
790 else
791 if (S_2(j)=='E');
792 R2(j)=0.714e-9;
793 else
794 if (S_2(j)=='G');
795 R2(j)=0.537e-9;
796 else
797 if (S_2(j)=='H');
798 R2(j)=0.732e-9;
799 else
800 if (S_2(j)=='I');
801 R2(j)=0.735e-9;
802 else
803 if(S_2(j)=='L');
804 R2(j)=0.734e-9;
805 else
806 if (S_2(j)=='K')
807 R2(j)=0.737e-9;
808 else
809 if (S_2(j)=='M')
810 R2(j)=0.741e-9;
811 else
812 if (S_2(j)=='F')
813 R2(j)=0.781e-9;
814 else
815 if (S_2(j)=='P');
816 R2(j)=0.672e-9;
817 else
818 if (S_2(j)=='S');
819 R2(j)=0.615e-9;
820 else
821 if (S_2(j)=='T');
822 R2(j)=0.659e-9;
823 else
824 if (S_2(j)=='W');
825 R2(j)=0.826e-9;
826 else
827 if (S_2(j)=='Y');
828 R2(j)=0.781e-9;
829 else
830 if (S_2(j)=='V');
831 R2(j)=0.694e-9;
832 end
833 end
834 end
835 end
836 end
```

```
837 end
838 end
839 end
840 end
841 end
842 end
843 end
844 end
845 end
846 end
847 end
848 end
849 end
850 end
851 end
852 end
853 end
854  Ra=0.6e-9;
855  Rr=0.809e-9;
856  Rn=0.682e-9;
857  Rd=0.665e-9;
858  Rc=0.629e-9;
859  Rq=0.725e-9;
860  Re=0.714e-9;
861  Rg=0.725e-9;
862  Rh=0.732e-9;
863  Ri=0.735e-9;
864  Rl=0.734e-9;
865  Rk=0.737e-9;
866  Rm=0.741e-9;
867  Rf=0.781e-9;
868  Rp=0.672e-9;
869  Rs=0.615e-9;
870  Rt=0.659e-9;
871  Rw=0.826e-9;
872  Ry=0.781e-9;
873  Rv=0.694e-9;
874 for i=1:length(S_1);
875 for j=1:length(S_2);
876 if (S_1(i)=='R'& S_2(j)=='D');
877     h(i,j)=.15*10^(-9)+Rr+Rd;
878 else
879 if (S_1(i)=='R'& S_2(j)=='E');
880       h(i,j)=.15*10^(-9)+Rr+Re;
881 else
882 if (S_1(i)=='D'& S_2(j)=='R');
883       h(i,j)=.15*10^(-9)+Rd+Rr;
884 else
885 if (S_1(i)=='D'& S_2(j)=='H');
886     h(i,j)=.15*10^(-9)+Rd+Rh;
887 else
888 if (S_1(i)=='D'& S_2(j)=='R');
```

```
889        h(i,j)=.15*10^(-9)+Rd+Rr;
890    else
891    if (S_1(i)=='D'& S_2(j)=='H');
892        h(i,j)=.15*10^(-9)+Rd+Rh;
893    else
894    if (S_1(i)=='D'& S_2(j)=='K');
895        h(i,j)=.15*10^(-9)+Rd+Rk;
896    else
897    if (S_1(i)=='E')& (S_2(j)=='R');
898        h(i,j)=.15*10^(-9)+Re+Rr;
899    else
900    if (S_1(i)=='E'& S_2(j)=='H');
901       h(i,j)=.15*10^(-9)+Re+Rh;
902    else
903    if (S_1(i)=='E'& S_2(j)=='K');
904       h(i,j)=.15*10^(-9)+Re+Rk;
905    else
906    if (S_1(i)=='H'& S_2(j)=='D')
907        h(i,j)=.15*10^(-9)+Rh+Rd;
908    else
909    if (S_1(i)=='H'& S_2(j)=='E')
910        h(i,j)=.15*10^(-9)+Rh+Re;
911    else
912    if (S_1(i)=='R'& S_2(j)=='R')
913        h(i,j)=.4*10^(-9)+Rr+Rr;
914    else
915    if (S_1(i)=='R'& S_2(j)=='H')
916       h(i,j)=.4*10^(-9)+Rr+Rh;
917    else
918    if (S_1(i)=='R'& S_2(j)=='H')
919        h(i,j)=.4*10^(-9)+Rr+Rh;
920    else
921    if (S_1(i)=='R'& S_2(j)=='K')
922        h(i,j)=.4*10^(-9)+Rr+Rk;
923    else
924    if (S_1(i)=='D'& S_2(j)=='E');
925        h(i,j)=.4*10^(-9)+Rd+Re;
926    else
927    if (S_1(i)=='D'& S_2(j)=='D');
928        h(i,j)=.4*10^(-9)+Rd+Rd;
929    else
930    if (S_1(i)=='H'& S_2(j)=='R')
931        h(i,j)=.4*10^(-9)+Rh+Rr;
932    else
933    if (S_1(i)=='H'& S_2(j)=='H')
934        h(i,j)=.4*10^(-9)+Rh+Rh;
935    else
936    if (S_1(i)=='H'& S_2(j)=='K')
937          h(i,j)=.4*10^(-9)+Rh+Rk;
938    else
939    if (S_1(i)=='K'& S_2(j)=='R')
940       h(i,j)=.4*10^(-9)+Rk+Rr;
```

```
941 else
942 if (S_1(i)=='K'& S_2(j)=='H')
943     h(i,j)=.4*10^(-9)+Rk+Rh;
944 else
945 if (S_1(i)=='K'& S_2(j)=='K')
946     h(i,j)=.4*10^(-9)+Rk+Rk;
947 else
948 if (S_1(i)=='N'& S_2(j)=='Q')
949    h(i,j)=.25*10^(-9)+Rn+Rq;
950 else
951 if (S_1(i)=='N'& S_2(j)=='S')
952    h(i,j)=.25*10^(-9)+Rn+Rs;
953 else
954 if (S_1(i)=='N'& S_2(j)=='Y')
955     h(i,j)=.25*10^(-9)+Rn+Ry;
956 else
957 if (S_1(i)=='Q'& S_2(j)=='S')| (S_1(i)=='Q')& (S_2(j)=='Y');
958     h(i,j)=.25*10^(-9)+Rq+Rs;
959 else
960 if (S_1(i)=='Q')& (S_2(j)=='Y');
961      h(i,j)=.25*10^(-9)+Rq+Ry;
962 else
963 if (S_1(i)=='S'& S_2(j)=='Y');
964     h(i,j)=.25*10^(-9)+Rs+Ry;
965 else
966     h(i,j)=1.76*10^(-9);
967 end
968 end
969 end
970 end
971 end
972 end
973 end
974 end
975 end
976 end
977 end
978 end
979 end
980 end
981 end
982 end
983 end
984 end
985 end
986 end
987 end
988 end
989 end
990 end
991 end
992 end
```

```
993  end
994  end
995  end
996  end
997  end
998  end
999
1000 function[A]=electrostatic(Q1,Q2,R1,R2,h,M,N,N1,epsilon)
1001 for i=1:N
1002     for j=1:M
1003         if R1(i)>R2(j)
1004             gamma(i,j)=R1(i)/R2(j);
1005         else
1006             if  R1(i)<R2(j)
1007                 gamma(i,j)=R2(j)/R1(i);
1008               else if R1(i)==R2(j);
1009      gamma(i,j)=R2(j)/R1(i);
1010          end
1011             end
1012         end
1013         if h(i,j)>(R1(i)+R2(j))
1014             r(i,j)=h(i,j)/(R1(i)+R2(j));
1015         else if  h(i,j)<=(R1(i)+R2(j))
1016             r(i,j)=(R1(i)+R2(j))/h(i,j);
1017         end
1018         end
1019     y(i,j)=(((r(i,j)^2*(1+gamma(i,j))^2)-...
1020     (1+(gamma(i,j))^2))/(2*gamma(i,j)));
1021     beta(i,j)=acosh(y(i,j));
1022     z(i,j)=exp(-beta(i,j));
1023     S12=0;
1024     S22=0;
1025     S11=0;
1026     for k=1:N1
1027         S_1(k)=(z(i,j)^k)/(((1-z(i,j)^(2*k)))*((gamma(i,j)+...
1028         y(i,j))-(y(i,j)^2-1)^(1/2)*...
1029         (1+z(i,j)^(2*k))/(1-z(i,j)^(2*k))));
1030         S11=S11+S_1(k);
1031         S_2(k)=(z(i,j)^(2*k))/(1-(z(i,j)^(2*k)));
1032         S12=S12+S_2(k);
1033         S_3(k)=(z(i,j)^k)/(((1-z(i,j)^(2*k)))*...
1034         ((1-gamma(i,j)*y(i,j))-...
1035         gamma(i,j)*(y(i,j)^2-1)^(1/2)*...
1036         (1+z(i,j)^(2*k))/(1-z(i,j)^(2*k))));
1037         S22=S22+S_3(k);
1038     end
1039     epsilon0=8.85418781762*10^(-12);
1040     c11(i,j)=(2*gamma(i,j)*((y(i,j)^2-1)^(1/2))).*S11;
1041     c22(i,j)=(2*gamma(i,j)*((y(i,j)^2-1)^(1/2))).*S22;
1042     c12(i,j)=-((2*gamma(i,j)*...
1043     ((y(i,j)^2-1))^(1/2))/(r(i,j)*...
1044     (1+gamma(i,j)))).*S12;
```

```
1045      delta(i,j)=((c11(i,j)*c22(i,j)-c12(i,j)^2));
1046       k=1/(4*pi*epsilon0);
1047      k1=1/(4*pi*epsilon* epsilon0);
1048          alpha(i,j)=Q2(i,j)/Q1(i,j);
1049      if R1(i)>R2(j)
1050          gamma(i,j)=R1(i)/R2(j);
1051    W1(i,j)=((1/k1)*R2(j)*gamma(i,j))*...
1052    ((1+gamma(i,j))/(2*alpha(i,j)))*...
1053    ((alpha(i,j)^2*c11(i,j)-...
1054    2*alpha(i,j)*c12(i,j)+c22(i,j))/delta(i,j));
1055          else if (R1(i)<R2(j))
1056              gamma(i,j)=R2(j)/R1(i);
1057              W1(i,j)=((1/k1)*R1(i)*gamma(i,j))*...
1058              ((1+gamma(i,j))/(2*alpha(i,j)))*...
1059              ((alpha(i,j)^2*c11(i,j)...
1060              -2*alpha(i,j)*c12(i,j)+c22(i,j))/delta(i,j));
1061       else if R1(i)==R2(j);
1062        W1(i,j)=((1/k1)*R1(i)*gamma(i,j))*...
1063        ((1+gamma(i,j))/(2*alpha(i,j)))*...
1064        ((alpha(i,j)^2*c11(i,j)-...
1065        2*alpha(i,j)*c12(i,j)+c22(i,j))/delta(i,j));
1066              end
1067              end
1068      end
1069      W2(i,j)=(k*(Q1(i,j)*Q2(i,j)))/(R1(i)+R2(j));
1070      A1(i,j)=W1(i,j);
1071      A2(i,j)=W2(i,j);
1072      A(i,j)=A1(i,j)/A2(i,j);
1073       end
1074  end
1075  return
```

References

1. S. D'Arcy, K.W. Martin, T. Panchenko, X. Chen, S. Bergeron, L.A. Stargell, B.E. Black, K. Luger, Mol. Cell. **51**(5) (2013)
2. L. Marino-Ramirez, M.G. Kann, B.A. Shoemaker, D. Landsman, J. Expert Rev. Proteomics **2**(5) (2005)
3. A.J. Andrew, G. Downing, K. Brown, Y.-J. Park, K. Luger, J. Biol. Chem. **283**(47) (2008)
4. W. Iwasaki, Y. Miya, N. Horikoshi, A. Osakabe, H. Taguchi, H. Tachiwana, T. Shibata, W. Kagawa, H. Kurumizaka, FEBS Open Bio **3** (2013)
5. R. Dias, B. Lindman, *DNA Interactions with Polymers and Surfactants* (Wiley, New York, 2007), pp. 135–172
6. Y.J. Park, K. Luger, The structure of nucleosome assembly protein
7. Y.J. Park, J.V. Chodaparambil, Y. Bao, S.J. McBryant, K. Luger, J. Biol. Chem. **280**(3), 1817–1825 (2005)
8. https://www.rcsb.org/structure/5G2E
9. T.T. Berezov, *Biological Chemistry* (Medicine, 1998)
10. The Universal Protein Resource, http://www.uniprot.org/
11. A.T. Fenley, D.A. Adams, A.V. Onufriev, Biophys. J. **99**, 1577–1585 (2010)
12. Yu.D. Semchikov, *Macromolecular Compounds* (Academy, New York, 2010)

13. M. Gerstein, F.M. Richards, Protein geometry: volumes, areas, and distances. Yale University (1977), http://papers.gersteinlab.org/e-print/geom-inttab/geom-inttab.pdf
14. J.C. Biro, Amino acid size, charge, hydropathy indices and matrices for protein structure analysis. 2006
15. Ya.I. Ryskin, Hydrogen bond and structure of hydrosilicates. Science (1972)
16. K.G. Kulikov, T.V. Koshlan, Mathematical simulation of interactions of protein molecules and prediction of their reactivity. Tech. Phys. **61**(10), 1572–1579 (2016)
17. V.A. Saranin, On the interaction of two electrically charged conducting balls. UFN **169**(4), 453–458 (1999)
18. V.A. Saranin, Electric field strength of charged conducting balls and the breakdown of the air gap between them. UFN. **172**(12), 1449–1454 (2002)
19. V.A. Saranin, Interaction of two charged conducting balls: theory and experiment. UFN **180**(10), 1109–1117 (2010)
20. V. Smythe, *Electrostatics and Electrodynamics* (Translation from the second American edition of A.V. Gaponova and M.A. Miller. Ed. IL, 1954)
21. C.C. Paige, P. Van Dooren, On the quadratic convergence of Kogbetliantr's algorithm for computing the singular value decomposition. Linear Algebra Appl. **77**, 301–313 (1985)
22. V.I. Gorbachenko, *Computational Linear Algebra with Examples MATLAB* (BHV Petersburg, St. Petersburg, 2011)
23. V.M. Verzhbitsky, *Computational Linear Algebra* (High School, 2009)
24. M. Huš, T. Urbic, Strength of hydrogen bonds of water depends on local environment. J. Chem. Phys. **136**, 144305-1–144305-7 (2012)
25. L. Marino-Ramirez, M.G. Kann, B.A. Shoemaker, D. Landsman, Expert Rev. Proteomics **2**(5) (2005)

Chapter 3
Mathematical Modelling of the Temperature Effect on the Character of Linking Between Monomeric Proteins in Aqueous Solutions

Abstract The mathematical model was developed for taking into account the influence of the temperature of an aqueous salt-free solution on the character of dimer formation by different sections of small proteins: H2A–H2B and H3–H4 dimers, and the effect of temperature on various sections of the Bcl-xl protein were studied. The analysis of the numerical calculations obtained in the course of the developed mathematical model revealed a different behavior of the histone dimers H2B-H2A and H3–H4, as well as the Bcl-xl$_{(1-212)}$–Bcl-xl$_{(1-212)}$, Bcl-xl–Bcl-xl, Bcl-xl$_{(1-212)}$–Bcl-xl$_{(213-233)}$, Bcl-xl$_{(213-233)}$–Bcl-xl$_{(213-233)}$ with an increase of temperature from 20 °C to 40 °C, as well as the contribution of different sections of the Bcl-xl protein to the formation of the biological complex.

3.1 Introduction

This chapter presents a developed theoretical method that allows one to analyze the effect of the temperature of an aqueous solution on the character of the formation of protein dimers, as well as to determine the regions of protein molecules that make the greatest contribution to the stabilization of dimers. The maximum protein size was 233 a.a. (Bcl-xl). The behavior of protein complexes was studied in water at various temperatures, without the addition of salts. Note that for the analysis of longer proteins by the number of amino acid residues, it is necessary to cut such proteins into domains, as was done with the Bcl-xl protein, dividing it into two sections Bcl-xl$_{(1-212)}$ and Bcl-xl$_{(213-233)}$ in order to determine the interaction of these domains and their role in the formation of the Bcl-xl–Bcl-xl homodimer.

Note that we have developed a mathematical model to take into account the influence of the temperature of the aqueous solution on the nature of the binding of protein molecules.

The physical model developed by us makes it possible to determine the stability of protein complexes, as well as to predict the possible aggregation of proteins when the temperature of the aqueous solution changes.

T. Koshlan and K. Kulikov, *Mathematical Modeling of Protein Complexes*,
Biological and Medical Physics, Biomedical Engineering,
https://doi.org/10.1007/978-3-319-98304-2_3

Let us note that the study of protein aggregation is a topical trend in contemporary molecular biophysics. The aggregation of proteins is one of the processes that occur continuously in a cell. Each protein is characterized by its native conformation, which allows it to fulfill its prescribed biological functions. However, genetic mutations and errors in the synthesis of proteins on a ribosome may lead to the formation of misfolded protein structures. There is always the probability of a partial distortion in the native structure, even for native proteins, especially under stress conditions (thermal, oxidative, or osmotic). When the native structure is distorted, proteins cease to fulfill their biological functions, become less stable, and may exhibit a tendency to aggregate, which may lead to a broad spectrum of pathological states in a cell and the organism as a whole.

Hence, the approach developed in this work will make it possible to study and explain the pathological aspects associated with the structural transformation of proteins in the process of aggregation.

The chapter consists of several parts. In the first part, we describe the principal properties of proteins and the character of their behavior at an increasing temperature. The second part shows the physical characteristics of the formation of dimers, H2A–H2B, H3–H4 and protein complexes Bcl-xl–Bcl-xl in previously conducted in vitro experiments. The third part gives a detailed description of the physical model of accounting for the effect of the temperature of an aqueous solution on protein complexes. The fourth part is devoted to numerical calculations and analysis of the data obtained on the example of the formation of dimers H2B-H2A and H3-H4, Bcl-xl$_{(1-212)}$–Bcl-xl$_{(1-212)}$, Bcl-xl–Bcl-xl, Bcl-xl$_{(1-212)}$–Bcl-xl$_{(213-233)}$, Bcl-xl$_{(213-233)}$–Bcl-xl$_{(213-233)}$, the major conclusions drawn in this work are given.

3.2 The Main Properties of Proteins and the Nature of Their Behavior with Increasing Temperature

It is known that a slight increase in temperature may lead to both the aggregation of some proteins and the dissolution of others. Let us consider the hydrophobic interactions [1–3] responsible for the aggregation of proteins in more detail. It has been hypothesized previously that interactions between hydrophobic amino-acid residues intensify with increasing temperature and abate with decreasing temperature in a short range of values [2]. While an amino-acid sequence acquires a native structure, hydrophobic residues are generally located inside the globular structure of a protein far from the water environment. When the temperature is increased, the native structure of a protein sustains distortion exhibited as a loss of protein functions. Hydrophobic residues become exposed on the surface of a molecule and may begin to interact with the hydrophobic residues of other proteins. There are several interactions of different natures that govern the structure of a protein, such as hydrogen and hydrophobic interactions, interactions between charged amino acids, and covalent bonds between cysteine residues. In its native state, a protein molecule is usually closely packed in such a way that the side groups located in the inner part of a molecule have restricted freedom of motion. The motion of the side groups that form the hydrophobic core of a molecule is especially restricted.

By increasing the temperature, it is possible to create ambient conditions under which the small-scale fluctuations of some groups of atoms become more intensive.

Since the hydrophobic residues are mainly inside the globular structure in the native state of the molecule, away from the water environment, then the temperature increase may lead to a violation of the native structure of protein [4], which is expressed in the loss of protein functions. Hydrophobic residues reach the surface of the molecule and can begin to interact with the hydrophobic residues of other proteins.

After hydrophobic amino-acid residues become exposed on the surface and begin to interact with other hydrophobic amino-acid residues, under certain conditions, a protein molecule is denaturated to form aggregates, e.g., of hydrophobic molecule parts.

Thus, the denaturation and rearrangement of the interacting amino acid residues with increasing temperature can lead to aggregation [5].

Let us note that denaturation is the destruction of the native special structure of a protein with the resulting loss of its bioactivity [3]. It is noteworthy that protein molecules do not necessarily form aggregates with increasing temperature, but may lose their native structure without forming any aggregates. The solubility of proteins of different kind is varied within broad ranges.

The solubility of a protein depends on the ratio between its polar and non-polar groups, their mutual arrangement, and the resulting dipole moment. A large number of polar groups must increase both the affinity of proteins to water and their solubility. However, ionic groups may have an inverse effect when they are bonded with oppositely charged groups and form intermolecular salt-like bonds. The formation of these intermolecular bonds always leads to dehydration and promotes the appearance of coarse insoluble aggregates, so the denaturation of proteins is also a function of the protein concentration in a solution. Electrostatic forces in water are reduced due to the high dielectric permeability, and interactions occur between the polar groups of a molecule and water. If the interaction between a protein and a solvent is stronger than between the amino acids of this protein, the protein is dissolved. Let us note that the capacity for water retention, as well as the solubility, simultaneously depends on the degree of both proteinwater and proteinprotein interactions and the conformation of a protein and its degree of denaturation. For this reason, thermal treatment has a strong effect on these physical properties [6–8]. In many cases, thermal treatment decreases the solubility of proteins and may increase the water-retention function under certain conditions. However, it is difficult to distinguish any general properties here. Each type of proteins exhibits its properties in different ways depending on its composition, structure, and conformation. Hence, in each case, the temperature effect on a protein requires careful study.

3.3 The Physical Properties of the Studied Proteins H2A, H2B, H3, H4, Bcl-xl

In this section, we consider previously published experimental studies of the physical properties of proteins H2A, H2B, H3, H4, Bcl-xl in vitro solutions with different physical characteristics.

In this section, we consider previously published experimental studies of the physical properties of proteins H2A, H2B, H3, H4, Bcl-xl in vitro in solutions with different physical characteristics. In [9] the thermal stability of the core histone dimer H2A–H2B has been studied by high-sensitivity differential scanning calorimetry and circular dichroism spectroscopy. The unfolding transition temperature of the 28kDa H2A–H2B dimer increases as a function of both the ionic strength of the solvent and the total protein concentration. At neutral pH and physiological ionic strength, the thermal denaturation is centered at about 50 °C. Analysis of the data shows that at low ionic strength and pH values between 6.5 and 8.5, the H2A–H2B dimer behaves as a highly cooperative system. The self-associative behavior of the $(H3–H4)_2$ in the absence of the H2A-H2B dimer, makes it very difficult to analyze its thermodynamic properties under conditions where its interaction with H2A–H2B dimer is particularly sensitive. The $(H3–H4)_2$ tetramer is responsive to changes in ionic strength. Extensive aggregation is promoted at higher protein concentrations, especially at high levels of NaCl.

Aggregation of the $(H3–H4)_2$ tetramer is prevented by the addition of the histone H2A-H2B dimer which acts as a molecular cap and regulates the assembly pathway toward the formation of octamers [10]. In [9] the H3–H4 dimer is thermally more stable than the H2A–H2B dimer. Comparison of corresponding data for the two dimeric proteins reveals that the unfolding temperature of the H3–H4 dimer is approximately 20 °C higher than that of the H2A–H2B dimer under similar experimental conditions. In [11] reported that at temperatures greater than 32 °C get aggregation of tetramer H3–H4.

Let us turn to the physical properties of the Bcl-xl protein in solutions.

In [12] reported that Bcl-xl irreversible aggregation and assembles into highly-ordered rope-like homogeneous fibrils under elevated temperatures. In [13] provide evidence that acidic pH promotes the assembly of Bcl-xl into a megadalton oligomer. Bcl-xl displays the propensity to oligomeriation in solution and that such oligomerization is driven by the intramolecular binding of its C–terminal TM domain to the canonical hydrophobic groove in a domain-swapped trans-fashion, whereby the TM domain of one monomer occupies the canonical groove within the other monomer and vise verse [14]. Bcl-xl exists in various associative [12] and can formation at 20 °C, ranging from monomer and dimer to higher-order oligomers. At 40 °C, the dimer and multimer conformers appear to shift in the direction of the polymeric conformation. The truncation of C–terminal TM domain completely abolished oligomerization of Bcl-xl under low temperatures 20 °C to 40 °C. A key role of TM domain is driving the intermolecular association of Bcl-xl into large aggregates in agreement with previous studies.

3.4 Description of the Physical Model

The mathematical model developed in this work is based on studies [15–17] and describes the temperature effect on the character of linking in protein dimers in an aqueous solution.The selected histone dimers H2A–H2B and H3–H4 as the model system, various sections of the Bcl-xl: Bcl-xl, $\text{Bcl-xl}_{(1-212)}$, $\text{Bcl-xl}_{(213-233)}$ protein, which form the Bcl-xl–Bcl-xl, $\text{Bcl-xl}_{(1-212)}$–$\text{Bcl-xl}_{(1-212)}$, $\text{Bcl-xl}_{(1-212)}$–$\text{Bcl-xl}_{(223-233)}$, $\text{Bcl-xl}_{(213-233)}$–$\text{Bcl-xl}_{(213-233)}$.

To analyze the stability of dimers in an aqueous solution while the temperature changes from 20 °C to 40 °C, we performed calculations of the matrix condition number, the elements of which are the potential energies of electrostatic interaction between pairwise taken amino acid residues. Several assumptions were made.

1. Each amino acid residue interacts with all other amino acid residues with a specific charge. Note, that the charge of each amino acid residue was obtained using the data from [2].

We took the data from [18], which shows changes in the volume of a protein with known amino acid sequences and using these data and calculated the order of magnitude δ on which the radius of each amino acid residue can vary with a temperature change of 5 °C. In the attached program, δ is denoted by Rt1.

2. The radius of hydrophobic amino-acid residues (A, I, L, M, F, P, W, Y, V) decreases by the value of δ for every 5 °C of temperature increase.

Let us note that different proteins demonstrate a complicated dependence of change in their volume with increasing temperature, as the process of heating may lead to different distortions in their three-dimensional structure and have diverse characters for different proteins [18, 19].

3. The radius of the other amino acid residues increases by the value of δ for every 5 °C of temperature increase.

4. When the temperature reaches 40 °C, we assume that there is a violation of the linear change in the parameters of physical quantities and the value of δ increases two-fold. Below is the general case of calculating the radii of amino acid residues at different temperatures of the aqueous solution.

$$R_a = r_a \cdot 10^{-9} - n\delta; \quad R_r = r_r \cdot 10^{-9} + n\delta; \quad R_n = r_n \cdot 10^{-9} + n\delta; \quad R_d = r_d \cdot 10^{-9} + n\delta;$$

$$R_c = r_c \cdot 10^{-9} + n\delta; \quad R_q = r_q \cdot 10^{-9} + n\delta; \quad R_e = r_e \cdot 10^{-9} + n\delta; \quad R_h = r_h \cdot 10^{-9} + n\delta;$$

$$R_i = r_i \cdot 10^{-9} - n\delta; \quad R_l = r_l \cdot 10^{-9} - n\delta; \quad R_g = r_g \cdot 10^{-9} + n\delta; \quad R_k = r_k \cdot 10^{-9} + n\delta;$$

$$R_m = r_m \cdot 10^{-9} - n\delta; \quad R_f = r_f \cdot 10^{-9} - n\delta; \quad R_p = r_p \cdot 10^{-9} - n\delta; \quad R_s = r_s \cdot 10^{-9} + n\delta;$$

$$R_t = r_t \cdot 10^{-9} + n\delta; \quad R_w = r_w \cdot 10^{-9} - n\delta; \quad R_y = r_y \cdot 10^{-9} - n\delta; \quad R_v = r_v \cdot 10^{-9} - n\delta,$$

where $\delta = 10^{-4}$ nm, r is the initial radius of the amino acid residue, R is the finite radius of the amino acid residue, n is an integer depending on the temperature of the aqueous solution (see Table 3.1).

Table 3.1 The value of the number n at different temperatures of the aqueous solution

Water temperature, °C	20°	25°	30°	35°	40°
n	1	1	2	3	8

Table 3.2 Relative dielectric permeability of water at different temperatures [20]

Water temperature, °C	20°	25°	30°	35°	40°
Dielectric permeability of water	80.103	78.304	76.546	74.828	73.151

5. The initial radii of amino acid residues are and the distances between the different amino acid residues are defined in Chap. 2.

Thus, every temperature corresponds to a certain set of 20 radii of amino-acid residues.

6. The distance between differently charged amino-acid residues is 0.15 nm.

7. The distance between the hydrophobic residues was taken at 0.36 nm.

8. The distance between identically charged amino acid residues is 0.4 nm.

9. We believe that the tendency of the values of lg(cond(W)) to decrease with a change in temperature of the aqueous solution may serve as an indicator of the possible aggregation of protein complexes.

In this case, the distances considered increased by a multiple of δ every 5 °C. In the program, this distance is represented by the sum of the distances between the boundaries of the spheres and the sum of the two mean radii of the spheres.

Since every temperature of an aqueous solution corresponds to a certain electrical permeability, its values for all temperatures used in our calculations were compiled in Table 3.2.

To analyze the temperature effect on the character of linking in protein dimers in aqueous solutions, we use the concept of the condition number (see Chap. 2) which will characterize the degree of stability in the configuration of a biocomplex in different temperature regimes in this physical formulation.

3.5 Numerical Modelling of the Effect of Temperature on the Character of Binding of Monomeric Proteins to Aqueous Solutions. Conclusion

To numerically model of the temperature effect on the character of the linking of monomeric proteins into dimers in aqueous solutions, we selected small proteins. We took the sequences of histone proteins H2A, H2B, H3,H4, and Bcl-xl from the database [21], where their numbers were P04911, P02293, P61830, P02309 and Q07817, respectively. Let us point out that the thermal motion energy will grow with increasing temperature, and this may lead to the destruction of solvate shells and, correspondingly, the aggregation of the system. We assume that the electrostatic

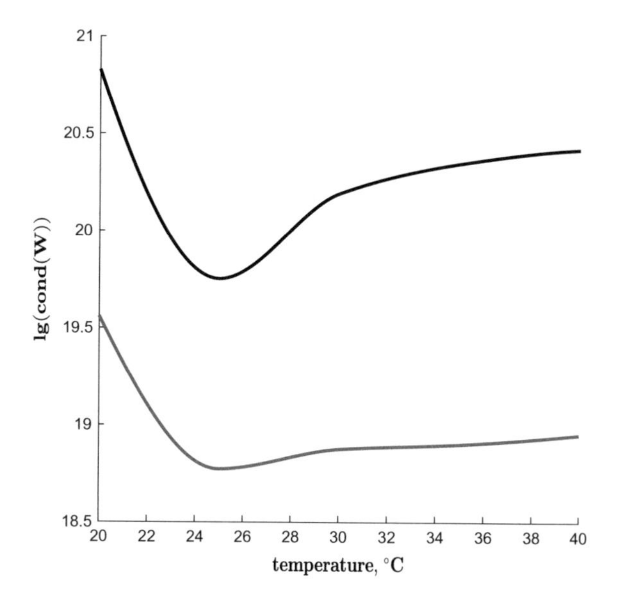

Fig. 3.1 lg(cond(W)) versus temperature for dimers H2A–H2B and H3–H4. The black color line for dimer H2A–H2B and red color line for dimer H3–H4

interaction between amino-acid residues in the process of aggregation is stronger, and this must lead to an decrease in lg(cond(W)), where cond(W)is the condition number of matrix of the potential energy of pairwise electrostatic interaction between the studied proteins (see expression 2.10). To select the most stable biochemical complex that links between proteins, we take the electrostatic potential energy matrix with the lowest condition number.

The results of numerical simulation of the interaction of histone dimers H2A–H2B and H3–H4 are presented in Fig. 3.1.

Numerical calculations were carried out for the whole Bcl-xl protein, as well as for the shortened Bcl-$xl_{(1-212)}$–Bcl-$xl_{(1-212)}$ protein, in which the TM domain was cut off from the C–terminus. A numerical calculation of the interaction of the shortened region of the protein Bcl-$xl_{(1-212)}$ with the TM domain of the protein Bcl-$xl_{(213-233)}$ was also performed. The results of numerical calculations for different regions of the Bcl-xl protein are shown in Fig. 3.2.

In Fig. 3.1 numerical calculations of the effect of temperature on the behavior of histone dimers in aqueous solutions in the range from 20° to 40° are presented. The values of lg(cond(W)) at 20° for the dimers H2A–H2B and H3–H4 were 20.828 and 19.560, respectively.

We also give the interaction values of the proteins H2A and H2B, H3 and H4, obtained in Chap. 2 at a temperature of 20°. In this case, the values of lg(cond(W)) corresponding to the formation of the dimers H2A–H2B and H3–H4 were 19.749 and 18.266, respectively.

We note that the developed thermal model makes it possible to analyze the stability of biological complexes in an aqueous solution when the temperature is changed by analyzing the curve lg(cond(W)). The numerical results obtained using the mathematical model from Chap. 2 allow one to directly compare the lg(cond(W))

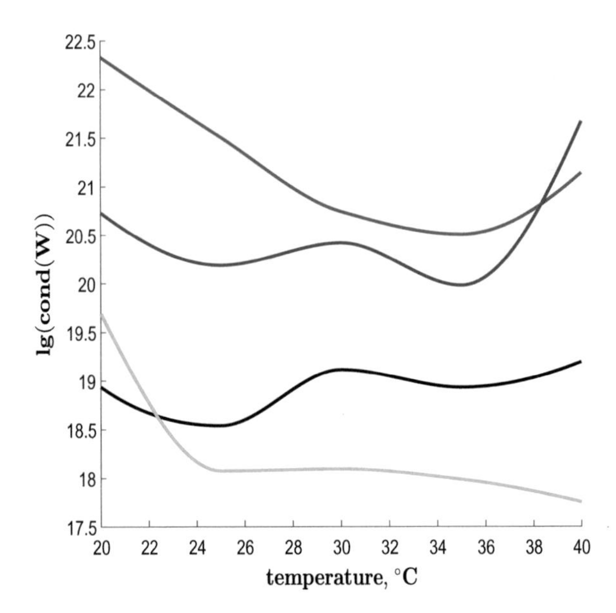

Fig. 3.2 lg(cond(W)) versus temperature $Bcl\text{-}xl_{(213-233)}$–$Bcl\text{-}xl_{(213-233)}$, Bcl-xl–Bcl-xl, $Bcl\text{-}xl_{(1-212)}$–$Bcl\text{-}xl_{(1-212)}$,$Bcl\text{-}xl_{(1-212)}$–$Bcl\text{-}xl_{(213-233)}$. The blue color line for $Bcl\text{-}xl_{(1-212)}$–$Bcl\text{-}xl_{(1-212)}$, red color line for Bcl-xl–Bcl-xl, black color line for $Bcl\text{-}xl_{(213-233)}$–$Bcl\text{-}xl_{(213-233)}$, green color line for $Bcl\text{-}xl_{(1-212)}$–$Bcl\text{-}xl_{(213-233)}$

values with the formation of different dimers for the analysis of the behavior of proteins.

Thus, from the obtained numerical data we see that the dimer H2A–H2B is less stable than the dimer H3–H4. This conclusion coincides with the numerical calculations obtained according to the thermal model.

The values of lg(cond(W)) in the interaction of H3 proteins with H4 are in a lower range of values than in the interaction of histones H2A and H2B. We assume that the effect of temperature in aqueous solutions leads to a more stable formation of a heterodimer involving H3 and H4 proteins than H2A and H2B, as well as to the possible aggregation of the H3–H4 dimer.

The formation of a pronounced minimum in the region 25 °C of the dimer H2A–H2B compared to the dimer H3–H4 can indicate structural changes in a given temperature range.

Let us now consider the numerical results of the interaction of different sections of the Bcl-xl protein. On the graph Fig. 3.2 are shown the results of numerical simulation of $Bcl\text{-}xl_{(213-233)}$–$Bcl\text{-}xl_{(213-233)}$, Bcl-xl–Bcl-xl, $Bcl\text{-}xl_{(1-212)}$–$Bcl\text{-}xl_{(1-212)}$,$Bcl\text{-}xl_{(1-212)}$–$Bcl\text{-}xl_{(213-233)}$.

We present the lg(cond(W)) values (see Table 3.3) obtained for these compounds using a mathematical model of the interactions between protein molecules and study of their propensity to form complex biological complexes (see Chap. 2).

Note that we will do the analysis of this table together with the obtained graphs for the selected pairs of proteins. We investigated the interaction of different regions of the Bcl-xl protein in order to determine the role of the selected sites in the formation of the whole homodimer Bcl-xl–Bcl-xl.

Table 3.3 Numerical results of the interaction of dimers Bcl-xl$_{(213-233)}$–Bcl-xl$_{(213-233)}$, Bcl-xl–Bcl-xl, Bcl-xl$_{(1-212)}$–Bcl-xl$_{(1-212)}$,Bcl-xl$_{(1-212)}$–Bcl-xl$_{(213-233)}$

Dimers	Bcl-xl–Bcl-xl	Bcl-xl$_{(1-212)}$– –Bcl-xl$_{(1-212)}$	Bcl-xl$_{(213-233)}$– –Bcl-xl$_{(213-233)}$	Bcl-xl$_{(1-212)}$– –Bcl-xl$_{(213-233)}$
lg(cond(w))	20.622	20.748	19.071	17.253

lg(cond(W)) is common logarithm of condition number

The results are shown in Fig. 3.2 and are described by the red and blue curves, respectively. As can be seen from Fig. 3.2, the homodimer (Bcl-xl)$_2$ is more stable than the homodimer of the cut off proteins Bcl-xl$_{(1-212)}$)$_2$ as the temperature changes.

In this case, the homodimer (Bcl-xl)$_2$ demonstrates a gradual decrease in lg(cond(W)) values as the temperature changes from 20° to 40°, in contrast to Bcl-xl$_{(1-212)}$)$_2$.

The curve of the values of lg(cond(W)) for the homodimer Bcl-xl$_{(1-212)}$)$_2$ increases from 20.730 to 21.668 at a temperature from 20° to 40°, and the curve of the values of lg(cond(W)) for the homodimer Bcl-xl$_2$ decreases from 22.328 to 21.140, respectively. A similar change in lg(cond(W)) within the framework of the constructed model is interpreted as an indicator of a possible aggregation of the protein complex. Analysis of the numerical values of Table 3.3 obtained on the basis of the mathematical model from Chap. 2 allows us to assume that the dimer (Bcl-xl)$_2$ is initially a more stable complex than the truncated dimer Bcl-xl$_{(1-212)}$)$_2$, since the values of lg(cond(W)) were 20.622 and 20.748, respectively.

Such differences in the behavior of protein dimers when the temperature varies from 20° to 40° are explained by the role of the TM domain of the protein Bcl-xl$_{(213-233)}$, which can take part in the formation of the dimer (Bcl-xl)$_2$.

To test this assumption, the interaction of the truncated Bcl-xl$_{(1-212)}$ protein with the TM domain of Bcl-xl$_{(213-233)}$ was studied. The results are shown in Fig. 3.2 and are described by a green curve. The numerical value of this interaction at a temperature of 20° is given in Table 3.3 and amounted to 17.253. Since this value is significantly smaller than the interaction of (Bcl-xl)$_2$ and Bcl-xl$_{(1-212)}$)$_2$ according to Table 3.3, we can conclude that the formed biological complex (Bcl-xl)$_{(1-212)}$-Bcl-xl$_{(213-233)}$ is more stable than (Bcl-xl)$_2$ and Bcl-xl$_{(1-212)}$)$_2$. This result correlates with the pattern of the curves in the Fig. 3.2.

Thus, the TM domain can play an essential role in the formation of the protein homodimer (Bcl-xl)$_2$, and its absence in the dimer Bcl-xl$_{(1-212)}$)$_2$ can lead to destabilization of this truncated homodimer.

The interaction of the TM domain of Bcl-xl$_{(213-233)}$ with itself was investigated separately. The results shown in the graph Fig. 3.2 are described by a black curve.

The numerical results (see Table 3.3) demonstrate that the values of lg(cond(W)) for the formation of the complex Bcl-$xl_{(213-233)}$ are higher than the values of lg(cond(W)) corresponding to the interaction (Bcl-xl)$_{(1-212)}$- Bcl-$xl_{(213-233)}$ and are lower than the values of lg(cond(W)) that correspond to the interactions of (Bcl-xl)$_2$ and Bcl-$xl_{(1-212)}$)$_2$. This result qualitatively coincides with the value of lg(cond(W)), which was obtained for a mathematical model of the temperature effect on the character of linking between monomeric proteins in aqueous solutions.

As can be seen from the presented graph Fig. 3.2, the values of lg(cond(W)) of the homodimer of Bcl-$xl_{(213-233)}$)$_2$ proteins lie in a lower range than the values of lg(cond(W)) of the protein homodimer (Bcl-xl)$_2$.

The result obtained for Bcl-$xl_{(213-233)}$)$_2$ can be interpreted as the formation of a biological complex, but with a temperature change it behaves less unstably than Bcl-$xl_{(1-212)}$–Bcl-$xl_{(213-233)}$.

The curve Bcl-$xl_{(213-233)}$)$_2$ tends to increase the values of lg(cond(W)) from 20° to 40°, varying from the values 18.937–19.195, respectively.

Thus, we assume that the TM domain of the protein Bcl-$xl_{(213-233)}$ can form a complex with the same site Bcl-$xl_{(213-233)}$, but the interaction with the globular part of the protein Bcl-xl is preferable for the TM domain.

The constructed mathematical model allows to determine:

– the influence of temperature on the character of binding of monomeric proteins in aqueous solutions

– the contribution of different sections of the examined proteins to the formation of a stable biological complex in a given temperature range.

3.6 Matlab Script for Mathematical Modelling of the Temperature Effect on the Character of Linking Between Monomeric Proteins in Aqueous Solutions

Input parameters:

1. S_1, S_{20} are amino acid sequences of biological omplexes ($S_1 \geq S_{20}$)
2. epsilon$_1$ is the dielectric constant of the medium
3. t is the temperature of the medium

Output parameters:

lg(cond(W) is the common logarithm of the condition number of the matrix W, where its elements are composed of the electrostatic potential energy which is created based on the interaction between pair of amino acid residues of biological complexes.

Compute:

lg(cond(W) is the common logarithm of the condition number of the matrix W, which will allow a prediction the reactivity of the studied biological complexes.

```
1  clear all
2  clc
3  format long e
4  %H2B
5  S_20=['M'  'S'  'A'  'K'  'A'  'E'  'K' 'K'...
6  'P'  'A'  'S'  'K'  'A'  'P'  'A'  'E'  'K'  'K'  'P'...
7  'A' 'A'  'K'  'K'  'T'  'S'  'T'  'S'  'T'  'D'  'G' ...
8  'K'  'K'  'R'  'S'  'K'  'A'  'R'  'K'  'E'  'T'  'Y' ...
9  'S'  'S'  'Y'  'I'  'Y'  'K'  'V'  'L'  'K'  'Q'  'T' ...
10 'H'  'P'  'D'  'T'  'G'  'I'  'S'  'Q'  'K'  'S'  'M' 'S'...
11 'M'  'S'  'I'  'L'  'N'  'S'  'F'  'V'  'N'  'D'  'I' 'F' ...
12 'E'  'R'  'I'  'A'  'T'  'E'  'A'  'S'  'K'  'L'  'A'...
13 'A'  'Y'  'N'  'K'  'K'  'S'  'T'  'I'  'S'  'A'  'R' ...
14 'E'  'I'  'Q'  'T'  'A'  'V'  'R'  'L'  'I'  'L'  'P' ...
15 'G'  'E'  'L'  'A'  'K'  'H'  'A'  'V'  'S'  'E'  'G'...
16 'T'  'R'  'A'  'V'  'T'  'K'  'Y'  'S'  'S'  'S'  'T' ...
17 'Q' 'A' ]
18 %H2A
19 S_1=['M'  'S'  'G'  'G'  'K'  'G'  'G'  'K'  ...
20 'A'  'G'  'S'  'A'  'A'  'K'  'A'  'S'  'Q'  'S'  'R' 'S'...
21 'A'  'K'  'A'  'G'  'L'  'T'  'F'  'P'  'V'  'G' 'R'  'V' ...
22 'H'  'R'  'L'  'L'  'R'  'R'  'G'  'N' 'Y'  'A'  'Q'  'R' ...
23 'I'  'G'  'S'  'G'  'A'  'P' 'V'  'Y'  'L'  'T'  'A'  'V' ...
24 'L'  'E'  'Y'  'L' 'A'  'A'  'E'  'I'  'L'  'E'  'L'  'A' ...
25 'G'  'N' 'A'  'A'  'R'  'D'  'N'  'K'  'K'  'T'  'R'  'I' ...
26 'I'  'P'  'R'  'H'  'L'  'Q'  'L'  'A'  'I'  'R' 'N'  'D'...
27 'D'  'E'  'L'  'N'  'K'  'L'  'L'  'G' 'N'  'V'  'T'  'I' ...
28 'A'  'Q'  'G'  'G'  'V'  'L' 'P'  'N'  'I'  'H'  'Q'  'N'
29 'L'  'L'  'P'  'K' 'K'  'S'  'A'  'K'  'A'  'T'  'K'  'A' ...
30  'S'  'Q' 'E'  'L' ]
31 t=20;
32 epsilon1=80.103;
33 rtt=0;
34 N1=300;
35 [S_1,S_2,Q1,Q2,R1,R2,h,M,N]=...
36 potential_20(t,epsilon1,rtt,S_1,S_20);
37 [A]=electrostatic(Q1,Q2, R1,R2,h,M,N,N1,epsilon1);
38 [R20]=condmy(A)
39 t=25;
40 epsilon1=78.304;
41 [S_1,S_2,Q1,Q2,R1,R2,h,M,N]=...
42 potential_25(t,epsilon1,rtt,S_1,S_20);
43 [A]=electrostatic(Q1,Q2, R1,R2,h,M,N,N1,epsilon1);
44 [R25]=condmy(A)
45 t=30;
46 epsilon1=76.546;
47 [S_1,S_2,Q1,Q2,R1,R2,h,M,N]=...
48 potential_30(t,epsilon1,rtt,S_1,S_20);
49 [A]=electrostatic(Q1,Q2, R1,R2,h,M,N,N1,epsilon1);
50 [R30]=condmy(A)
```

```
51  t=35;
52  epsilon1=74.828;
53  [S_1,S_2,Q1,Q2,R1,R2,h,M,N]=...
54  potential_35(t,epsilon1,rtt,S_1,S_20);
55  [A]=electrostatic(Q1,Q2, R1,R2,h,M,N,N1,epsilon1);
56  [R35]=condmy(A)
57  t=40;
58  epsilon1=73.151;
59  [S_1,S_2,Q1,Q2,R1,R2,h,M,N]=...
60  potential_40(t,epsilon1,rtt,S_1,S_20);
61  [A]=electrostatic(Q1,Q2, R1,R2,h,M,N,N1,epsilon1);
62  [R40]=condmy(A)
63  %-----------------------------------------------
64  %H3
65  S_1=['M'  'A'  'R'  'T'  'K'  'Q'  'T'  'A'  'R'  'K' ...
66  'S'  'T'  'G'  'G'  'K'  'A'  'P'  'R'  'K'  'Q'  'L'...
67  'A'  'S'  'K'  'A'  'A'  'R'  'K'  'S'  'A'  'P'  'S'  'T'...
68  'G'  'G'  'V'  'K'  'K'  'P'  'H'  'R'  'Y'  'K'  'P'  'G'...
69  'T'  'V'  'A'  'L'  'R'  'E'  'I'  'R'  'R'  'F'  'Q'  'K'...
70  'S'  'T'  'E'  'L'  'L'  'I'  'R'  'K'  'L'  'P'  'F'  'Q'...
71  'R'  'L'  'V'  'R'  'E'  'I'  'A'  'Q'  'D'  'F'  'K'  'T'...
72  'D'  'L'  'R'  'F'  'Q'  'S'  'S'  'A'  'I'  'G'  'A'  'L'...
73  'Q'  'E'  'S'  'V'  'E'  'A'  'Y'  'L'  'V'  'S'  'L'  'F'...
74  'E'  'D'  'T'  'N'  'L'  'A'  'A'  'I'  'H'  'A'  'K'  'R'...
75  'V'  'T'  'I'  'Q'  'K'  'K'  'D'  'I'  'K'  'L'  'A'  'R'...
76  'R' 'L'  'R'  'G'  'E'  'R'  'S' ]
77  %H4
78  S_20= ['M'  'S'  'G'  'R'  'G'  'K'  'G'  'G'...
79  'K'  'G'  'L'  'G'  'K'  'G'  'G'  'A'  'K'  'R'  'H'...
80  'R'  'K'  'I'  'L'  'R'  'D'  'N'  'I'  'Q'  'G'  'I'...
81  'T'  'K'  'P'  'A'  'I'  'R'  'R'  'L'  'A'  'R'  'R'  'G'...
82  'G'  'V'  'K'  'R'  'I'  'S'  'G'  'L'  'I'  'Y'  'E'  'E'...
83  'V'  'R'  'A'  'V'  'L'  'K'  'S'  'F'  'L'  'E'  'S'  'V'...
84  'I'  'R'  'D'  'S'  'V'  'T'  'Y'  'T'  'E'  'H'  'A'  'K'...
85  'R'  'K'  'T'  'V'  'T'  'S'  'L'  'D'  'V'  'V'  'Y'  'A'
86  'L'  'K'  'R'  'Q'  'G'  'R'  'T'  'L'  'Y'  'G'  'F'...
87    'G' 'G']
88  t=20;
89  epsilon1=80.103;
90  [S_1,S_2,Q1,Q2,R1,R2,h,M,N]=...
91  potential_20(t,epsilon1,rtt,S_1,S_20);
92  [A]=electrostatic(Q1,Q2, R1,R2,h,M,N,N1,epsilon1);
93  [R_20]=condmy(A)
94  t=25;
95  epsilon1=78.304;
96  [S_1,S_2,Q1,Q2,R1,R2,h,M,N]=...
97  potential_25(t,epsilon1,rtt,S_1,S_20);
98  [A]=electrostatic(Q1,Q2, R1,R2,h,M,N,N1,epsilon1);
99  [R_25]=condmy(A)
100 t=30;
101 epsilon1=76.546;
```

```
102 [S_1,S_2,Q1,Q2,R1,R2,h,M,N]=...
103 potential_30(t,epsilon1,rtt,S_1,S_20);
104 [A]=electrostatic(Q1,Q2, R1,R2,h,M,N,N1,epsilon1);
105 [R_30]=condmy(A)
106 t=35;
107 epsilon1=74.828;
108 [S_1,S_2,Q1,Q2,R1,R2,h,M,N]=...
109 potential_35(t,epsilon1,rtt,S_1,S_20);
110 [A]=electrostatic(Q1,Q2, R1,R2,h,M,N,N1,epsilon1);
111 [R_35]=condmy(A)
112 t=40;
113 epsilon1=73.151;
114 [S_1,S_2,Q1,Q2,R1,R2,h,M,N]=...
115 potential_40(t,epsilon1,rtt,S_1,S_20);
116 [A]=electrostatic(Q1,Q2, R1,R2,h,M,N,N1,epsilon1);
117 [R_40]=condmy(A)
118 T=  [20 25 30 35 40 ];
119 R_1=[R20 R25 R30 R35 R40 ];
120 R_2=[R_20 R_25 R_30 R_35 R_40 ];
121 h = .1;
122 xi = 20:h:40;
123 yi1 = interp1(T,R_1, xi, 'cubic');
124 yi2 = interp1(T,R_2, xi, 'cubic');
125 N5=14;
126 set(0,'DefaultTextInterpreter', 'latex');
127 figure
128 hold on
129 plot(xi, yi1, 'k', 'LineWidth' ,2)
130 plot(xi, yi2, 'r', 'LineWidth' ,2);
131 legend('H2A-H2B', 'H3-H4')
132 set(0,'DefaultTextFontSize',N5,...
133 'DefaultTextFontName','Arial Cyr');
134 xlabel('temperature, °C');
135 set(0,'DefaultTextFontSize',N5, ...
136 'DefaultTextFontName','Arial Cyr');
137 ylabel('lg(cond(W))');
138
139 %----------------------------------------------------------
140 clear all
141 clc
142 format long e
143 %Bcl-xl(213-233)
144 S_1=['W'  'F'  'L'  'T'  'G'  'M'  'T'  'V' 'A'  'G'  ....
145 'V'  'V'  'L'  'L'  'G'  'S'  'L'  'F' 'S'  'R'  'K'] ;
146 %Bcl-xl(213-233)
147 S_20=['W'  'F'  'L'  'T'  'G'  'M'  'T'  'V' 'A' ....
148 'G'  'V'  'V'  'L'  'L'  'G'  'S'  'L'  'F' 'S' ...
149 'R'  'K'] ;
150 t=20;
151 epsilon1=80.103;
152 rtt=0;
153 N1=300;
```

```
154 [S_1,S_2,Q1,Q2,R1,R2,h,M,N]=...
155 potential_20(t,epsilon1,rtt,S_1,S_20);
156 [A]=electrostatic(Q1,Q2, R1,R2,h,M,N,N1,epsilon1);
157 [R20]=condmy(A)
158 t=25;
159 epsilon1=78.304;
160 [S_1,S_2,Q1,Q2,R1,R2,h,M,N]=...
161 potential_25(t,epsilon1,rtt,S_1,S_20);
162 [A]=electrostatic(Q1,Q2, R1,R2,h,M,N,N1,epsilon1);
163 [R25]=condmy(A)
164 t=30;
165 epsilon1=76.546;
166 [S_1,S_2,Q1,Q2,R1,R2,h,M,N]=...
167 potential_30(t,epsilon1,rtt,S_1,S_20);
168 [A]=electrostatic(Q1,Q2, R1,R2,h,M,N,N1,epsilon1);
169 [R30]=condmy(A)
170 t=35;
171 epsilon1=74.828;
172 [S_1,S_2,Q1,Q2,R1,R2,h,M,N]=...
173 potential_35(t,epsilon1,rtt,S_1,S_20);
174 [A]=electrostatic(Q1,Q2, R1,R2,h,M,N,N1,epsilon1);
175 [R35]=condmy(A)
176 t=40;
177 epsilon1=73.151;
178 [S_1,S_2,Q1,Q2,R1,R2,h,M,N]=...
179 potential_40(t,epsilon1,rtt,S_1,S_20);
180 [A]=electrostatic(Q1,Q2, R1,R2,h,M,N,N1,epsilon1);
181 [R40]=condmy(A)
182 %----------------------------------------------
183 %Bcl-xl
184 S_1=['M' 'S' 'Q' 'S' 'N' 'R' 'E' 'L' 'V' 'V'...
185 'D' 'F' 'L' 'S' 'Y' 'K' 'L' 'S' 'Q' 'K' 'G'...
186 'Y' 'S' 'W' 'S' 'Q' 'F' 'S' 'D' 'V' 'E' 'E'...
187 'N' 'R' 'T' 'E' 'A' 'P' 'E' 'G' 'T' 'E' 'S' ...
188 'E' 'M' 'E' 'T' 'P' 'S' 'A' 'I' 'N' 'G' 'N' ...
189 'P' 'S' 'W' 'H' 'L' 'A' 'D' 'S' 'P' 'A' 'V'...
190 'N' 'G' 'A' 'T' 'G' 'H' 'S' 'S' 'S' 'L' 'D' ...
191 'A' 'R' 'E' 'V' 'I' 'P' 'M' 'A' 'A' 'V' 'K' ...
192 'Q' 'A' 'L' 'R' 'E' 'A' 'G' 'D' 'E' 'F' 'E'...
193 'L' 'R' 'Y' 'R' 'R' 'A' 'F' 'S' 'D' 'L' 'T'...
194 'S' 'Q' 'L' 'H' 'I' 'T' 'P' 'G' 'T' 'A' 'Y' 'Q'...
195 'S' 'F' 'E' 'Q' 'V' 'V' 'N' 'E' 'L' 'F' 'R' 'D' ...
196 'G' 'V' 'N' 'W' 'G' 'R' 'I' 'V' 'A' 'F' 'F' 'S' ...
197 'F' 'G' 'G' 'A' 'L' 'C' 'V' 'E' 'S' 'V' 'D' 'K'...
198 'E' 'M' 'Q' 'V' 'L' 'V' 'S' 'R' 'I' 'A' 'A' 'W'...
199 'M' 'A' 'T' 'Y' 'L' 'N' 'D' 'H' 'L' 'E' 'P' 'W' ...
200 'I' 'Q' 'E' 'N' 'G' 'G' 'W' 'D' 'T' 'F' 'V' 'E' ...
201 'L' 'Y' 'G' 'N' 'N' 'A' 'A' 'A' 'E' 'S' 'R' 'K'...
202 'G' 'Q' 'E' 'R' 'F' 'N' 'R' 'W' 'F' 'L' 'T' 'G' ...
203 'M' 'T' 'V' 'A' 'G' 'V' 'V' 'L' 'L' 'G' 'S' 'L' ...
204 'F' 'S' 'R' 'K'] ;
205 %Bcl-xl
```

```
206 S_20=['M'  'S'  'Q'  'S'  'N'  'R'  'E'  'L'  'V'  'V'...
207 'D'  'F'  'L'  'S'  'Y'  'K'  'L'  'S'  'Q'  'K' 'G'...
208 'Y'  'S'  'W'  'S'  'Q'  'F'  'S'  'D'  'V' 'E'  'E'...
209 'N'  'R'  'T'  'E'  'A'  'P'  'E'  'G' 'T'  'E'  'S' ...
210 'E'  'M'  'E'  'T'  'P'  'S'  'A'  'I'  'N'  'G'  'N' ...
211 'P'  'S'  'W'  'H'  'L'  'A' 'D'  'S'  'P'  'A'  'V'...
212 'N'  'G'  'A'  'T'  'G' 'H'  'S'  'S'  'S'  'L'  'D' ...
213 'A'  'R'  'E'  'V' 'I'  'P'  'M'  'A'  'A'  'V'  'K' ...
214 'Q'  'A'  'L' 'R'  'E'  'A'  'G'  'D'  'E'  'F'  'E'...
215 'L'  'R'  'Y'  'R'  'R'  'A'  'F'  'S'  'D'  'L'  'T'...
216 'S'  'Q'  'L'  'H'  'I'  'T'  'P'  'G'  'T'  'A' 'Y' 'Q'...
217 'S'  'F'  'E'  'Q'  'V'  'V'  'N'  'E'  'L' 'F'  'R'  'D' ...
218 'G' 'V'  'N'  'W'  'G'  'R'  'I' 'V'  'A'  'F'  'F'  'S' ...
219 'F'  'G'  'G'  'A'  'L'  'C' 'V'  'E'  'S'  'V'  'D'  'K'...
220 'E'  'M'  'Q' 'V'  'L'  'V'  'S'  'R'  'I'  'A'  'A'  'W'...
221 'M' 'A'  'T'  'Y'  'L'  'N'  'D'  'H'  'L'  'E'  'P' 'W' ...
222 'I'  'Q'  'E'  'N'  'G'  'G'  'W'  'D'  'T' 'F'  'V'  'E' ...
223 'L'  'Y'  'G'  'N'  'N'  'A'  'A'  'A'  'E'  'S'  'R'  'K'...
224 'G'  'Q'  'E'  'R'  'F' 'N'  'R'  'W'  'F'  'L'  'T'  'G' ...
225 'M'  'T'  'V' 'A'  'G'  'V'  'V'  'L'  'L'  'G'  'S'  'L' ...
226 'F' 'S'  'R'  'K'] ;
227 t=20;
228 epsilon1=80.103;
229 [S_1,S_2,Q1,Q2,R1,R2,h,M,N]=...
230 potential_20(t,epsilon1,rtt,S_1,S_20);
231 [A]=electrostatic(Q1,Q2, R1,R2,h,M,N,N1,epsilon1);
232 [R_20]=condmy(A)
233 t=25;
234 epsilon1=78.304;
235 [S_1,S_2,Q1,Q2,R1,R2,h,M,N]=...
236 potential_25(t,epsilon1,rtt,S_1,S_20);
237 [A]=electrostatic(Q1,Q2, R1,R2,h,M,N,N1,epsilon1);
238 [R_25]=condmy(A)
239 t=30;
240 epsilon1=76.546;
241 [S_1,S_2,Q1,Q2,R1,R2,h,M,N]=...
242 potential_30(t,epsilon1,rtt,S_1,S_20);
243 [A]=electrostatic(Q1,Q2, R1,R2,h,M,N,N1,epsilon1);
244 [R_30]=condmy(A)
245 t=35;
246 epsilon1=74.828;
247 [S_1,S_2,Q1,Q2,R1,R2,h,M,N]=...
248 potential_35(t,epsilon1,rtt,S_1,S_20);
249 [A]=electrostatic(Q1,Q2, R1,R2,h,M,N,N1,epsilon1);
250 [R_35]=condmy(A)
251 t=40;
252 epsilon1=73.151;
253 [S_1,S_2,Q1,Q2,R1,R2,h,M,N]=...
254 potential_40(t,epsilon1,rtt,S_1,S_20);
255 [A]=electrostatic(Q1,Q2, R1,R2,h,M,N,N1,epsilon1);
256 [R_40]=condmy(A)
257 %-----------------------------------------------
```

```
258 %Bcl-xl (1-212)
259 S_1=['M'   'S'   'Q'   'S'   'N'   'R'   'E'   'L'   'V'   'V'  ...
260 'D'   'F'   'L'   'S'   'Y'   'K'   'L'   'S'   'Q'   'K'  'G'  ...
261 'Y'   'S'   'W'   'S'   'Q'   'F'   'S'   'D'   'V'  'E'   'E'...
262 'N'   'R'   'T'   'E'   'A'   'P'   'E'   'G'  'T'   'E'   'S'...
263 'E'   'M'   'E'   'T'   'P'   'S'   'A'   'I'   'N'   'G'   'N'...
264 'P'   'S'   'W'   'H'   'L'   'A'  'D'   'S'   'P'   'A'   'V'...
265 'N'   'G'   'A'   'T'   'G'  'H'   'S'   'S'   'S'   'L'   'D'  ...
266 'A'   'R'   'E'   'V'  'I'   'P'   'M'   'A'   'A'   'V'   'K'...
267 'Q'   'A'   'L'  'R'   'E'   'A'   'G'   'D'   'E'   'F'   'E'  ...
268 'L'   'R'    'Y'   'R'   'R'   'A'   'F'   'S'   'D'   'L'   'T'...
269 'S'  'Q'   'L'   'H'   'I'   'T'   'P'   'G'   'T'   'A'   'Y'  ...
270 'Q'   'S'   'F'   'E'   'Q'   'V'   'V'   'N'   'E'   'L'  'F'...
271 'R'   'D'   'G'   'V'   'N'   'W'   'G'   'R'   'I'  'V'   'A'...
272 'F'   'F'   'S'   'F'   'G'   'G'   'A'   'L'   'C'  'V'   'E'  ...
273 'S'   'V'   'D'   'K'   'E'   'M'   'Q'  'V'   'L'   'V'   'S'  ...
274 'R'   'I'   'A'   'A'   'W'   'M'  'A'   'T'   'Y'   'L'   'N'...
275 'D'   'H'   'L'   'E'   'P'  'W'   'I'   'Q'   'E'   'N'   'G'  ...
276 'G'   'W'   'D'   'T'  'F'   'V'   'E'   'L'   'Y'   'G'   'N'...
277 'N'   'A'   'A'   'A'   'E'   'S'   'R'   'K'   'G'   'Q'   'E'...
278 'R'   'F'  'N'   'R' ] ;
279 %Bcl-xl (1-212)
280 S_20=['M'   'S'   'Q'   'S'   'N'   'R'   'E'   'L'   'V'   'V'  ...
281 'D'   'F'   'L'   'S'   'Y'   'K'   'L'   'S'   'Q'   'K'  'G'  ...
282 'Y'   'S'   'W'   'S'   'Q'   'F'   'S'   'D'   'V'  'E'   'E'...
283 'N'   'R'   'T'   'E'   'A'   'P'   'E'   'G'  'T'   'E'   'S'...
284 'E'   'M'   'E'   'T'   'P'   'S'   'A'   'I'   'N'   'G'   'N'...
285 'P'   'S'   'W'   'H'   'L'   'A'  'D'   'S'   'P'   'A'   'V'...
286 'N'   'G'   'A'   'T'   'G'  'H'   'S'   'S'   'S'   'L'   'D'  ...
287 'A'   'R'   'E'   'V'  'I'   'P'   'M'   'A'   'A'   'V'   'K'...
288 'Q'   'A'   'L'  'R'   'E'   'A'   'G'   'D'   'E'   'F'   'E'  ...
289 'L'   'R'    'Y'   'R'   'R'   'A'   'F'   'S'   'D'   'L'   'T'...
290 'S'  'Q'   'L'   'H'   'I'   'T'   'P'   'G'   'T'   'A'   'Y'  ...
291 'Q'   'S'   'F'   'E'   'Q'   'V'   'V'   'N'   'E'   'L'  'F'...
292 'R'   'D'   'G'   'V'   'N'   'W'   'G'   'R'   'I'  'V'   'A'...
293 'F'   'F'   'S'   'F'   'G'   'G'   'A'   'L'   'C'  'V'   'E'  ...
294 'S'   'V'   'D'   'K'   'E'   'M'   'Q'  'V'   'L'   'V'   'S'  ...
295 'R'   'I'   'A'   'A'   'W'   'M'  'A'   'T'   'Y'   'L'   'N'...
296 'D'   'H'   'L'   'E'   'P'  'W'   'I'   'Q'   'E'   'N'   'G'  ...
297 'G'   'W'   'D'   'T'  'F'   'V'   'E'   'L'   'Y'   'G'   'N'...
298 'N'   'A'   'A'   'A'   'E'   'S'   'R'   'K'   'G'   'Q'   'E'...
299 'R'   'F'  'N'   'R' ] ;
300 t=20;
301 epsilon1=80.103;
302 [S_1,S_2,Q1,Q2,R1,R2,h,M,N]=...
303 potential_20(t,epsilon1,rtt,S_1,S_20);
304 [A]=electrostatic(Q1,Q2, R1,R2,h,M,N,N1,epsilon1);
305 [R_20_1]=condmy(A)
306 t=25;
307 epsilon1=78.304;
308 [S_1,S_2,Q1,Q2,R1,R2,h,M,N]=...
309 potential_25(t,epsilon1,rtt,S_1,S_20);
```

```
310 [A]=electrostatic(Q1,Q2, R1,R2,h,M,N,N1,epsilon1);
311 [R_25_1]=condmy(A)
312 t=30;
313 epsilon1=76.546;
314 [S_1,S_2,Q1,Q2,R1,R2,h,M,N]=...
315 potential_30(t,epsilon1,rtt,S_1,S_20);
316 [A]=electrostatic(Q1,Q2, R1,R2,h,M,N,N1,epsilon1);
317 [R_30_1]=condmy(A)
318 t=35;
319 epsilon1=74.828;
320 [S_1,S_2,Q1,Q2,R1,R2,h,M,N]=...
321 potential_35(t,epsilon1,rtt,S_1,S_20);
322 [A]=electrostatic(Q1,Q2, R1,R2,h,M,N,N1,epsilon1);
323 [R_35_1]=condmy(A)
324 t=40;
325 epsilon1=73.151;
326 [S_1,S_2,Q1,Q2,R1,R2,h,M,N]=...
327 potential_40(t,epsilon1,rtt,S_1,S_20);
328 [A]=electrostatic(Q1,Q2, R1,R2,h,M,N,N1,epsilon1);
329 [R_40_1]=condmy(A)
330 %---------------------------------------------------
331 %Bcl-xl (1-212)
332 S_1=['M' 'S' 'Q' 'S' 'N' 'R' 'E' 'L' 'V' 'V' ...
333 'D' 'F' 'L' 'S' 'Y' 'K' 'L' 'S' 'Q' 'K' 'G' ...
334 'Y' 'S' 'W' 'S' 'Q' 'F' 'S' 'D' 'V' 'E' 'E'...
335 'N' 'R' 'T' 'E' 'A' 'P' 'E' 'G' 'T' 'E' 'S'...
336 'E' 'M' 'E' 'T' 'P' 'S' 'A' 'I' 'N' 'G' 'N'...
337 'P' 'S' 'W' 'H' 'L' 'A' 'D' 'S' 'P' 'A' 'V'...
338 'N' 'G' 'A' 'T' 'G' 'H' 'S' 'S' 'S' 'L' 'D' ...
339 'A' 'R' 'E' 'V' 'I' 'P' 'M' 'A' 'A' 'V' 'K'...
340 'Q' 'A' 'L' 'R' 'E' 'A' 'G' 'D' 'E' 'F' 'E' ...
341 'L' 'R' 'Y' 'R' 'R' 'A' 'F' 'S' 'D' 'L' 'T'...
342 'S' 'Q' 'L' 'H' 'I' 'T' 'P' 'G' 'T' 'A' 'Y' ...
343 'Q' 'S' 'F' 'E' 'Q' 'V' 'V' 'N' 'E' 'L' 'F'...
344 'R' 'D' 'G' 'V' 'N' 'W' 'G' 'R' 'I' 'V' 'A'...
345 'F' 'F' 'S' 'F' 'G' 'G' 'A' 'L' 'C' 'V' 'E' ...
346 'S' 'V' 'D' 'K' 'E' 'M' 'Q' 'V' 'L' 'V' 'S' ...
347 'R' 'I' 'A' 'A' 'W' 'M' 'A' 'T' 'Y' 'L' 'N'...
348 'D' 'H' 'L' 'E' 'P' 'W' 'I' 'Q' 'E' 'N' 'G' ...
349 'G' 'W' 'D' 'T' 'F' 'V' 'E' 'L' 'Y' 'G' 'N'...
350 'N' 'A' 'A' 'A' 'E' 'S' 'R' 'K' 'G' 'Q' 'E'...
351 'R' 'F' 'N' 'R' ] ;
352 %Bcl-xl(213-233)
353 S_20=['W' 'F' 'L' 'T' 'G' 'M' ...
354 'T' 'V' 'A' 'G' 'V' 'V' 'L' 'L' 'G' 'S' 'L'...
355 'F' 'S' 'R' 'K'] ;
356 t=20;
357 epsilon1=80.103;
358 [S_1,S_2,Q1,Q2,R1,R2,h,M,N]=...
359 potential_20(t,epsilon1,rtt,S_1,S_20);
360 [A]=electrostatic(Q1,Q2, R1,R2,h,M,N,N1,epsilon1);
361 [R_20_2]=condmy(A)
```

```
362 t=25;
363 epsilon1=78.304;
364 [S_1,S_2,Q1,Q2,R1,R2,h,M,N]=...
365 potential_25(t,epsilon1,rtt,S_1,S_20);
366 [A]=electrostatic(Q1,Q2, R1,R2,h,M,N,N1,epsilon1);
367 [R_25_2]=condmy(A)
368 t=30;
369 epsilon1=76.546;
370 [S_1,S_2,Q1,Q2,R1,R2,h,M,N]=...
371 potential_30(t,epsilon1,rtt,S_1,S_20);
372 [A]=electrostatic(Q1,Q2, R1,R2,h,M,N,N1,epsilon1);
373 [R_30_2]=condmy(A)
374 t=35;
375 epsilon1=74.828;
376 [S_1,S_2,Q1,Q2,R1,R2,h,M,N]=...
377 potential_35(t,epsilon1,rtt,S_1,S_20);
378 [A]=electrostatic(Q1,Q2, R1,R2,h,M,N,N1,epsilon1);
379 [R_35_2]=condmy(A)
380 t=40;
381 epsilon1=73.151;
382 [S_1,S_2,Q1,Q2,R1,R2,h,M,N]=...
383 potential_40(t,epsilon1,rtt,S_1,S_20);
384 [A]=electrostatic(Q1,Q2, R1,R2,h,M,N,N1,epsilon1);
385 [R_40_2]=condmy(A)
386 T=  [20 25 30 35 40 ];
387 R_1=[R20 R25 R30 R35 R40 ];
388 R_2=[R_20 R_25 R_30 R_35 R_40 ];
389 R_3=[R_20_1 R_25_1 R_30_1 R_35_1 R_40_1 ];
390 R_4=[R_20_2 R_25_2 R_30_2 R_35_2 R_40_2 ];
391 h = .1;
392 xi = 20:h:40;
393 yi1 = interp1(T,R_1, xi, 'cubic');
394 yi2 = interp1(T,R_2, xi, 'cubic');
395 yi3 = interp1(T,R_3, xi, 'cubic');
396 yi4 = interp1(T,R_4, xi, 'cubic');
397 N5=14;
398 set(0,'DefaultTextInterpreter', 'latex');
399 hold on
400 plot(xi,yi1, 'k', 'LineWidth' ,2)
401 plot(xi,yi2, 'r', 'LineWidth' ,2);
402 plot(xi,yi3, 'b', 'LineWidth' ,2);
403 plot(xi,yi4, 'g', 'LineWidth' ,2);
404 legend('Bcl-xl(213-233)-Bcl-xl(213-233)','Bcl-xl-Bcl-xl',...
405 'Bcl-xl(1-212)-Bcl-xl(1-212)','Bcl-xl(1-212)-Bcl-xl(212-233)')
406 set(0,'DefaultTextFontSize',N5,...
407 'DefaultTextFontName','Arial Cyr');
408 xlabel('temperature, °C');
409 set(0,'DefaultTextFontSize',N5,...
410 'DefaultTextFontName','Arial Cyr');
411 ylabel('lg(cond(W))');
412 %-----------------------------------------------
413
414 function [S_1,S_2,Q1,Q2,R1,R2,h,M,N]=...
```

```
415 potential_20(t,epsilon1,rtt,S_1,S_20);
416 Hhidro=0;
417 Rt1=(4.6*1E-13);
418 epsilon0=8.85418781762*10^(-12);
419 k=1/(4*pi*epsilon0);
420 B=(1.38064852*10^(-23))/(1.6021766208*10^(-19));
421 Ea=1.8;
422 ra=0.6;
423 Ra=ra*1E-9-Rt1;
424 pha=Ea*(t+273)*B;
425 qA=(pha*Ra*epsilon1)*k^(-1);
426 Er=-0.9;
427 rr=0.8;
428 Rr=rr*1E-9+Rt1;
429 phr=Er*(t+273)*B;
430 qR=(phr*Rr*epsilon1)*k^(-1);
431 En=0.2;
432 rn=0.682;
433 Rn=rn*1E-9+Rt1;
434 phn=En*(t+273)*B;
435 qN=(phn*Rn*epsilon1)*k^(-1);
436 Ed=-0.01;
437 rd=0.666;
438 Rd=rd*1E-9+Rt1;
439 phd=Ed*(t+273)*B;
440 qD=(phd*Rd*epsilon1)*k^(-1);
441 Ec=2.45;
442 rc=0.629;
443 phc=Ec*(t+273)*B;
444 Rc=rc*1E-9+Rt1;
445 qC=(phc*Rc*epsilon1)*k^(-1);
446 Eq=-0.65;
447 rq=0.725;
448 Rq=rq*1E-9+Rt1;
449 phq=Eq*(t+273)*B;
450 qQ=(phq*Rq*epsilon1)*k^(-1);
451 Ee=-0.1;
452 re=0.714;
453 Re=re*1E-9+Rt1;
454 phe=Ee*(t+273)*B;
455 qE=(phe*Re*epsilon1)*k^(-1);
456 Eg=1.05;
457 rg=0.725;
458 Rg=rg*1E-9+Rt1;
459 phg=Eg*(t+273)*B;
460 qG=(phg*Rg*epsilon1)*k^(-1);
461 Eh=0.05;
462 rh=0.725;
463 Rh=rh*1E-9+Rt1;
464 phh=Eh*(t+273)*B;
465 qH=(phh*Rh*epsilon1)*k^(-1);
466 Ei=0.92;
```

```
467 ri=0.735;
468 Ri=ri*1E-9-Rt1;
469 phi=Ei*(t+273)*B;
470 qI=(phi*Ri*epsilon1)*k^(-1);
471 El=0.75;
472 rl=0.734;
473 Rl=rl*1E-9-Rt1;
474 phl=El*(t+273)*B;
475 qL=(phl*Rl*epsilon1)*k^(-1);
476 Ek=-1.2;
477 rk=0.737;
478 Rk=rk*1E-9+Rt1;
479 phk=Ek*(t+273)*B;
480 qK=(phk*Rk*epsilon1)*k^(-1);
481 Em=0.25;
482 rm=0.741;
483 Rm=rm*1E-9-Rt1;
484 phm=Em*(t+273)*B;
485 qM=(phm*Rm*epsilon1)*k^(-1);
486 Ef=0.72;
487 rf=0.781;
488 Rf=rf*1E-9-Rt1;
489 phf=Ef*(t+273)*B;
490 qF=(phf*Rf*epsilon1)*k^(-1);
491 Ep=0.3;
492 rp=0.672;
493 Rp=rp*1E-9-Rt1;
494 php=Ep*(t+273)*B;
495 qP=(php*Rp*epsilon1)*k^(-1);
496 Es=0.55;
497 rs=0.615;
498 Rs=rs*1E-9+Rt1;
499 phs=Es*(t+273)*B;
500 qS=(phs*Rs*epsilon1)*k^(-1);
501 Et=0.85;
502 rt=0.659;
503 Rt=rt*1E-9+Rt1;
504 pht=Et*(t+273)*B;
505 qT=(pht*Rt*epsilon1)*k^(-1);
506 Ew=0.67;
507 rw=0.826;
508 Rw=rw*1E-9-Rt1;
509 phw=Ew*(t+273)*B;
510 qW=(phw*Rw*epsilon1)*k^(-1);
511 Ey=0.5;
512 ry=0.781;
513 Ry=ry*1E-9-Rt1;
514 phy=Ey*(t+273)*B;
515 qY=(phy*Ry*epsilon1)*k^(-1);
516 Ev=0.8;
517 rv=0.694;
518 Rv=rv*1E-9-Rt1;
```

```
519 phv=Ev*(t+273)*B;
520 qV=(phv*Rv*epsilon1)*k^(-1);
521 N=length(S_1);
522 M=length(S_20);
523 S_2=S_20;
524 Q1=[];
525 Q2=[];
526 Q3=[];
527 Q4=[];
528 R1=[];
529 R2=[];
530 h=[];
531 for i=1:length(S_1);
532 if (S_1(i)=='A')
533 Q1(i)=qA;
534 else
535    if (S_1(i)=='R')
536 Q1(i)=qR;
537    else
538   if (S_1(i)=='N')
539 Q1(i)=qN;
540   else
541    if (S_1(i)=='D')
542 Q1(i)=qD;
543   else
544   if (S_1(i)=='C')
545 Q1(i)=qC;
546   else
547    if (S_1(i)=='Q')
548 Q1(i)=qQ;
549    else
550    if (S_1(i)=='E')
551 Q1(i)=qE;
552    else
553     if (S_1(i)=='G')
554 Q1(i)=qG;
555     else
556   if (S_1(i)=='K')
557 Q1(i)=qK;
558   else
559    if (S_1(i)=='P')
560 Q1(i)=qP;
561    else
562   if (S_1(i)=='S')
563 Q1(i)=qS;
564   else
565    if (S_1(i)=='T')
566 Q1(i)=qT;
567    else
568   if (S_1(i)=='I')
569 Q1(i)=qI;
570   else
```

```
571     if (S_1(i)=='V')
572 Q1(i)=qV;
573    else
574     if (S_1(i)=='L')
575 Q1(i)=qL;
576     else
577     if (S_1(i)=='F')
578 Q1(i)=qF;
579    else
580    if (S_1(i)=='W')
581 Q1(i)=qW;
582    else
583    if (S_1(i)=='Y')
584 Q1(i)=qY;
585    else
586    if (S_1(i)=='M')
587 Q1(i)=qM;
588    else
589    if (S_1(i)=='H')
590 Q1(i)=qH;
591 end
592 end
593 end
594 end
595 end
596 end
597 end
598 end
599 end
600 end
601 end
602 end
603 end
604 end
605 end
606 end
607 end
608 end
609 end
610 end
611 end
612 for j=1:length(S_2);
613 if (S_2(j)=='A')
614 Q2(j)=qA;
615 else
616     if (S_2(j)=='R')
617 Q2(j)=qR;
618     else
619    if (S_2(j)=='N')
620 Q2(j)=qN;
621    else
622    if (S_2(j)=='D')
```

```
623 Q2(j)=qD;
624   else
625   if (S_2(j)=='C')
626 Q2(j)=qC;
627   else
628    if (S_2(j)=='Q')
629 Q2(j)=qQ;
630    else
631    if (S_2(j)=='E')
632 Q2(j)=qE;
633    else
634     if (S_2(j)=='G')
635 Q2(j)=qG;
636     else
637   if (S_2(j)=='K')
638 Q2(j)=qK;
639   else
640    if (S_2(j)=='P')
641 Q2(j)=qP;
642    else
643   if (S_2(j)=='S')
644 Q2(j)=qS;
645   else
646    if (S_2(j)=='T')
647 Q2(j)=qT;
648    else
649   if (S_2(j)=='I')
650 Q2(j)=qI;
651   else
652    if (S_2(j)=='V')
653 Q2(j)=qV;
654   else
655    if (S_2(j)=='L')
656 Q2(j)=qL;
657    else
658    if (S_2(j)=='F')
659 Q2(j)=qF;
660   else
661   if (S_2(j)=='W')
662 Q2(j)=qW;
663   else
664   if (S_2(j)=='Y')
665 Q2(j)=qY;
666   else
667   if (S_2(j)=='M')
668 Q2(j)=qM;
669   else
670   if (S_2(j)=='H')
671 Q2(j)=qH;
672 end
673 end
674 end
```

```
675  end
676  end
677  end
678  end
679  end
680  end
681  end
682  end
683  end
684  end
685  end
686  end
687  end
688  end
689  end
690  end
691  end
692  end
693  for i=1:length(S_1);
694      for j=1:length(S_2);
695       if (S_1(i)=='A')|(S_2(j)=='A');
696              R1(i)=Ra;
697              R2(j)=Ra;
698          else
699          if (S_1(i)=='R')|(S_2(j)=='R');
700              R1(i)=Rr;
701              R2(j)=Rr;
702          else
703  if (S_1(i)=='N')|(S_2(j)=='N');
704              R1(i)=Rn;
705              R2(j)=Rn;
706  else
707  if (S_1(i)=='D')|(S_2(j)=='D');
708      R1(i)=Rd;
709      R2(j)=Rd;
710  else
711        if (S_1(i)=='C')|(S_2(j)=='C');
712              R1(i)=Rc;
713              R2(j)=Rc;
714          else
715  if (S_1(i)=='Q')|(S_2(j)=='Q');
716              R1(i)=Rc;
717              R2(j)=Rc;
718          else
719          if (S_1(i)=='E')|(S_2(j)=='E');
720              R1(i)=Re;
721              R2(j)=Re;
722          else
723            if (S_1(i)=='G')|(S_2(j)=='G');
724              R1(i)=Rg;
725              R2(j)=Rg;
726          else
```

```
727  if (S_1(i)=='H')|(S_2(j)=='H');
728       R1(i)=Rh;
729       R2(j)=Rh;
730  else
731      if (S_1(i)=='I')|(S_2(j)=='I');
732           R1(i)=0.735E-9-Rt1;
733           R2(j)=0.735E-9-Rt1;
734         else
735      if (S_1(i)=='L')|(S_2(j)=='L');
736           R1(i)=Rl;
737           R2(j)=Rl;
738         else
739      if (S_1(i)=='K')|(S_2(j)=='K')
740          R1(i)=Rk;
741           R2(j)=Rk;
742         else
743      if (S_1(i)=='M')|(S_2(j)=='M')
744           R1(i)=Rm;
745           R2(j)=Rm;
746         else
747      if (S_1(i)=='F')|(S_2(j)=='F')
748           R1(i)=Rf;
749           R2(j)=Rf;
750      else
751        if (S_1(i)=='P')|(S_2(j)=='P');
752           R1(i)=Rp;
753           R2(j)=Rp;
754         else
755          if (S_1(i)=='S')|(S_2(j)=='S');
756           R1(i)=Rs;
757           R2(j)=Rs;
758         else
759          if (S_1(i)=='T')|(S_2(j)=='T');
760           R1(i)=Rt;
761           R2(j)=Rt;
762         else
763          if (S_1(i)=='W')|(S_2(j)=='W');
764           R1(i)=Rw;
765           R2(j)=Rw;
766         else
767          if (S_1(i)=='Y')|(S_2(j)=='Y');
768           R1(i)=Ry;
769           R2(j)=Ry;
770          else
771             if (S_1(i)=='V')|(S_2(j)=='V');
772             R1(i)=Rv;
773             R2(j)=Rv;
774             else
775
776 end
777 end
778 end
```

```
779 end
780 end
781 end
782 end
783 end
784 end
785 end
786 end
787 end
788 end
789 end
790 end
791 end
792 end
793 end
794 end
795 end
796 end
797 end
798 for i=1:length(S_1);
799   for j=1:length(S_2);
800
801 if (S_1(i)=='R'& S_2(j)=='D');
802     h(i,j)=.15*10^(-9)+Rr+Rd+2*Rt1;
803 else
804 if (S_1(i)=='R'& S_2(j)=='E');
805         h(i,j)=.15*10^(-9)+Rr+Re+2*Rt1;
806           else
807 if (S_1(i)=='D'& S_2(j)=='R');
808 h(i,j)=.15*10^(-9)+Rd+Rr+2*Rt1;
809 else
810 if (S_1(i)=='D'& S_2(j)=='H');
811 h(i,j)=.15*10^(-9)+Rd+Rh+2*Rt1;
812   else
813 if (S_1(i)=='D'& S_2(j)=='R');
814 h(i,j)=.15*10^(-9)+Rd+Rr+2*Rt1;
815 else
816     if (S_1(i)=='D'& S_2(j)=='H');
817     h(i,j)=.15*10^(-9)+Rd+Rh+2*Rt1;
818   else
819 if (S_1(i)=='D'& S_2(j)=='K');
820 h(i,j)=.15*10^(-9)+Rd+Rk+2*Rt1;
821  else
822 if (S_1(i)=='E')& (S_2(j)=='R');
823 h(i,j)=.15*10^(-9)+Re+Rr+2*Rt1;
824           else
825 if (S_1(i)=='E'& S_2(j)=='H');
826 h(i,j)=.15*10^(-9)+Re+Rh+2*Rt1;
827       else
828 if (S_1(i)=='E'& S_2(j)=='K');
829 h(i,j)=.15*10^(-9)+Re+Rk+2*Rt1;
830   else
```

```
831 if (S_1(i)=='H'& S_2(j)=='D')
832 h(i,j)=.15*10^(-9)+Rh+Rd+2*Rt1;
833 else
834 if (S_1(i)=='H'& S_2(j)=='E')
835 h(i,j)=.15*10^(-9)+Rh+Re+2*Rt1;
836 else
837
838 if (S_1(i)=='R'& S_2(j)=='R')
839     h(i,j)=.4*10^(-9)+Rr+Rr;
840    else
841   if (S_1(i)=='R'& S_2(j)=='H')
842    h(i,j)=.4*10^(-9)+Rr+Rh;
843   else
844 if (S_1(i)=='R'& S_2(j)=='H')
845     h(i,j)=.4*10^(-9)+Rr+Rh;
846 else
847   if (S_1(i)=='R'& S_2(j)=='K')
848     h(i,j)=.4*10^(-9)+Rr+Rk;
849  else
850 if (S_1(i)=='D'& S_2(j)=='E');
851         h(i,j)=.4*10^(-9)+Rd+Re;
852 else
853    if (S_1(i)=='D'& S_2(j)=='D');
854    h(i,j)=.4*10^(-9)+Rd+Rd;
855   else
856 if (S_1(i)=='H'& S_2(j)=='R')
857     h(i,j)=.4*10^(-9)+Rh+Rr;
858 else
859   if (S_1(i)=='H'& S_2(j)=='H')
860      h(i,j)=.4*10^(-9)+Rh+Rh;
861  else
862    if (S_1(i)=='H'& S_2(j)=='K')
863      h(i,j)=.4*10^(-9)+Rh+Rk;
864    else
865 if (S_1(i)=='K'& S_2(j)=='R')
866         h(i,j)=.4*10^(-9)+Rk+Rr;
867 else
868  if (S_1(i)=='K'& S_2(j)=='H')
869       h(i,j)=.4*10^(-9)+Rk+Rh;
870  else
871   if (S_1(i)=='K'& S_2(j)=='K')
872     h(i,j)=.4*10^(-9)+Rk+Rk;
873  else
874 if (S_1(i)=='N'& S_2(j)=='Q')
875         h(i,j)=.25*10^(-9)+Rn+Rq;
876 else
877  if (S_1(i)=='N'& S_2(j)=='S')
878    h(i,j)=.25*10^(-9)+Rn+Rs;
879   else
880   if (S_1(i)=='N'& S_2(j)=='Y')
881     h(i,j)=.25*10^(-9)+Rn+Ry;
882  else
```

```
883 if (S_1(i)=='Q'& S_2(j)=='S')| (S_1(i)=='Q')& (S_2(j)=='Y');
884         h(i,j)=.25*10^(-9)+Rq+Rs;
885 else
886  if  (S_1(i)=='Q')& (S_2(j)=='Y');
887       h(i,j)=.25*10^(-9)+Rq+Ry;
888 else
889 if (S_1(i)=='S'& S_2(j)=='Y');
890         h(i,j)=.25*10^(-9)+Rs+Ry;
891 else
892 if (S_1(i)=='I'& S_2(j)=='V')|(S_1(i)=='I'& S_2(j)=='L')|...
893 (S_1(i)=='I'& S_2(j)=='F')|(S_1(i)=='I'& S_2(j)=='W')|...
894 (S_1(i)=='I'& S_2(j)=='Y')|(S_1(i)=='I'& S_2(j)=='M')|...
895 (S_1(i)=='I'& S_2(j)=='H')|(S_1(i)=='V'& S_2(j)=='V')|...
896 (S_1(i)=='V'& S_2(j)=='L')|(S_1(i)=='V'& S_2(j)=='F')|...
897 (S_1(i)=='V'& S_2(j)=='W')|(S_1(i)=='V'& S_2(j)=='M')|...
898 (S_1(i)=='V'& S_2(j)=='H')|(S_1(i)=='L'& S_2(j)=='F')|...
899 (S_1(i)=='L'& S_2(j)=='W')|(S_1(i)=='L'& S_2(j)=='Y')|...
900 (S_1(i)=='L'& S_2(j)=='M')|(S_1(i)=='L'& S_2(j)=='H')|...
901 (S_1(i)=='F'& S_2(j)=='W')|(S_1(i)=='F'& S_2(j)=='F')|...
902 (S_1(i)=='F'& S_2(j)=='Y')|(S_1(i)=='F'& S_2(j)=='M')|...
903 (S_1(i)=='F'& S_2(j)=='H')|(S_1(i)=='W'& S_2(j)=='W')|...
904 (S_1(i)=='W'& S_2(j)=='Y')|(S_1(i)=='W'& S_2(j)=='M')|...
905 (S_1(i)=='W'& S_2(j)=='H')|(S_1(i)=='Y'& S_2(j)=='Y')|...
906 (S_1(i)=='Y'& S_2(j)=='M')|(S_1(i)=='Y'& S_2(j)=='H')|...
907 (S_1(i)=='M'& S_2(j)=='M')|(S_1(i)=='M'& S_2(j)=='H')|...
908 (S_1(i)=='H'& S_2(j)=='H')|(S_1(i)=='I'& S_2(j)=='I')|...
909 (S_1(i)=='V'& S_2(j)=='V')|(S_1(i)=='L'& S_2(j)=='L')|...
910 (S_1(i)=='F'& S_2(j)=='F')|(S_1(i)=='W'& S_2(j)=='W')
911        h(i,j)=.36*10^(-9)+(0.736*10^(-9))*2;
912  else
913   if (S_2(j)=='I'& S_1(i)=='V')|(S_2(j)=='I'& S_1(i)=='L')|...
914 (S_2(j)=='I'& S_1(i)=='F')|(S_2(j)=='I'& S_1(i)=='W')|...
915 (S_2(j)=='I'& S_1(i)=='Y')|(S_2(j)=='I'& S_1(i)=='M')|...
916 (S_2(j)=='I'& S_1(i)=='H')|(S_2(j)=='V'& S_1(i)=='V')|...
917 (S_2(j)=='V'& S_1(i)=='L')|(S_2(j)=='V'& S_1(i)=='F')|...
918 (S_2(j)=='V'& S_1(i)=='W')|(S_2(j)=='V'& S_1(i)=='M')|...
919 (S_2(j)=='V'& S_1(i)=='H')|(S_2(j)=='L'& S_1(i)=='F')|...
920 (S_2(j)=='L'& S_1(i)=='W')||(S_2(j)=='L'& S_1(i)=='Y')|...
921 (S_2(j)=='L'&  S_1(i)=='M')|(S_2(j)=='L'&  S_1(i)=='H')|...
922 (S_2(j)=='F'&  S_1(i)=='W')|(S_2(j)=='F'& S_1(i)=='F')|...
923 (S_2(j)=='F'& S_1(i)=='Y')|(S_2(j)=='F'& S_1(i)=='M')|...
924 (S_2(j)=='F'& S_1(i)=='H')|(S_2(j)=='W'& S_1(i)=='W')|...
925 (S_2(j)=='W'& S_1(i)=='Y')|(S_2(j)=='W'& S_1(i)=='M')|...
926 (S_2(j)=='W'& S_1(i)=='H')|(S_2(j)=='Y'& S_1(i)=='Y')|...
927 (S_2(j)=='Y'& S_1(i)=='M')|(S_2(j)=='Y'& S_1(i)=='H')|...
928 (S_2(j)=='M'& S_1(i)=='M')|(S_2(j)=='M'& S_1(i)=='H')|...
929 (S_2(j)=='H'& S_1(i)=='H')|(S_2(j)=='I'& S_1(i)=='I')|...
930 (S_2(j)=='V'& S_1(i)=='V')|(S_2(j)=='L'& S_1(i)=='L')|...
931 (S_2(j)=='F'& S_1(i)=='F')|(S_2(j)=='W'& S_1(i)=='W')
932         h(i,j)=.36*10^(-9)+(0.736*10^(-9))*2;
933 else
934         h(i,j)=(0.71286*10^(-9))*2+2*rtt+0.3*10^(-9);
```

```
935 end
936 end
937 end
938 end
939 end
940 end
941 end
942 end
943 end
944 end
945 end
946 end
947 end
948 end
949 end
950 end
951 end
952 end
953 end
954 end
955 end
956 end
957 end
958 end
959 end
960 end
961 end
962 end
963 end
964 end
965 end
966 end
967 end
968 end
969
970 function[S_1,S_2,Q1,Q2,R1,R2,h,M,N]=...
971 potential_25(t,epsilon1,rtt,S_1,S_20);
972 Hhidro=0;
973 Rt1=(4.6*1E-13);
974 epsilon0=8.85418781762*10^(-12);
975 k=1/(4*pi*epsilon0);
976 B=(1.38064852*10^(-23))/(1.6021766208*10^(-19));
977 Ea=1.9;
978 ra=0.6;
979 Ra=ra*1E-9-Rt1;
980 pha=Ea*(t+273)*B;
981 qA=(pha*Ra*epsilon1)*k^(-1);
982 Er=-0.82;
983 rr=0.8;
984 Rr=rr*1E-9+Rt1;
985 phr=Er*(t+273)*B;
986 qR=(phr*Rr*epsilon1)*k^(-1);
```

```
987  En=0.11;
988  rn=0.682;
989  Rn=rn*1E-9+Rt1;
990  phn=En*(t+273)*B;
991  qN=(phn*Rn*epsilon1)*k^(-1);
992  Ed=-0.09;
993  rd=0.666;
994  Rd=rd*1E-9+Rt1;
995  phd=Ed*(t+273)*B;
996  qD=(phd*Rd*epsilon1)*k^(-1);
997  Ec=2.55;
998  rc=0.629;
999  phc=Ec*(t+273)*B;
1000 Rc=rc*1E-9+Rt1;
1001 qC=(phc*Rc*epsilon1)*k^(-1);
1002 Eq=-0.69;
1003 rq=0.725;
1004 Rq=rq*1E-9+Rt1;
1005 phq=Eq*(t+273)*B;
1006 qQ=(phq*Rq*epsilon1)*k^(-1);
1007 Ee=-0.15;
1008 re=0.714;
1009 Re=re*1E-9+Rt1;
1010 phe=Ee*(t+273)*B;
1011 qE=(phe*Re*epsilon1)*k^(-1);
1012 Eg=1.05;
1013 rg=0.725;
1014 Rg=rg*1E-9+Rt1;
1015 phg=Eg*(t+273)*B;
1016 qG=(phg*Rg*epsilon1)*k^(-1);
1017 Eh=0.1;
1018 rh=0.725;
1019 Rh=rh*1E-9+Rt1;
1020 phh=Eh*(t+273)*B;
1021 qH=(phh*Rh*epsilon1)*k^(-1);
1022 Ei=0.93;
1023 ri=0.735;
1024 Ri=ri*1E-9-Rt1;
1025 phi=Ei*(t+273)*B;
1026 qI=(phi*Ri*epsilon1)*k^(-1);
1027 El=0.76;
1028 rl=0.734;
1029 Rl=rl*1E-9-Rt1;
1030 phl=El*(t+273)*B;
1031 qL=(phl*Rl*epsilon1)*k^(-1);
1032 Ek=-1.05;
1033 rk=0.737;
1034 Rk=rk*1E-9+Rt1;
1035 phk=Ek*(t+273)*B;
1036 qK=(phk*Rk*epsilon1)*k^(-1);
1037 Em=0.29;
1038 rm=0.741;
```

```
1039 Rm=rm*1E-9-Rt1;
1040 phm=Em*(t+273)*B;
1041 qM=(phm*Rm*epsilon1)*k^(-1);
1042 Ef=0.76;
1043 rf=0.781;
1044 Rf=rf*1E-9-Rt1;
1045 phf=Ef*(t+273)*B;
1046 qF=(phf*Rf*epsilon1)*k^(-1);
1047 Ep=0.45;
1048 rp=0.672;
1049 Rp=rp*1E-9-Rt1;
1050 php=Ep*(t+273)*B;
1051 qP=(php*Rp*epsilon1)*k^(-1);
1052 Es=0.6;
1053 rs=0.615;
1054 Rs=rs*1E-9+Rt1;
1055 phs=Es*(t+273)*B;
1056 qS=(phs*Rs*epsilon1)*k^(-1);
1057 Et=0.9;
1058 rt=0.659;
1059 Rt=rt*1E-9+Rt1;
1060 pht=Et*(t+273)*B;
1061 qT=(pht*Rt*epsilon1)*k^(-1);
1062 Ew=0.68;
1063 rw=0.826;
1064 Rw=rw*1E-9-Rt1;
1065 phw=Ew*(t+273)*B;
1066 qW=(phw*Rw*epsilon1)*k^(-1);
1067 Ey=0.48;
1068 ry=0.781;
1069 Ry=ry*1E-9-Rt1;
1070 phy=Ey*(t+273)*B;
1071 qY=(phy*Ry*epsilon1)*k^(-1);
1072 Ev=0.84;
1073 rv=0.694;
1074 Rv=rv*1E-9-Rt1;
1075 phv=Ev*(t+273)*B;
1076 qV=(phv*Rv*epsilon1)*k^(-1);
1077 N=length(S_1);
1078 M=length(S_20);
1079 S_2=S_20;
1080 Q1=[];
1081 Q2=[];
1082 Q3=[];
1083 Q4=[];
1084 R1=[];
1085 R2=[];
1086 h=[];
1087 for i=1:length(S_1);
1088 if (S_1(i)=='A')
1089 Q1(i)=qA;
1090 else
```

```
1091     if (S_1(i)=='R')
1092 Q1(i)=qR;
1093     else
1094    if (S_1(i)=='N')
1095 Q1(i)=qN;
1096    else
1097     if (S_1(i)=='D')
1098 Q1(i)=qD;
1099    else
1100    if (S_1(i)=='C')
1101 Q1(i)=qC;
1102    else
1103     if (S_1(i)=='Q')
1104 Q1(i)=qQ;
1105     else
1106     if (S_1(i)=='E')
1107 Q1(i)=qE;
1108     else
1109      if (S_1(i)=='G')
1110 Q1(i)=qG;
1111      else
1112    if (S_1(i)=='K')
1113 Q1(i)=qK;
1114    else
1115     if (S_1(i)=='P')
1116 Q1(i)=qP;
1117     else
1118    if (S_1(i)=='S')
1119 Q1(i)=qS;
1120    else
1121     if (S_1(i)=='T')
1122 Q1(i)=qT;
1123     else
1124    if (S_1(i)=='I')
1125 Q1(i)=qI;
1126    else
1127     if (S_1(i)=='V')
1128 Q1(i)=qV;
1129    else
1130     if (S_1(i)=='L')
1131 Q1(i)=qL;
1132     else
1133     if (S_1(i)=='F')
1134 Q1(i)=qF;
1135    else
1136    if (S_1(i)=='W')
1137 Q1(i)=qW;
1138    else
1139    if (S_1(i)=='Y')
1140 Q1(i)=qY;
1141    else
1142    if (S_1(i)=='M')
```

```
1143 Q1(i)=qM;
1144   else
1145   if (S_1(i)=='H')
1146 Q1(i)=qH;
1147 end
1148 end
1149 end
1150 end
1151 end
1152 end
1153 end
1154 end
1155 end
1156 end
1157 end
1158 end
1159 end
1160 end
1161 end
1162 end
1163 end
1164 end
1165 end
1166 end
1167 end
1168 for j=1:length(S_2);
1169 if (S_2(j)=='A')
1170 Q2(j)=qA;
1171 else
1172    if (S_2(j)=='R')
1173 Q2(j)=qR;
1174    else
1175   if (S_2(j)=='N')
1176 Q2(j)=qN;
1177   else
1178   if (S_2(j)=='D')
1179 Q2(j)=qD;
1180   else
1181   if (S_2(j)=='C')
1182 Q2(j)=qC;
1183   else
1184    if (S_2(j)=='Q')
1185 Q2(j)=qQ;
1186    else
1187    if (S_2(j)=='E')
1188 Q2(j)=qE;
1189    else
1190     if (S_2(j)=='G')
1191 Q2(j)=qG;
1192     else
1193   if (S_2(j)=='K')
```

```
1194 Q2(j)=qK;
1195   else
1196    if (S_2(j)=='P')
1197 Q2(j)=qP;
1198    else
1199   if (S_2(j)=='S')
1200 Q2(j)=qS;
1201   else
1202    if (S_2(j)=='T')
1203 Q2(j)=qT;
1204    else
1205   if (S_2(j)=='I')
1206 Q2(j)=qI;
1207   else
1208    if (S_2(j)=='V')
1209 Q2(j)=qV;
1210   else
1211    if (S_2(j)=='L')
1212 Q2(j)=qL;
1213    else
1214    if (S_2(j)=='F')
1215 Q2(j)=qF;
1216   else
1217   if (S_2(j)=='W')
1218 Q2(j)=qW;
1219   else
1220   if (S_2(j)=='Y')
1221 Q2(j)=qY;
1222   else
1223   if (S_2(j)=='M')
1224 Q2(j)=qM;
1225   else
1226   if (S_2(j)=='H')
1227 Q2(j)=qH;
1228 end
1229 end
1230 end
1231 end
1232 end
1233 end
1234 end
1235 end
1236 end
1237 end
1238 end
1239 end
1240 end
1241 end
1242 end
1243 end
1244 end
1245 end
```

```
1246  end
1247  end
1248  end
1249  for i=1:length(S_1);
1250      for j=1:length(S_2);
1251       if (S_1(i)=='A')|(S_2(j)=='A');
1252              R1(i)=Ra;
1253              R2(j)=Ra;
1254          else
1255          if (S_1(i)=='R')|(S_2(j)=='R');
1256              R1(i)=Rr;
1257              R2(j)=Rr;
1258          else
1259  if (S_1(i)=='N')|(S_2(j)=='N');
1260              R1(i)=Rn;
1261              R2(j)=Rn;
1262  else
1263  if (S_1(i)=='D')|(S_2(j)=='D');
1264      R1(i)=Rd;
1265      R2(j)=Rd;
1266  else
1267       if (S_1(i)=='C')|(S_2(j)=='C');
1268              R1(i)=Rc;
1269              R2(j)=Rc;
1270          else
1271  if (S_1(i)=='Q')|(S_2(j)=='Q');
1272              R1(i)=Rq;
1273              R2(j)=Rq;
1274          else
1275          if (S_1(i)=='E')|(S_2(j)=='E');
1276              R1(i)=Re;
1277              R2(j)=Re;
1278          else
1279            if (S_1(i)=='G')|(S_2(j)=='G');
1280              R1(i)=Rg;
1281              R2(j)=Rg;
1282          else
1283   if (S_1(i)=='H')|(S_2(j)=='H');
1284        R1(i)=Rh;
1285        R2(j)=Rh;
1286   else
1287       if (S_1(i)=='I')|(S_2(j)=='I');
1288              R1(i)=Ri;
1289              R2(j)=Ri;
1290          else
1291       if (S_1(i)=='L')|(S_2(j)=='L');
1292              R1(i)=Rl;
1293              R2(j)=Rl;
1294          else
1295       if (S_1(i)=='K')|(S_2(j)=='K')
1296             R1(i)=Rk;
1297              R2(j)=Rk;
```

```
1298            else
1299         if (S_1(i)=='M')|(S_2(j)=='M')
1300                 R1(i)=Rm;
1301                 R2(j)=Rm;
1302            else
1303         if (S_1(i)=='F')|(S_2(j)=='F')
1304                 R1(i)=Rf;
1305                 R2(j)=Rf;
1306         else
1307           if (S_1(i)=='P')|(S_2(j)=='P');
1308                 R1(i)=Rp;
1309                 R2(j)=Rp;
1310            else
1311             if (S_1(i)=='S')|(S_2(j)=='S');
1312                 R1(i)=Rs;
1313                 R2(j)=Rs;
1314            else
1315             if (S_1(i)=='T')|(S_2(j)=='T');
1316                 R1(i)=Rt;
1317                 R2(j)=Rt;
1318            else
1319             if (S_1(i)=='W')|(S_2(j)=='W');
1320                 R1(i)=Rw;
1321                 R2(j)=Rw;
1322            else
1323             if (S_1(i)=='Y')|(S_2(j)=='Y');
1324                 R1(i)=Ry;
1325                 R2(j)=Ry;
1326             else
1327                 if (S_1(i)=='V')|(S_2(j)=='V');
1328                 R1(i)=Rv;
1329                 R2(j)=Rv;
1330                 else
1331                 if (S_1(i)=='X')|(S_2(j)=='X')
1332                  R1(i)=0.194E-9;
1333                  R2(j)=0.994E-9;
1334    end
1335    end
1336    end
1337    end
1338    end
1339    end
1340    end
1341    end
1342    end
1343    end
1344    end
1345    end
1346    end
1347    end
1348    end
1349    end
```

```
1350 end
1351 end
1352 end
1353 end
1354 end
1355 end
1356 end
1357 for i=1:length(S_1);
1358 for j=1:length(S_2);
1359 if (S_1(i)=='R'& S_2(j)=='D');
1360     h(i,j)=.15*10^(-9)+Rr+Rd+Rt1;
1361 else
1362 if (S_1(i)=='R'& S_2(j)=='E');
1363         h(i,j)=.15*10^(-9)+Rr+Re+Rt1;
1364          else
1365 if (S_1(i)=='D'& S_2(j)=='R');
1366 h(i,j)=.15*10^(-9)+Rd+Rr+Rt1;
1367 else
1368 if (S_1(i)=='D'& S_2(j)=='H');
1369 h(i,j)=.15*10^(-9)+Rd+Rh+Rt1;
1370   else
1371 if (S_1(i)=='D'& S_2(j)=='R');
1372 h(i,j)=.15*10^(-9)+Rd+Rr+Rt1;
1373 else
1374     if (S_1(i)=='D'& S_2(j)=='H');
1375     h(i,j)=.15*10^(-9)+Rd+Rh+Rt1;
1376   else
1377 if (S_1(i)=='D'& S_2(j)=='K');
1378 h(i,j)=.15*10^(-9)+Rd+Rk+Rt1;
1379  else
1380 if (S_1(i)=='E')& (S_2(j)=='R');
1381 h(i,j)=.15*10^(-9)+Re+Rr+Rt1;
1382          else
1383 if (S_1(i)=='E'& S_2(j)=='H');
1384 h(i,j)=.15*10^(-9)+Re+Rh+Rt1;
1385       else
1386 if (S_1(i)=='E'& S_2(j)=='K');
1387 h(i,j)=.15*10^(-9)+Re+Rk+Rt1;
1388   else
1389 if (S_1(i)=='H'& S_2(j)=='D')
1390 h(i,j)=.15*10^(-9)+Rh+Rd+Rt1;
1391 else
1392 if (S_1(i)=='H'& S_2(j)=='E')
1393 h(i,j)=.15*10^(-9)+Rh+Re+Rt1;
1394 else
1395 if (S_1(i)=='R'& S_2(j)=='R')
1396     h(i,j)=.4*10^(-9)+Rr+Rr+Rt1;
1397     else
1398   if (S_1(i)=='R'& S_2(j)=='H')
1399    h(i,j)=.4*10^(-9)+Rr+Rh+Rt1;
1400   else
1401 if (S_1(i)=='R'& S_2(j)=='H')
```

```
1402      h(i,j)=.4*10^(-9)+Rr+Rh+Rt1;
1403  else
1404     if (S_1(i)=='R'& S_2(j)=='K')
1405       h(i,j)=.4*10^(-9)+Rr+Rk+Rt1;
1406    else
1407  if (S_1(i)=='D'& S_2(j)=='E');
1408             h(i,j)=.4*10^(-9)+Rd+Re+Rt1;
1409  else
1410      if (S_1(i)=='D'& S_2(j)=='D');
1411      h(i,j)=.4*10^(-9)+Rd+Rd+Rt1;
1412     else
1413  if (S_1(i)=='H'& S_2(j)=='R')
1414       h(i,j)=.4*10^(-9)+Rh+Rr+Rt1;
1415  else
1416     if (S_1(i)=='H'& S_2(j)=='H')
1417         h(i,j)=.4*10^(-9)+Rh+Rh+Rt1;
1418    else
1419      if (S_1(i)=='H'& S_2(j)=='K')
1420         h(i,j)=.4*10^(-9)+Rh+Rk+Rt1;
1421      else
1422  if (S_1(i)=='K'& S_2(j)=='R')
1423             h(i,j)=.4*10^(-9)+Rk+Rr+Rt1;
1424  else
1425    if (S_1(i)=='K'& S_2(j)=='H')
1426           h(i,j)=.4*10^(-9)+Rk+Rh+Rt1;
1427    else
1428     if (S_1(i)=='K'& S_2(j)=='K')
1429       h(i,j)=.4*10^(-9)+Rk+Rk+Rt1;
1430    else
1431  if (S_1(i)=='N'& S_2(j)=='Q')
1432             h(i,j)=.25*10^(-9)+Rn+Rq+Rt1;
1433  else
1434    if (S_1(i)=='N'& S_2(j)=='S')
1435      h(i,j)=.25*10^(-9)+Rn+Rs+Rt1;
1436     else
1437     if (S_1(i)=='N'& S_2(j)=='Y')
1438       h(i,j)=.25*10^(-9)+Rn+Ry+Rt1;
1439    else
1440  if (S_1(i)=='Q'& S_2(j)=='S')| (S_1(i)=='Q')& (S_2(j)=='Y');
1441             h(i,j)=.25*10^(-9)+Rq+Rs+Rt1;
1442  else
1443    if  (S_1(i)=='Q')& (S_2(j)=='Y');
1444         h(i,j)=.25*10^(-9)+Rq+Ry+Rt1;
1445  else
1446  if (S_1(i)=='S'& S_2(j)=='Y');
1447             h(i,j)=.25*10^(-9)+Rs+Ry+Rt1;
1448  else
1449  if (S_1(i)=='I'& S_2(j)=='V')|(S_1(i)=='I'& S_2(j)=='L')|...
1450   (S_1(i)=='I'& S_2(j)=='F')|(S_1(i)=='I'& S_2(j)=='W')|...
1451   (S_1(i)=='I'& S_2(j)=='Y')|(S_1(i)=='I'& S_2(j)=='M')|...
1452   (S_1(i)=='I'& S_2(j)=='A')|(S_1(i)=='V'& S_2(j)=='V')|...
1453   (S_1(i)=='V'& S_2(j)=='L')|(S_1(i)=='V'& S_2(j)=='F')|...
```

```
1454 (S_1(i)=='V'& S_2(j)=='W')|(S_1(i)=='V'& S_2(j)=='M')|...
1455 (S_1(i)=='V'& S_2(j)=='A')|(S_1(i)=='L'& S_2(j)=='F')|...
1456 (S_1(i)=='L'& S_2(j)=='W')|(S_1(i)=='L'& S_2(j)=='Y')|...
1457 (S_1(i)=='L'& S_2(j)=='M')|(S_1(i)=='L'& S_2(j)=='A')|...
1458 (S_1(i)=='F'& S_2(j)=='W')|(S_1(i)=='F'& S_2(j)=='F')|...
1459 (S_1(i)=='F'& S_2(j)=='Y')|(S_1(i)=='F'& S_2(j)=='M')|...
1460 (S_1(i)=='F'& S_2(j)=='A')|(S_1(i)=='W'& S_2(j)=='W')|...
1461 (S_1(i)=='W'& S_2(j)=='Y')|(S_1(i)=='W'& S_2(j)=='M')|...
1462 (S_1(i)=='W'& S_2(j)=='A')|(S_1(i)=='Y'& S_2(j)=='Y')|...
1463 (S_1(i)=='Y'& S_2(j)=='M')|(S_1(i)=='Y'& S_2(j)=='A')|...
1464 (S_1(i)=='M'& S_2(j)=='M')|(S_1(i)=='M'& S_2(j)=='A')|...
1465 (S_1(i)=='A'& S_2(j)=='A')|(S_1(i)=='I'& S_2(j)=='I')|...
1466 (S_1(i)=='V'& S_2(j)=='V')|(S_1(i)=='L'& S_2(j)=='L')|...
1467 (S_1(i)=='F'& S_2(j)=='F')|(S_1(i)=='W'& S_2(j)=='W')|...
1468 (S_1(i)=='P'& S_2(j)=='I')| (S_1(i)=='P'& S_2(j)=='V')|...
1469 (S_1(i)=='P'& S_2(j)=='L')|(S_1(i)=='P'& S_2(j)=='F')|...
1470 (S_1(i)=='P'& S_2(j)=='W')| (S_1(i)=='P'& S_2(j)=='Y')|...
1471 (S_1(i)=='P'& S_2(j)=='M')| (S_1(i)=='P'& S_2(j)=='A');
1472 h(i,j)=.36*10^(-9)+(0.736*10^(-9))*2;
1473  else
1474   if (S_2(j)=='I'& S_1(i)=='V')|(S_2(j)=='I'& S_1(i)=='L')|...
1475 (S_2(j)=='I'& S_1(i)=='F')|(S_2(j)=='I'& S_1(i)=='W')|...
1476 (S_2(j)=='I'& S_1(i)=='Y')|(S_2(j)=='I'& S_1(i)=='M')|...
1477 (S_2(j)=='I'& S_1(i)=='A')|(S_2(j)=='V'& S_1(i)=='V')|...
1478 (S_2(j)=='V'& S_1(i)=='L')|(S_2(j)=='V'& S_1(i)=='F')|...
1479 (S_2(j)=='V'& S_1(i)=='W')|(S_2(j)=='V'& S_1(i)=='M')|...
1480 (S_2(j)=='V'& S_1(i)=='A')|(S_2(j)=='L'& S_1(i)=='F')|...
1481 (S_2(j)=='L'& S_1(i)=='W')|(S_2(j)=='L'& S_1(i)=='Y')|...
1482 (S_2(j)=='L'& S_1(i)=='M')|(S_2(j)=='L'& S_1(i)=='A')|...
1483 (S_2(j)=='F'& S_1(i)=='W')|(S_2(j)=='F'& S_1(i)=='F')|...
1484 (S_2(j)=='F'& S_1(i)=='Y')|(S_2(j)=='F'& S_1(i)=='M')|...
1485 (S_2(j)=='F'& S_1(i)=='A')|(S_2(j)=='W'& S_1(i)=='W')|...
1486 (S_2(j)=='W'& S_1(i)=='Y')|(S_2(j)=='W'& S_1(i)=='M')|...
1487 (S_2(j)=='W'& S_1(i)=='A')|(S_2(j)=='Y'& S_1(i)=='Y')|...
1488 (S_2(j)=='Y'& S_1(i)=='M')|(S_2(j)=='Y'& S_1(i)=='A')|...
1489 (S_2(j)=='M'& S_1(i)=='M')|(S_2(j)=='M'& S_1(i)=='A')|...
1490 (S_2(j)=='A'& S_1(i)=='A')|(S_2(j)=='I'& S_1(i)=='I')|...
1491 (S_2(j)=='V'& S_1(i)=='V')|(S_2(j)=='L'& S_1(i)=='L')|...
1492 (S_2(j)=='F'& S_1(i)=='F')|(S_2(j)=='W'& S_1(i)=='W')|...
1493 (S_2(j)=='P'& S_1(i)=='I')|(S_2(j)=='P'& S_1(i)=='V')|...
1494 (S_2(j)=='P'& S_1(i)=='L')|(S_2(j)=='P'& S_1(i)=='F')|...
1495 (S_2(j)=='P'& S_1(i)=='W')|(S_2(j)=='P'& S_1(i)=='Y')|...
1496 (S_2(j)=='P'& S_1(i)=='M')|(S_2(j)=='P'& S_1(i)=='A');
1497 h(i,j)=.36*10^(-9)+(0.736*10^(-9))*2;
1498 else
1499          h(i,j)=(0.71286*10^(-9))*2+0.3*10^(-9)+Rt1;
1500 end
1501 end
1502 end
1503 end
1504 end
1505 end
```

```
1506 end
1507 end
1508 end
1509 end
1510 end
1511 end
1512 end
1513 end
1514 end
1515 end
1516 end
1517 end
1518 end
1519 end
1520 end
1521 end
1522 end
1523 end
1524 end
1525 end
1526 end
1527 end
1528 end
1529 end
1530 end
1531 end
1532 end
1533 end
1534
1535 function [S_1,S_2,Q1,Q2,R1,R2,h,M,N]=...
1536 potential_30(t,epsilon1,rtt,S_1,S_20)
1537 Hhidro=0;
1538 Rt1=(4.6*1E-13);
1539 epsilon0=8.85418781762*10^(-12);
1540 k=1/(4*pi*epsilon0);
1541 B=(1.38064852*10^(-23))/(1.6021766208*10^(-19));
1542 Ea=1.88;
1543 ra=0.6;
1544 Ra=ra*1E-9-2*Rt1;
1545 pha=Ea*(t+273)*B;
1546 qA=(pha*Ra*epsilon1)*k^(-1);
1547 Er=-0.81;
1548 rr=0.8;
1549 Rr=rr*1E-9+2*Rt1;
1550 phr=Er*(t+273)*B;
1551 qR=(phr*Rr*epsilon1)*k^(-1);
1552 En=0.08;
1553 rn=0.682;
1554 Rn=rn*1E-9+2*Rt1;
1555 phn=En*(t+273)*B;
1556 qN=(phn*Rn*epsilon1)*k^(-1);
1557 Ed=-0.11;
```

```
1558 rd=0.666;
1559 Rd=rd*1E-9+2*Rt1;
1560 phd=Ed*(t+273)*B;
1561 qD=(phd*Rd*epsilon1)*k^(-1);
1562 Ec=2.53;
1563 rc=0.629;
1564 phc=Ec*(t+273)*B;
1565 Rc=rc*1E-9+2*Rt1;
1566 qC=(phc*Rc*epsilon1)*k^(-1);
1567 Eq=-0.69;
1568 rq=0.725;
1569 Rq=rq*1E-9+2*Rt1;
1570 phq=Eq*(t+273)*B;
1571 qQ=(phq*Rq*epsilon1)*k^(-1);
1572 Ee=-0.15;
1573 re=0.714;
1574 Re=re*1E-9+2*Rt1;
1575 phe=Ee*(t+273)*B;
1576 qE=(phe*Re*epsilon1)*k^(-1);
1577 Eg=1.1;
1578 rg=0.725;
1579 Rg=rg*1E-9+2*Rt1;
1580 phg=Eg*(t+273)*B;
1581 qG=(phg*Rg*epsilon1)*k^(-1);
1582 Eh=0.12;
1583 rh=0.725;
1584 Rh=rh*1E-9+2*Rt1;
1585 phh=Eh*(t+273)*B;
1586 qH=(phh*Rh*epsilon1)*k^(-1);
1587 Ei=0.92;
1588 ri=0.735;
1589 Ri=ri*1E-9-2*Rt1;
1590 phi=Ei*(t+273)*B;
1591 qI=(phi*Ri*epsilon1)*k^(-1);
1592 El=0.76;
1593 rl=0.734;
1594 Rl=rl*1E-9-2*Rt1;
1595 phl=El*(t+273)*B;
1596 qL=(phl*Rl*epsilon1)*k^(-1);
1597 Ek=-1.12;
1598 rk=0.737;
1599 Rk=rk*1E-9+2*Rt1;
1600 phk=Ek*(t+273)*B;
1601 qK=(phk*Rk*epsilon1)*k^(-1);
1602 Em=0.31;
1603 rm=0.741;
1604 Rm=rm*1E-9-2*Rt1;
1605 phm=Em*(t+273)*B;
1606 qM=(phm*Rm*epsilon1)*k^(-1);
1607 Ef=0.69;
1608 rf=0.781;
1609 Rf=rf*1E-9-2*Rt1;
```

```
1610 phf=Ef*(t+273)*B;
1611 qF=(phf*Rf*epsilon1)*k^(-1);
1612 Ep=0.2;
1613 rp=0.672;
1614 Rp=rp*1E-9-2*Rt1;
1615 php=Ep*(t+273)*B;
1616 qP=(php*Rp*epsilon1)*k^(-1);
1617 Es=0.6;
1618 rs=0.615;
1619 Rs=rs*1E-9+2*Rt1;
1620 phs=Es*(t+273)*B;
1621 qS=(phs*Rs*epsilon1)*k^(-1);
1622 Et=0.89;
1623 rt=0.659;
1624 Rt=rt*1E-9+2*Rt1;
1625 pht=Et*(t+273)*B;
1626 qT=(pht*Rt*epsilon1)*k^(-1);
1627 Ew=0.63;
1628 rw=0.826;
1629 Rw=rw*1E-9-2*Rt1;
1630 phw=Ew*(t+273)*B;
1631 qW=(phw*Rw*epsilon1)*k^(-1);
1632 Ey=0.5;
1633 ry=0.781;
1634 Ry=ry*1E-9-2*Rt1;
1635 phy=Ey*(t+273)*B;
1636 qY=(phy*Ry*epsilon1)*k^(-1);
1637 Ev=0.84;
1638 rv=0.694;
1639 Rv=rv*1E-9-2*Rt1;
1640 phv=Ev*(t+273)*B;
1641 qV=(phv*Rv*epsilon1)*k^(-1);
1642 N=length(S_1);
1643 M=length(S_20);
1644 S_2=S_20;
1645 Q1=[];
1646 Q2=[];
1647 Q3=[];
1648 Q4=[];
1649 R1=[];
1650 R2=[];
1651 h=[];
1652 for i=1:length(S_1);
1653 if (S_1(i)=='A')
1654 Q1(i)=qA;
1655 else
1656    if (S_1(i)=='R')
1657 Q1(i)=qR;
1658    else
1659   if (S_1(i)=='N')
1660 Q1(i)=qN;
1661   else
```

```
1662      if (S_1(i)=='D')
1663   Q1(i)=qD;
1664     else
1665     if (S_1(i)=='C')
1666   Q1(i)=qC;
1667     else
1668      if (S_1(i)=='Q')
1669   Q1(i)=qQ;
1670      else
1671      if (S_1(i)=='E')
1672   Q1(i)=qE;
1673      else
1674       if (S_1(i)=='G')
1675   Q1(i)=qG;
1676       else
1677     if (S_1(i)=='K')
1678   Q1(i)=qK;
1679     else
1680      if (S_1(i)=='P')
1681   Q1(i)=qP;
1682      else
1683     if (S_1(i)=='S')
1684   Q1(i)=qS;
1685     else
1686      if (S_1(i)=='T')
1687   Q1(i)=qT;
1688      else
1689     if (S_1(i)=='I')
1690   Q1(i)=qI;
1691     else
1692      if (S_1(i)=='V')
1693   Q1(i)=qV;
1694     else
1695      if (S_1(i)=='L')
1696   Q1(i)=qL;
1697      else
1698      if (S_1(i)=='F')
1699   Q1(i)=qF;
1700     else
1701     if (S_1(i)=='W')
1702   Q1(i)=qW;
1703     else
1704     if (S_1(i)=='Y')
1705   Q1(i)=qY;
1706     else
1707     if (S_1(i)=='M')
1708   Q1(i)=qM;
1709     else
1710     if (S_1(i)=='H')
1711   Q1(i)=qH;
1712   end
1713   end
```

```
1714 end
1715 end
1716 end
1717 end
1718 end
1719 end
1720 end
1721 end
1722 end
1723 end
1724 end
1725 end
1726 end
1727 end
1728 end
1729 end
1730 end
1731 end
1732 end
1733 for j=1:length(S_2);
1734 if (S_2(j)=='A')
1735 Q2(j)=qA;
1736 else
1737    if (S_2(j)=='R')
1738 Q2(j)=qR;
1739    else
1740   if (S_2(j)=='N')
1741 Q2(j)=qN;
1742   else
1743   if (S_2(j)=='D')
1744 Q2(j)=qD;
1745   else
1746   if (S_2(j)=='C')
1747 Q2(j)=qC;
1748   else
1749    if (S_2(j)=='Q')
1750 Q2(j)=qQ;
1751    else
1752    if (S_2(j)=='E')
1753 Q2(j)=qE;
1754    else
1755     if (S_2(j)=='G')
1756 Q2(j)=qG;
1757     else
1758   if (S_2(j)=='K')
1759 Q2(j)=qK;
1760   else
1761    if (S_2(j)=='P')
1762 Q2(j)=qP;
1763    else
1764   if (S_2(j)=='S')
1765 Q2(j)=qS;
```

```
1766     else
1767      if (S_2(j)=='T')
1768  Q2(j)=qT;
1769      else
1770     if (S_2(j)=='I')
1771  Q2(j)=qI;
1772     else
1773      if (S_2(j)=='V')
1774  Q2(j)=qV;
1775     else
1776      if (S_2(j)=='L')
1777  Q2(j)=qL;
1778      else
1779      if (S_2(j)=='F')
1780  Q2(j)=qF;
1781     else
1782     if (S_2(j)=='W')
1783  Q2(j)=qW;
1784     else
1785     if (S_2(j)=='Y')
1786  Q2(j)=qY;
1787     else
1788     if (S_2(j)=='M')
1789  Q2(j)=qM;
1790     else
1791     if (S_2(j)=='H')
1792  Q2(j)=qH;
1793  end
1794  end
1795  end
1796  end
1797  end
1798  end
1799  end
1800  end
1801  end
1802  end
1803  end
1804  end
1805  end
1806  end
1807  end
1808  end
1809  end
1810  end
1811  end
1812  end
1813  end
1814  for i=1:length(S_1);
1815      for j=1:length(S_2);
1816       if (S_1(i)=='A')|(S_2(j)=='A');
1817              R1(i)=Ra;
```

```
1818                R2(j)=Ra;
1819            else
1820            if (S_1(i)=='R')|(S_2(j)=='R');
1821                R1(i)=Rr;
1822                R2(j)=Rr;
1823            else
1824    if (S_1(i)=='N')|(S_2(j)=='N');
1825                R1(i)=Rn;
1826                R2(j)=Rn;
1827    else
1828    if (S_1(i)=='D')|(S_2(j)=='D');
1829        R1(i)=Rd;
1830        R2(j)=Rd;
1831    else
1832          if (S_1(i)=='C')|(S_2(j)=='C');
1833                R1(i)=Rc;
1834                R2(j)=Rc;
1835            else
1836    if (S_1(i)=='Q')|(S_2(j)=='Q');
1837                R1(i)=Rq;
1838                R2(j)=Rq;
1839            else
1840            if (S_1(i)=='E')|(S_2(j)=='E');
1841                R1(i)=Re;
1842                R2(j)=Re;
1843            else
1844              if (S_1(i)=='G')|(S_2(j)=='G');
1845                R1(i)=Rg;
1846                R2(j)=Rg;
1847            else
1848     if (S_1(i)=='H')|(S_2(j)=='H');
1849          R1(i)=Rh;
1850          R2(j)=Rh;
1851     else
1852         if (S_1(i)=='I')|(S_2(j)=='I');
1853                R1(i)=Ri;
1854                R2(j)=Ri;
1855            else
1856         if (S_1(i)=='L')|(S_2(j)=='L');
1857                R1(i)=Rl;
1858                R2(j)=Rl;
1859            else
1860         if (S_1(i)=='K')|(S_2(j)=='K')
1861               R1(i)=Rk;
1862                R2(j)=Rk;
1863            else
1864         if (S_1(i)=='M')|(S_2(j)=='M')
1865                R1(i)=Rm;
1866                R2(j)=Rm;
1867            else
1868         if (S_1(i)=='F')|(S_2(j)=='F')
1869                R1(i)=Rf;
```

```
1870                R2(j)=Rf;
1871        else
1872          if (S_1(i)=='P')|(S_2(j)=='P');
1873                R1(i)=Rp;
1874                R2(j)=Rp;
1875             else
1876              if (S_1(i)=='S')|(S_2(j)=='S');
1877                R1(i)=Rs;
1878                R2(j)=Rs;
1879             else
1880              if (S_1(i)=='T')|(S_2(j)=='T');
1881                R1(i)=Rt;
1882                R2(j)=Rt;
1883             else
1884              if (S_1(i)=='W')|(S_2(j)=='W');
1885                R1(i)=Rw;
1886                R2(j)=Rw;
1887             else
1888              if (S_1(i)=='Y')|(S_2(j)=='Y');
1889                R1(i)=Ry;
1890                R2(j)=Ry;
1891              else
1892                 if (S_1(i)=='V')|(S_2(j)=='V');
1893                 R1(i)=Rv;
1894                 R2(j)=Rv;
1895                 else
1896                 if (S_1(i)=='X')|(S_2(j)=='X')
1897                  R1(i)=0.194E-9;
1898                  R2(j)=0.994E-9;
1899 end
1900 end
1901 end
1902 end
1903 end
1904 end
1905 end
1906 end
1907 end
1908 end
1909 end
1910 end
1911 end
1912 end
1913 end
1914 end
1915 end
1916 end
1917 end
1918 end
1919 end
1920 end
1921 end
```

```
1922 for i=1:length(S_1);
1923   for j=1:length(S_2);
1924 if (S_1(i)=='R'& S_2(j)=='D');
1925     h(i,j)=.15*10^(-9)+Rr+Rd+2*Rt1;
1926 else
1927 if (S_1(i)=='R'& S_2(j)=='E');
1928         h(i,j)=.15*10^(-9)+Rr+Re+2*Rt1;
1929          else
1930 if (S_1(i)=='D'& S_2(j)=='R');
1931 h(i,j)=.15*10^(-9)+Rd+Rr+2*Rt1;
1932 else
1933 if (S_1(i)=='D'& S_2(j)=='H');
1934 h(i,j)=.15*10^(-9)+Rd+Rh+2*Rt1;
1935   else
1936 if (S_1(i)=='D'& S_2(j)=='R');
1937 h(i,j)=.15*10^(-9)+Rd+Rr+2*Rt1;
1938 else
1939     if (S_1(i)=='D'& S_2(j)=='H');
1940     h(i,j)=.15*10^(-9)+Rd+Rh+2*Rt1;
1941   else
1942 if (S_1(i)=='D'& S_2(j)=='K');
1943 h(i,j)=.15*10^(-9)+Rd+Rk+2*Rt1;
1944  else
1945 if (S_1(i)=='E')& (S_2(j)=='R');
1946 h(i,j)=.15*10^(-9)+Re+Rr+2*Rt1;
1947          else
1948 if (S_1(i)=='E'& S_2(j)=='H');
1949 h(i,j)=.15*10^(-9)+Re+Rh+2*Rt1;
1950        else
1951 if (S_1(i)=='E'& S_2(j)=='K');
1952 h(i,j)=.15*10^(-9)+Re+Rk+2*Rt1;
1953   else
1954 if (S_1(i)=='H'& S_2(j)=='D')
1955 h(i,j)=.15*10^(-9)+Rh+Rd+2*Rt1;
1956 else
1957 if (S_1(i)=='H'& S_2(j)=='E')
1958 h(i,j)=.15*10^(-9)+Rh+Re+2*Rt1;
1959 else
1960 if (S_1(i)=='R'& S_2(j)=='R')
1961     h(i,j)=.4*10^(-9)+Rr+Rr+2*Rt1;
1962    else
1963   if (S_1(i)=='R'& S_2(j)=='H')
1964    h(i,j)=.4*10^(-9)+Rr+Rh+2*Rt1;
1965   else
1966 if (S_1(i)=='R'& S_2(j)=='H')
1967     h(i,j)=.4*10^(-9)+Rr+Rh+2*Rt1;
1968 else
1969   if (S_1(i)=='R'& S_2(j)=='K')
1970     h(i,j)=.4*10^(-9)+Rr+Rk+2*Rt1;
1971  else
1972 if (S_1(i)=='D'& S_2(j)=='E');
1973         h(i,j)=.4*10^(-9)+Rd+Re+2*Rt1;
```

```
1974 else
1975     if (S_1(i)=='D'& S_2(j)=='D');
1976     h(i,j)=.4*10^(-9)+Rd+Rd+Rt1;
1977    else
1978 if (S_1(i)=='H'& S_2(j)=='R')
1979      h(i,j)=.4*10^(-9)+Rh+Rr+2*Rt1;
1980 else
1981    if (S_1(i)=='H'& S_2(j)=='H')
1982       h(i,j)=.4*10^(-9)+Rh+Rh+2*Rt1;
1983  else
1984     if (S_1(i)=='H'& S_2(j)=='K')
1985       h(i,j)=.4*10^(-9)+Rh+Rk+2*Rt1;
1986     else
1987 if (S_1(i)=='K'& S_2(j)=='R')
1988          h(i,j)=.4*10^(-9)+Rk+Rr+2*Rt1;
1989 else
1990  if (S_1(i)=='K'& S_2(j)=='H')
1991        h(i,j)=.4*10^(-9)+Rk+Rh+2*Rt1;
1992  else
1993   if (S_1(i)=='K'& S_2(j)=='K')
1994      h(i,j)=.4*10^(-9)+Rk+Rk+2*Rt1;
1995  else
1996 if (S_1(i)=='N'& S_2(j)=='Q')
1997          h(i,j)=.25*10^(-9)+Rn+Rq+2*Rt1;
1998 else
1999  if (S_1(i)=='N'& S_2(j)=='S')
2000     h(i,j)=.25*10^(-9)+Rn+Rs+2*Rt1;
2001    else
2002    if (S_1(i)=='N'& S_2(j)=='Y')
2003      h(i,j)=.25*10^(-9)+Rn+Ry+2*Rt1;
2004  else
2005 if (S_1(i)=='Q'& S_2(j)=='S')|(S_1(i)=='Q')& (S_2(j)=='Y');
2006          h(i,j)=.25*10^(-9)+Rq+Rs+2*Rt1;
2007 else
2008  if  (S_1(i)=='Q')& (S_2(j)=='Y');
2009      h(i,j)=.25*10^(-9)+Rq+Ry+2*Rt1;
2010 else
2011 if (S_1(i)=='S'& S_2(j)=='Y');
2012          h(i,j)=.25*10^(-9)+Rs+Ry+2*Rt1;
2013 else
2014 if (S_1(i)=='I'& S_2(j)=='V')|(S_1(i)=='I'& S_2(j)=='L')|...
2015 (S_1(i)=='I'& S_2(j)=='F')|(S_1(i)=='I'& S_2(j)=='W')|...
2016 (S_1(i)=='I'& S_2(j)=='Y')|(S_1(i)=='I'& S_2(j)=='M')|...
2017 (S_1(i)=='I'& S_2(j)=='A')|(S_1(i)=='V'& S_2(j)=='V')|...
2018 (S_1(i)=='V'& S_2(j)=='L')|(S_1(i)=='V'& S_2(j)=='F')|...
2019 (S_1(i)=='V'& S_2(j)=='W')|(S_1(i)=='V'& S_2(j)=='M')|...
2020 (S_1(i)=='V'& S_2(j)=='A')|(S_1(i)=='L'& S_2(j)=='F')|...
2021 (S_1(i)=='L'& S_2(j)=='W')|(S_1(i)=='L'& S_2(j)=='Y')|...
2022 (S_1(i)=='L'& S_2(j)=='M')|(S_1(i)=='L'& S_2(j)=='A')|...
2023 (S_1(i)=='F'& S_2(j)=='W')|(S_1(i)=='F'& S_2(j)=='F')|...
2024 (S_1(i)=='F'& S_2(j)=='Y')|(S_1(i)=='F'& S_2(j)=='M')|...
2025 (S_1(i)=='F'& S_2(j)=='A')|(S_1(i)=='W'& S_2(j)=='W')|...
```

```
2026 (S_1(i)=='W'& S_2(j)=='Y')|(S_1(i)=='W'& S_2(j)=='M')|...
2027 (S_1(i)=='W'& S_2(j)=='A')|(S_1(i)=='Y'& S_2(j)=='Y')|...
2028 (S_1(i)=='Y'& S_2(j)=='M')|(S_1(i)=='Y'& S_2(j)=='A')|...
2029 (S_1(i)=='M'& S_2(j)=='M')|(S_1(i)=='M'& S_2(j)=='A')|...
2030 (S_1(i)=='A'& S_2(j)=='A')|(S_1(i)=='I'& S_2(j)=='I')|...
2031 (S_1(i)=='V'& S_2(j)=='V')|(S_1(i)=='L'& S_2(j)=='L')|...
2032 (S_1(i)=='F'& S_2(j)=='F')|(S_1(i)=='W'& S_2(j)=='W')|...
2033 (S_1(i)=='P'& S_2(j)=='I')|(S_1(i)=='P'& S_2(j)=='V')|...
2034 (S_1(i)=='P'& S_2(j)=='L')|(S_1(i)=='P'& S_2(j)=='F')|...
2035 (S_1(i)=='P'& S_2(j)=='W')|(S_1(i)=='P'& S_2(j)=='Y')|...
2036 (S_1(i)=='P'& S_2(j)=='M')|(S_1(i)=='P'& S_2(j)=='A');
2037 h(i,j)=.36*10^(-9)+(0.736*10^(-9))*2;
2038  else
2039   if (S_2(j)=='I'& S_1(i)=='V')|(S_2(j)=='I'& S_1(i)=='L')|...
2040 (S_2(j)=='I'& S_1(i)=='F')|(S_2(j)=='I'& S_1(i)=='W')|...
2041 (S_2(j)=='I'& S_1(i)=='Y')|(S_2(j)=='I'& S_1(i)=='M')|...
2042 (S_2(j)=='I'& S_1(i)=='A')|(S_2(j)=='V'& S_1(i)=='V')|...
2043 (S_2(j)=='V'& S_1(i)=='L')|(S_2(j)=='V'& S_1(i)=='F')|...
2044 (S_2(j)=='V'& S_1(i)=='W')|(S_2(j)=='V'& S_1(i)=='M')|...
2045 (S_2(j)=='V'& S_1(i)=='A')|(S_2(j)=='L'& S_1(i)=='F')|...
2046 (S_2(j)=='L'& S_1(i)=='W')|(S_2(j)=='L'& S_1(i)=='Y')|...
2047 (S_2(j)=='L'& S_1(i)=='M')|(S_2(j)=='L'&S_1(i)=='A')|...
2048 (S_2(j)=='F'& S_1(i)=='W')|(S_2(j)=='F'& S_1(i)=='F')|...
2049 (S_2(j)=='F'& S_1(i)=='Y')|(S_2(j)=='F'& S_1(i)=='M')|...
2050 (S_2(j)=='F'& S_1(i)=='A')|(S_2(j)=='W'& S_1(i)=='W')|...
2051 (S_2(j)=='W'& S_1(i)=='Y')|(S_2(j)=='W'& S_1(i)=='M')|...
2052 (S_2(j)=='W'& S_1(i)=='A')|(S_2(j)=='Y'& S_1(i)=='Y')|...
2053 (S_2(j)=='Y'& S_1(i)=='M')|(S_2(j)=='Y'& S_1(i)=='A')|...
2054 (S_2(j)=='M'& S_1(i)=='M')|(S_2(j)=='M'& S_1(i)=='A')|...
2055 (S_2(j)=='A'& S_1(i)=='A')|(S_2(j)=='I'& S_1(i)=='I')|...
2056 (S_2(j)=='V'& S_1(i)=='V')|(S_2(j)=='L'& S_1(i)=='L')|...
2057 (S_2(j)=='F'& S_1(i)=='F')|(S_2(j)=='W'& S_1(i)=='W')|...
2058 (S_2(j)=='P'& S_1(i)=='I')|(S_2(j)=='P'& S_1(i)=='V')|...
2059 (S_2(j)=='P'& S_1(i)=='L')|(S_2(j)=='P'& S_1(i)=='F')|...
2060 (S_2(j)=='P'& S_1(i)=='W')|(S_2(j)=='P'& S_1(i)=='Y')|...
2061 (S_2(j)=='P'& S_1(i)=='M')|(S_2(j)=='P'& S_1(i)=='A');
2062 h(i,j)=.36*10^(-9)+(0.736*10^(-9))*2;
2063 else
2064          h(i,j)=(0.71286*10^(-9))*2+0.3*10^(-9)+2*Rt1;
2065 end
2066 end
2067 end
2068 end
2069 end
2070 end
2071 end
2072 end
2073 end
2074 end
2075 end
2076 end
2077 end
```

```
2078 end
2079 end
2080 end
2081 end
2082 end
2083 end
2084 end
2085 end
2086 end
2087 end
2088 end
2089 end
2090 end
2091 end
2092 end
2093 end
2094 end
2095 end
2096 end
2097 end
2098 end
2099
2100 function [S_1,S_2,Q1,Q2,R1,R2,h,M,N]=...
2101 potential_35(t,epsilon1,rtt,S_1,S_20);
2102 Hhidro=0;
2103 Rt1=(4.6*1E-13);
2104 epsilon0=8.85418781762*10^(-12);
2105 k=1/(4*pi*epsilon0);
2106 B=(1.38064852*10^(-23))/(1.6021766208*10^(-19));
2107 Ea=1.85;
2108 ra=0.6;
2109 Ra=ra*1E-9-3*Rt1;
2110 pha=Ea*(t+273)*B;
2111 qA=(pha*Ra*epsilon1)*k^(-1);
2112 Er=-0.8;
2113 rr=0.8;
2114 Rr=rr*1E-9+3*Rt1;
2115 phr=Er*(t+273)*B;
2116 qR=(phr*Rr*epsilon1)*k^(-1);
2117 En=0.09;
2118 rn=0.682;
2119 Rn=rn*1E-9+3*Rt1;
2120 phn=En*(t+273)*B;
2121 qN=(phn*Rn*epsilon1)*k^(-1);
2122 Ed=-0.15;
2123 rd=0.666;
2124 Rd=rd*1E-9+3*Rt1;
2125 phd=Ed*(t+273)*B;
2126 qD=(phd*Rd*epsilon1)*k^(-1);
2127 Ec=2.48;
2128 rc=0.629;
2129 phc=Ec*(t+273)*B;
```

```
2130 Rc=rc*1E-9+3*Rt1;
2131 qC=(phc*Rc*epsilon1)*k^(-1);
2132 Eq=-0.65;
2133 rq=0.725;
2134 Rq=rq*1E-9+3*Rt1;
2135 phq=Eq*(t+273)*B;
2136 qQ=(phq*Rq*epsilon1)*k^(-1);
2137 Ee=-0.2;
2138 re=0.714;
2139 Re=re*1E-9+3*Rt1;
2140 phe=Ee*(t+273)*B;
2141 qE=(phe*Re*epsilon1)*k^(-1);
2142 Eg=1.05;
2143 rg=0.725;
2144 Rg=rg*1E-9+3*Rt1;
2145 phg=Eg*(t+273)*B;
2146 qG=(phg*Rg*epsilon1)*k^(-1);
2147 Eh=0.12;
2148 rh=0.725;
2149 Rh=rh*1E-9+3*Rt1;
2150 phh=Eh*(t+273)*B;
2151 qH=(phh*Rh*epsilon1)*k^(-1);
2152 Ei=0.9;
2153 ri=0.735;
2154 Ri=ri*1E-9-3*Rt1;
2155 phi=Ei*(t+273)*B;
2156 qI=(phi*Ri*epsilon1)*k^(-1);
2157 El=0.74;
2158 rl=0.734;
2159 Rl=rl*1E-9-3*Rt1;
2160 phl=El*(t+273)*B;
2161 qL=(phl*Rl*epsilon1)*k^(-1);
2162 Ek=-1.1;
2163 rk=0.737;
2164 Rk=rk*1E-9+3*Rt1;
2165 phk=Ek*(t+273)*B;
2166 qK=(phk*Rk*epsilon1)*k^(-1);
2167 Em=0.27;
2168 rm=0.741;
2169 Rm=rm*1E-9-3*Rt1;
2170 phm=Em*(t+273)*B;
2171 qM=(phm*Rm*epsilon1)*k^(-1);
2172 Ef=0.72;
2173 rf=0.781;
2174 Rf=rf*1E-9-3*Rt1;
2175 phf=Ef*(t+273)*B;
2176 qF=(phf*Rf*epsilon1)*k^(-1);
2177 Ep=0.18;
2178 rp=0.672;
2179 Rp=rp*1E-9-3*Rt1;
2180 php=Ep*(t+273)*B;
2181 qP=(php*Rp*epsilon1)*k^(-1);
```

```
2182 Es=0.65;
2183 rs=0.615;
2184 Rs=rs*1E-9+3*Rt1;
2185 phs=Es*(t+273)*B;
2186 qS=(phs*Rs*epsilon1)*k^(-1);
2187 Et=0.88;
2188 rt=0.659;
2189 Rt=rt*1E-9+3*Rt1;
2190 pht=Et*(t+273)*B;
2191 qT=(pht*Rt*epsilon1)*k^(-1);
2192 Ew=0.55;
2193 rw=0.826;
2194 Rw=rw*1E-9-3*Rt1;
2195 phw=Ew*(t+273)*B;
2196 qW=(phw*Rw*epsilon1)*k^(-1);
2197 Ey=0.44;
2198 ry=0.781;
2199 Ry=ry*1E-9-3*Rt1;
2200 phy=Ey*(t+273)*B;
2201 qY=(phy*Ry*epsilon1)*k^(-1);
2202 Ev=0.83;
2203 rv=0.694;
2204 Rv=rv*1E-9-3*Rt1;
2205 phv=Ev*(t+273)*B;
2206 qV=(phv*Rv*epsilon1)*k^(-1);
2207 N=length(S_1);
2208 M=length(S_20);
2209 S_2=S_20;
2210 Q1=[];
2211 Q2=[];
2212 Q3=[];
2213 Q4=[];
2214 R1=[];
2215 R2=[];
2216 h=[];
2217 for i=1:length(S_1);
2218 if (S_1(i)=='A')
2219 Q1(i)=qA;
2220 else
2221    if (S_1(i)=='R')
2222 Q1(i)=qR;
2223    else
2224   if (S_1(i)=='N')
2225 Q1(i)=qN;
2226   else
2227    if (S_1(i)=='D')
2228 Q1(i)=qD;
2229   else
2230   if (S_1(i)=='C')
2231 Q1(i)=qC;
2232   else
2233    if (S_1(i)=='Q')
```

```
2234 Q1(i)=qQ;
2235    else
2236    if (S_1(i)=='E')
2237 Q1(i)=qE;
2238    else
2239     if (S_1(i)=='G')
2240 Q1(i)=qG;
2241     else
2242   if (S_1(i)=='K')
2243 Q1(i)=qK;
2244   else
2245    if (S_1(i)=='P')
2246 Q1(i)=qP;
2247    else
2248   if (S_1(i)=='S')
2249 Q1(i)=qS;
2250   else
2251    if (S_1(i)=='T')
2252 Q1(i)=qT;
2253    else
2254   if (S_1(i)=='I')
2255 Q1(i)=qI;
2256   else
2257    if (S_1(i)=='V')
2258 Q1(i)=qV;
2259   else
2260    if (S_1(i)=='L')
2261 Q1(i)=qL;
2262    else
2263    if (S_1(i)=='F')
2264 Q1(i)=qF;
2265   else
2266   if (S_1(i)=='W')
2267 Q1(i)=qW;
2268   else
2269   if (S_1(i)=='Y')
2270 Q1(i)=qY;
2271   else
2272   if (S_1(i)=='M')
2273 Q1(i)=qM;
2274   else
2275   if (S_1(i)=='H')
2276 Q1(i)=qH;
2277 end
2278 end
2279 end
2280 end
2281 end
2282 end
2283 end
2284 end
2285 end
```

```
2286 end
2287 end
2288 end
2289 end
2290 end
2291 end
2292 end
2293 end
2294 end
2295 end
2296 end
2297 end
2298 for j=1:length(S_2);
2299 if (S_2(j)=='A')
2300 Q2(j)=qA;
2301 else
2302    if (S_2(j)=='R')
2303 Q2(j)=qR;
2304    else
2305   if (S_2(j)=='N')
2306 Q2(j)=qN;
2307   else
2308   if (S_2(j)=='D')
2309 Q2(j)=qD;
2310   else
2311   if (S_2(j)=='C')
2312 Q2(j)=qC;
2313   else
2314    if (S_2(j)=='Q')
2315 Q2(j)=qQ;
2316    else
2317    if (S_2(j)=='E')
2318 Q2(j)=qE;
2319    else
2320     if (S_2(j)=='G')
2321 Q2(j)=qG;
2322     else
2323   if (S_2(j)=='K')
2324 Q2(j)=qK;
2325   else
2326    if (S_2(j)=='P')
2327 Q2(j)=qP;
2328    else
2329   if (S_2(j)=='S')
2330 Q2(j)=qS;
2331   else
2332    if (S_2(j)=='T')
2333 Q2(j)=qT;
2334    else
2335   if (S_2(j)=='I')
2336 Q2(j)=qI;
2337   else
```

```
2338     if (S_2(j)=='V')
2339 Q2(j)=qV;
2340    else
2341     if (S_2(j)=='L')
2342 Q2(j)=qL;
2343     else
2344     if (S_2(j)=='F')
2345 Q2(j)=qF;
2346    else
2347    if (S_2(j)=='W')
2348 Q2(j)=qW;
2349    else
2350    if (S_2(j)=='Y')
2351 Q2(j)=qY;
2352    else
2353    if (S_2(j)=='M')
2354 Q2(j)=qM;
2355    else
2356    if (S_2(j)=='H')
2357 Q2(j)=qH;
2358 end
2359 end
2360 end
2361 end
2362 end
2363 end
2364 end
2365 end
2366 end
2367 end
2368 end
2369 end
2370 end
2371 end
2372 end
2373 end
2374 end
2375 end
2376 end
2377 end
2378 end
2379 for i=1:length(S_1);
2380     for j=1:length(S_2);
2381      if (S_1(i)=='A')|(S_2(j)=='A');
2382             R1(i)=Ra;
2383            R2(j)=Ra;
2384        else
2385        if (S_1(i)=='R')|(S_2(j)=='R');
2386            R1(i)=Rr;
2387            R2(j)=Rr;
2388        else
2389 if (S_1(i)=='N')|(S_2(j)=='N');
```

```
2390                R1(i)=Rn;
2391                R2(j)=Rn;
2392  else
2393  if (S_1(i)=='D')|(S_2(j)=='D');
2394      R1(i)=Rd;
2395      R2(j)=Rd;
2396  else
2397        if (S_1(i)=='C')|(S_2(j)=='C');
2398              R1(i)=Rc;
2399              R2(j)=Rc;
2400          else
2401  if (S_1(i)=='Q')|(S_2(j)=='Q');
2402              R1(i)=Rq;
2403              R2(j)=Rq;
2404          else
2405          if (S_1(i)=='E')|(S_2(j)=='E');
2406              R1(i)=Re;
2407              R2(j)=Re;
2408          else
2409            if (S_1(i)=='G')|(S_2(j)=='G');
2410              R1(i)=Rg;
2411              R2(j)=Rg;
2412          else
2413   if (S_1(i)=='H')|(S_2(j)=='H');
2414        R1(i)=Rh;
2415        R2(j)=Rh;
2416   else
2417       if (S_1(i)=='I')|(S_2(j)=='I');
2418              R1(i)=Ri;
2419              R2(j)=Ri;
2420          else
2421       if (S_1(i)=='L')|(S_2(j)=='L');
2422              R1(i)=Rl;
2423              R2(j)=Rl;
2424          else
2425       if (S_1(i)=='K')|(S_2(j)=='K')
2426             R1(i)=Rk;
2427              R2(j)=Rk;
2428          else
2429       if (S_1(i)=='M')|(S_2(j)=='M')
2430              R1(i)=Rm;
2431              R2(j)=Rm;
2432          else
2433       if (S_1(i)=='F')|(S_2(j)=='F')
2434              R1(i)=Rf;
2435              R2(j)=Rf;
2436       else
2437         if (S_1(i)=='P')|(S_2(j)=='P');
2438              R1(i)=Rp;
2439              R2(j)=Rp;
2440          else
2441           if (S_1(i)=='S')|(S_2(j)=='S');
```

```
2442                R1(i)=Rs;
2443                R2(j)=Rs;
2444            else
2445             if (S_1(i)=='T')|(S_2(j)=='T');
2446                R1(i)=Rt;
2447                R2(j)=Rt;
2448            else
2449             if (S_1(i)=='W')|(S_2(j)=='W');
2450                R1(i)=Rw;
2451                R2(j)=Rw;
2452            else
2453             if (S_1(i)=='Y')|(S_2(j)=='Y');
2454                R1(i)=Ry;
2455                R2(j)=Ry;
2456             else
2457                if (S_1(i)=='V')|(S_2(j)=='V');
2458                R1(i)=Rv;
2459                R2(j)=Rv;
2460                else
2461                if (S_1(i)=='X')|(S_2(j)=='X')
2462                 R1(i)=0.194E-9;
2463                 R2(j)=0.994E-9;
2464 end
2465 end
2466 end
2467 end
2468 end
2469 end
2470 end
2471 end
2472 end
2473 end
2474 end
2475 end
2476 end
2477 end
2478 end
2479 end
2480 end
2481 end
2482 end
2483 end
2484 end
2485 end
2486 end
2487 for i=1:length(S_1);
2488 for j=1:length(S_2);
2489 if (S_1(i)=='R'& S_2(j)=='D');
2490     h(i,j)=.15*10^(-9)+Rr+Rd+3*Rt1;
2491 else
2492 if (S_1(i)=='R'& S_2(j)=='E');
2493        h(i,j)=.15*10^(-9)+Rr+Re+3*Rt1;
```

```
2494            else
2495 if (S_1(i)=='D'& S_2(j)=='R');
2496 h(i,j)=.15*10^(-9)+Rd+Rr+3*Rt1;
2497 else
2498 if (S_1(i)=='D'& S_2(j)=='H');
2499 h(i,j)=.15*10^(-9)+Rd+Rh+3*Rt1;
2500   else
2501 if (S_1(i)=='D'& S_2(j)=='R');
2502 h(i,j)=.15*10^(-9)+Rd+Rr+3*Rt1;
2503 else
2504     if (S_1(i)=='D'& S_2(j)=='H');
2505     h(i,j)=.15*10^(-9)+Rd+Rh+3*Rt1;
2506   else
2507 if (S_1(i)=='D'& S_2(j)=='K');
2508 h(i,j)=.15*10^(-9)+Rd+Rk+3*Rt1;
2509  else
2510 if (S_1(i)=='E')& (S_2(j)=='R');
2511 h(i,j)=.15*10^(-9)+Re+Rr+3*Rt1;
2512          else
2513 if (S_1(i)=='E'& S_2(j)=='H');
2514 h(i,j)=.15*10^(-9)+Re+Rh+3*Rt1;
2515       else
2516 if (S_1(i)=='E'& S_2(j)=='K');
2517 h(i,j)=.15*10^(-9)+Re+Rk+3*Rt1;
2518   else
2519 if (S_1(i)=='H'& S_2(j)=='D')
2520 h(i,j)=.15*10^(-9)+Rh+Rd+3*Rt1;
2521 else
2522 if (S_1(i)=='H'& S_2(j)=='E')
2523 h(i,j)=.15*10^(-9)+Rh+Re+3*Rt1;
2524 else
2525 if (S_1(i)=='R'& S_2(j)=='R')
2526     h(i,j)=.4*10^(-9)+Rr+Rr+3*Rt1;
2527    else
2528   if (S_1(i)=='R'& S_2(j)=='H')
2529    h(i,j)=.4*10^(-9)+Rr+Rh+3*Rt1;
2530   else
2531 if (S_1(i)=='R'& S_2(j)=='H')
2532     h(i,j)=.4*10^(-9)+Rr+Rh+2*Rt1;
2533 else
2534   if (S_1(i)=='R'& S_2(j)=='K')
2535     h(i,j)=.4*10^(-9)+Rr+Rk+3*Rt1;
2536  else
2537 if (S_1(i)=='D'& S_2(j)=='E');
2538         h(i,j)=.4*10^(-9)+Rd+Re+3*Rt1;
2539 else
2540    if (S_1(i)=='D'& S_2(j)=='D');
2541    h(i,j)=.4*10^(-9)+Rd+Rd+3*Rt1;
2542   else
2543 if (S_1(i)=='H'& S_2(j)=='R')
2544     h(i,j)=.4*10^(-9)+Rh+Rr+3*Rt1;
2545 else
```

```
2546     if (S_1(i)=='H'& S_2(j)=='H')
2547        h(i,j)=.4*10^(-9)+Rh+Rh+3*Rt1;
2548    else
2549      if (S_1(i)=='H'& S_2(j)=='K')
2550        h(i,j)=.4*10^(-9)+Rh+Rk+3*Rt1;
2551      else
2552   if (S_1(i)=='K'& S_2(j)=='R')
2553           h(i,j)=.4*10^(-9)+Rk+Rr+3*Rt1;
2554   else
2555    if (S_1(i)=='K'& S_2(j)=='H')
2556         h(i,j)=.4*10^(-9)+Rk+Rh+3*Rt1;
2557    else
2558     if (S_1(i)=='K'& S_2(j)=='K')
2559       h(i,j)=.4*10^(-9)+Rk+Rk+3*Rt1;
2560    else
2561   if (S_1(i)=='N'& S_2(j)=='Q')
2562           h(i,j)=.25*10^(-9)+Rn+Rq+3*Rt1;
2563   else
2564    if (S_1(i)=='N'& S_2(j)=='S')
2565      h(i,j)=.25*10^(-9)+Rn+Rs+3*Rt1;
2566     else
2567     if (S_1(i)=='N'& S_2(j)=='Y')
2568       h(i,j)=.25*10^(-9)+Rn+Ry+3*Rt1;
2569    else
2570   if (S_1(i)=='Q'& S_2(j)=='S')| (S_1(i)=='Q')& (S_2(j)=='Y');
2571           h(i,j)=.25*10^(-9)+Rq+Rs+3*Rt1;
2572   else
2573    if  (S_1(i)=='Q')& (S_2(j)=='Y');
2574        h(i,j)=.25*10^(-9)+Rq+Ry+3*Rt1;
2575   else
2576   if (S_1(i)=='S'& S_2(j)=='Y');
2577           h(i,j)=.25*10^(-9)+Rs+Ry+3*Rt1;
2578   else
2579   if (S_1(i)=='I'& S_2(j)=='V')|(S_1(i)=='I'& S_2(j)=='L')|...
2580   (S_1(i)=='I'& S_2(j)=='F')|(S_1(i)=='I'& S_2(j)=='W')|...
2581   (S_1(i)=='I'& S_2(j)=='Y')|(S_1(i)=='I'& S_2(j)=='M')|...
2582   (S_1(i)=='I'& S_2(j)=='A')|(S_1(i)=='V'& S_2(j)=='V')|...
2583   (S_1(i)=='V'& S_2(j)=='L')|(S_1(i)=='V'& S_2(j)=='F')|...
2584   (S_1(i)=='V'& S_2(j)=='W')|(S_1(i)=='V'& S_2(j)=='M')|...
2585   (S_1(i)=='V'& S_2(j)=='A')|(S_1(i)=='L'& S_2(j)=='F')|...
2586   (S_1(i)=='L'& S_2(j)=='W')|(S_1(i)=='L'& S_2(j)=='Y')|...
2587   (S_1(i)=='L'& S_2(j)=='M')|(S_1(i)=='L'& S_2(j)=='A')|...
2588   (S_1(i)=='F'& S_2(j)=='W')|(S_1(i)=='F'& S_2(j)=='F')|...
2589   (S_1(i)=='F'& S_2(j)=='Y')|(S_1(i)=='F'& S_2(j)=='M')|...
2590   (S_1(i)=='F'& S_2(j)=='A')|(S_1(i)=='W'& S_2(j)=='W')|...
2591   (S_1(i)=='W'& S_2(j)=='Y')|(S_1(i)=='W'& S_2(j)=='M')|...
2592   (S_1(i)=='W'& S_2(j)=='A')|(S_1(i)=='Y'& S_2(j)=='Y')|...
2593   (S_1(i)=='Y'& S_2(j)=='M')|(S_1(i)=='Y'& S_2(j)=='A')|...
2594   (S_1(i)=='M'& S_2(j)=='M')|(S_1(i)=='M'& S_2(j)=='A')|...
2595   (S_1(i)=='A'& S_2(j)=='A')|(S_1(i)=='I'& S_2(j)=='I')|...
2596   (S_1(i)=='V'& S_2(j)=='V')|(S_1(i)=='L'& S_2(j)=='L')|...
2597   (S_1(i)=='F'& S_2(j)=='F')|(S_1(i)=='W'& S_2(j)=='W')|...
```

```
2598 (S_1(i)=='P'& S_2(j)=='I')|(S_1(i)=='P'& S_2(j)=='V')|...
2599 (S_1(i)=='P'& S_2(j)=='L')|(S_1(i)=='P'& S_2(j)=='F')|...
2600 (S_1(i)=='P'& S_2(j)=='W')|(S_1(i)=='P'& S_2(j)=='Y')|...
2601 (S_1(i)=='P'& S_2(j)=='M')|(S_1(i)=='P'& S_2(j)=='A');
2602 h(i,j)=.36*10^(-9)+(0.736*10^(-9))*2;
2603  else
2604 if (S_2(j)=='I'& S_1(i)=='V')|(S_2(j)=='I'& S_1(i)=='L')|...
2605 (S_2(j)=='I'& S_1(i)=='F')|(S_2(j)=='I'& S_1(i)=='W')|...
2606 (S_2(j)=='I'& S_1(i)=='Y')|(S_2(j)=='I'& S_1(i)=='M')|...
2607 (S_2(j)=='I'& S_1(i)=='A')|(S_2(j)=='V'& S_1(i)=='V')|...
2608 (S_2(j)=='V'& S_1(i)=='L')|(S_2(j)=='V'& S_1(i)=='F')|...
2609 (S_2(j)=='V'& S_1(i)=='W')|(S_2(j)=='V'& S_1(i)=='M')|...
2610 (S_2(j)=='V'& S_1(i)=='A')|(S_2(j)=='L'& S_1(i)=='F')|...
2611 (S_2(j)=='L'& S_1(i)=='W')|(S_2(j)=='L'& S_1(i)=='Y')|...
2612 (S_2(j)=='L'& S_1(i)=='M')|(S_2(j)=='L'& S_1(i)=='A')|...
2613 (S_2(j)=='F'& S_1(i)=='W')|(S_2(j)=='F'& S_1(i)=='F')|...
2614 (S_2(j)=='F'& S_1(i)=='Y')|(S_2(j)=='F'& S_1(i)=='M')|...
2615 (S_2(j)=='F'& S_1(i)=='A')|(S_2(j)=='W'& S_1(i)=='W')|...
2616 (S_2(j)=='W'& S_1(i)=='Y')|(S_2(j)=='W'& S_1(i)=='M')|...
2617 (S_2(j)=='W'& S_1(i)=='A')|(S_2(j)=='Y'& S_1(i)=='Y')|...
2618 (S_2(j)=='Y'& S_1(i)=='M')|(S_2(j)=='Y'& S_1(i)=='A')|...
2619 (S_2(j)=='M'& S_1(i)=='M')|(S_2(j)=='M'& S_1(i)=='A')|...
2620 (S_2(j)=='A'& S_1(i)=='A')|(S_2(j)=='I'& S_1(i)=='I')|...
2621 (S_2(j)=='V'& S_1(i)=='V')|(S_2(j)=='L'& S_1(i)=='L')|...
2622 (S_2(j)=='F'& S_1(i)=='F')|(S_2(j)=='W'& S_1(i)=='W')|...
2623 (S_2(j)=='P'& S_1(i)=='I')|(S_2(j)=='P'& S_1(i)=='V')|...
2624 (S_2(j)=='P'& S_1(i)=='L')|(S_2(j)=='P'& S_1(i)=='F')|...
2625 (S_2(j)=='P'& S_1(i)=='W')|(S_2(j)=='P'& S_1(i)=='Y')|...
2626 (S_2(j)=='P'& S_1(i)=='M')|(S_2(j)=='P'& S_1(i)=='A');
2627 h(i,j)=.36*10^(-9)+(0.736*10^(-9))*2;
2628 else
2629         h(i,j)=(0.71286*10^(-9))*2+0.3*10^(-9)+3*Rt1;
2630 end
2631 end
2632 end
2633 end
2634 end
2635 end
2636 end
2637 end
2638 end
2639 end
2640 end
2641 end
2642 end
2643 end
2644 end
2645 end
2646 end
2647 end
2648 end
2649 end
```

```
2650 end
2651 end
2652 end
2653 end
2654 end
2655 end
2656 end
2657 end
2658 end
2659 end
2660 end
2661 end
2662 end
2663 end
2664
2665 function [S_1,S_2,Q1,Q2,R1,R2,h,M,N]=...
2666 potential_40(t,epsilon1,rLL,S_1,S_20);
2667 Hhidro=0;
2668 Rt1=(4.6*1E-13)*2;
2669 epsilon0=8.85418781762*10^(-12);
2670 k=1/(4*pi*epsilon0);
2671 B=(1.38064852*10^(-23))/(1.6021766208*10^(-19));
2672 Ea=1.75;
2673 ra=0.6;
2674 Ra=ra*1E-9-4*Rt1;
2675 pha=Ea*(t+273)*B;
2676 qA=(pha*Ra*epsilon1)*k^(-1);
2677 Er=-0.79;
2678 rr=0.8;
2679 Rr=rr*1E-9+4*Rt1;
2680 phr=Er*(t+273)*B;
2681 qR=(phr*Rr*epsilon1)*k^(-1);
2682 En=0.11;
2683 rn=0.682;
2684 Rn=rn*1E-9+4*Rt1;
2685 phn=En*(t+273)*B;
2686 qN=(phn*Rn*epsilon1)*k^(-1);
2687 Ed=-0.2;
2688 rd=0.666;
2689 Rd=rd*1E-9+4*Rt1;
2690 phd=Ed*(t+273)*B;
2691 qD=(phd*Rd*epsilon1)*k^(-1);
2692 Ec=2.35;
2693 rc=0.629;
2694 phc=Ec*(t+273)*B;
2695 Rc=rc*1E-9+4*Rt1;
2696 qC=(phc*Rc*epsilon1)*k^(-1);
2697 Eq=-0.61;
2698 rq=0.725;
2699 Rq=rq*1E-9+4*Rt1;
2700 phq=Eq*(t+273)*B;
2701 qQ=(phq*Rq*epsilon1)*k^(-1);
```

```
2702 Ee=-0.25;
2703 re=0.714;
2704 Re=re*1E-9+4*Rt1;
2705 phe=Ee*(t+273)*B;
2706 qE=(phe*Re*epsilon1)*k^(-1);
2707 Eg=1;
2708 rg=0.725;
2709 Rg=rg*1E-9+4*Rt1;
2710 phg=Eg*(t+273)*B;
2711 qG=(phg*Rg*epsilon1)*k^(-1);
2712 Eh=0.11;
2713 rh=0.725;
2714 Rh=rh*1E-9+4*Rt1;
2715 phh=Eh*(t+273)*B;
2716 qH=(phh*Rh*epsilon1)*k^(-1);
2717 Ei=0.82;
2718 ri=0.735;
2719 Ri=ri*1E-9-4*Rt1;
2720 phi=Ei*(t+273)*B;
2721 qI=(phi*Ri*epsilon1)*k^(-1);
2722 El=0.70;
2723 rl=0.734;
2724 Rl=rl*1E-9-4*Rt1;
2725 phl=El*(t+273)*B;
2726 qL=(phl*Rl*epsilon1)*k^(-1);
2727 Ek=-1.1;
2728 rk=0.737;
2729 Rk=rk*1E-9+4*Rt1;
2730 phk=Ek*(t+273)*B;
2731 qK=(phk*Rk*epsilon1)*k^(-1);
2732 Em=0.24;
2733 rm=0.741;
2734 Rm=rm*1E-9-4*Rt1;
2735 phm=Em*(t+273)*B;
2736 qM=(phm*Rm*epsilon1)*k^(-1);
2737 Ef=0.65;
2738 rf=0.781;
2739 Rf=rf*1E-9-4*Rt1;
2740 phf=Ef*(t+273)*B;
2741 qF=(phf*Rf*epsilon1)*k^(-1);
2742 Ep=0.1;
2743 rp=0.672;
2744 Rp=rp*1E-9-4*Rt1;
2745 php=Ep*(t+273)*B;
2746 qP=(php*Rp*epsilon1)*k^(-1);
2747 Es=0.62;
2748 rs=0.615;
2749 Rs=rs*1E-9+4*Rt1;
2750 phs=Es*(t+273)*B;
2751 qS=(phs*Rs*epsilon1)*k^(-1);
2752 Et=0.85;
2753 rt=0.659;
```

```
2754 Rt=rt*1E-9+4*Rt1;
2755 pht=Et*(t+273)*B;
2756 qT=(pht*Rt*epsilon1)*k^(-1);
2757 Ew=0.43;
2758 rw=0.826;
2759 Rw=rw*1E-9-4*Rt1;
2760 phw=Ew*(t+273)*B;
2761 qW=(phw*Rw*epsilon1)*k^(-1);
2762 Ey=0.37;
2763 ry=0.781;
2764 Ry=ry*1E-9-4*Rt1;
2765 phy=Ey*(t+273)*B;
2766 qY=(phy*Ry*epsilon1)*k^(-1);
2767 Ev=0.79;
2768 rv=0.694;
2769 Rv=rv*1E-9-4*Rt1;
2770 phv=Ev*(t+273)*B;
2771 qV=(phv*Rv*epsilon1)*k^(-1);
2772 N=length(S_1);
2773 M=length(S_20);
2774 S_2=S_20;
2775 Q1=[];
2776 Q2=[];
2777 Q3=[];
2778 Q4=[];
2779 R1=[];
2780 R2=[];
2781 h=[];
2782 for i=1:length(S_1);
2783 if (S_1(i)=='A')
2784 Q1(i)=qA;
2785 else
2786    if (S_1(i)=='R')
2787 Q1(i)=qR;
2788    else
2789   if (S_1(i)=='N')
2790 Q1(i)=qN;
2791   else
2792    if (S_1(i)=='D')
2793 Q1(i)=qD;
2794   else
2795   if (S_1(i)=='C')
2796 Q1(i)=qC;
2797   else
2798    if (S_1(i)=='Q')
2799 Q1(i)=qQ;
2800    else
2801    if (S_1(i)=='E')
2802 Q1(i)=qE;
2803    else
2804     if (S_1(i)=='G')
2805 Q1(i)=qG;
```

```
2806        else
2807      if (S_1(i)=='K')
2808  Q1(i)=qK;
2809      else
2810       if (S_1(i)=='P')
2811  Q1(i)=qP;
2812       else
2813      if (S_1(i)=='S')
2814  Q1(i)=qS;
2815      else
2816       if (S_1(i)=='T')
2817  Q1(i)=qT;
2818       else
2819      if (S_1(i)=='I')
2820  Q1(i)=qI;
2821      else
2822       if (S_1(i)=='V')
2823  Q1(i)=qV;
2824      else
2825       if (S_1(i)=='L')
2826  Q1(i)=qL;
2827       else
2828       if (S_1(i)=='F')
2829  Q1(i)=qF;
2830      else
2831      if (S_1(i)=='W')
2832  Q1(i)=qW;
2833      else
2834      if (S_1(i)=='Y')
2835  Q1(i)=qY;
2836      else
2837      if (S_1(i)=='M')
2838  Q1(i)=qM;
2839      else
2840      if (S_1(i)=='H')
2841  Q1(i)=qH;
2842  end
2843  end
2844  end
2845  end
2846  end
2847  end
2848  end
2849  end
2850  end
2851  end
2852  end
2853  end
2854  end
2855  end
2856  end
2857  end
```

```
2858 end
2859 end
2860 end
2861 end
2862 end
2863 for j=1:length(S_2);
2864 if (S_2(j)=='A')
2865 Q2(j)=qA;
2866 else
2867    if (S_2(j)=='R')
2868 Q2(j)=qR;
2869    else
2870   if (S_2(j)=='N')
2871 Q2(j)=qN;
2872   else
2873   if (S_2(j)=='D')
2874 Q2(j)=qD;
2875   else
2876   if (S_2(j)=='C')
2877 Q2(j)=qC;
2878   else
2879    if (S_2(j)=='Q')
2880 Q2(j)=qQ;
2881    else
2882    if (S_2(j)=='E')
2883 Q2(j)=qE;
2884    else
2885     if (S_2(j)=='G')
2886 Q2(j)=qG;
2887     else
2888   if (S_2(j)=='K')
2889 Q2(j)=qK;
2890   else
2891    if (S_2(j)=='P')
2892 Q2(j)=qP;
2893    else
2894   if (S_2(j)=='S')
2895 Q2(j)=qS;
2896   else
2897    if (S_2(j)=='T')
2898 Q2(j)=qT;
2899    else
2900   if (S_2(j)=='I')
2901 Q2(j)=qI;
2902   else
2903    if (S_2(j)=='V')
2904 Q2(j)=qV;
2905   else
2906    if (S_2(j)=='L')
2907 Q2(j)=qL;
2908    else
2909    if (S_2(j)=='F')
```

```
2910 Q2(j)=qF;
2911   else
2912   if (S_2(j)=='W')
2913 Q2(j)=qW;
2914   else
2915   if (S_2(j)=='Y')
2916 Q2(j)=qY;
2917   else
2918   if (S_2(j)=='M')
2919 Q2(j)=qM;
2920   else
2921   if (S_2(j)=='H')
2922 Q2(j)=qH;
2923 end
2924 end
2925 end
2926 end
2927 end
2928 end
2929 end
2930 end
2931 end
2932 end
2933 end
2934 end
2935 end
2936 end
2937 end
2938 end
2939 end
2940 end
2941 end
2942 end
2943 end
2944 for i=1:length(S_1);
2945 for j=1:length(S_2);
2946      if (S_1(i)=='A')|(S_2(j)=='A');
2947             R1(i)=Ra;
2948             R2(j)=Ra;
2949         else
2950         if (S_1(i)=='R')|(S_2(j)=='R');
2951             R1(i)=Rr;
2952             R2(j)=Rr;
2953         else
2954 if (S_1(i)=='N')|(S_2(j)=='N');
2955             R1(i)=Rn;
2956             R2(j)=Rn;
2957 else
2958 if (S_1(i)=='D')|(S_2(j)=='D');
2959     R1(i)=Rd;
2960     R2(j)=Rd;
2961 else
```

```
2962        if (S_1(i)=='C')|(S_2(j)=='C');
2963              R1(i)=Rc;
2964              R2(j)=Rc;
2965          else
2966  if (S_1(i)=='Q')|(S_2(j)=='Q');
2967              R1(i)=Rq;
2968              R2(j)=Rq;
2969          else
2970          if (S_1(i)=='E')|(S_2(j)=='E');
2971              R1(i)=Re;
2972              R2(j)=Re;
2973          else
2974            if (S_1(i)=='G')|(S_2(j)=='G');
2975              R1(i)=Rg;
2976              R2(j)=Rg;
2977          else
2978   if (S_1(i)=='H')|(S_2(j)=='H');
2979        R1(i)=Rh;
2980        R2(j)=Rh;
2981   else
2982       if (S_1(i)=='I')|(S_2(j)=='I');
2983              R1(i)=0.735E-9-4*Rt1;
2984              R2(j)=0.735E-9-4*Rt1;
2985          else
2986       if (S_1(i)=='L')|(S_2(j)=='L');
2987              R1(i)=0.734E-9-4*Rt1;
2988              R2(j)=0.734E-9-4*Rt1;
2989          else
2990       if (S_1(i)=='K')|(S_2(j)=='K');
2991             R1(i)=Rk+4*Rt1;
2992              R2(j)=Rk+4*Rt1;
2993          else
2994       if (S_1(i)=='M')|(S_2(j)=='M')
2995              R1(i)=0.741E-9-4*Rt1;
2996              R2(j)=0.741E-9-4*Rt1;
2997       else
2998       if (S_1(i)=='F')|(S_2(j)=='F');
2999              R1(i)=0.781E-9-4*Rt1;
3000              R2(j)=0.781E-9-4*Rt1;
3001       else
3002         if (S_1(i)=='P')|(S_2(j)=='P');
3003              R1(i)=0.672E-9-4*Rt1;
3004              R2(j)=0.672E-9-4*Rt1;
3005          else
3006           if (S_1(i)=='S')|(S_2(j)=='S');
3007              R1(i)=0.615E-9+4*Rt1;
3008              R2(j)=0.615E-9+4*Rt1;
3009          else
3010           if (S_1(i)=='T')|(S_2(j)=='T');
3011              R1(i)=0.659E-9+4*Rt1;
3012              R2(j)=0.659E-9+4*Rt1;
3013          else
```

```
3014            if (S_1(i)=='W')|(S_2(j)=='W');
3015               R1(i)=0.826E-9-4*Rt1;
3016               R2(j)=0.826E-9-4*Rt1;
3017           else
3018            if (S_1(i)=='Y')|(S_2(j)=='Y');
3019               R1(i)=0.781E-9-4*Rt1;
3020               R2(j)=0.781E-9-4*Rt1;
3021            else
3022               if (S_1(i)=='V')|(S_2(j)=='V');
3023               R1(i)=0.694E-9-4*Rt1;
3024               R2(j)=0.694E-9-4*Rt1;
3025               else
3026               if (S_1(i)=='X')|(S_2(j)=='X')
3027                R1(i)=0.194E-9;
3028                R2(j)=0.994E-9;
3029 end
3030 end
3031 end
3032 end
3033 end
3034 end
3035 end
3036 end
3037 end
3038 end
3039 end
3040 end
3041 end
3042 end
3043 end
3044 end
3045 end
3046 end
3047 end
3048 end
3049 end
3050 end
3051 end
3052 for i=1:length(S_1);
3053 for j=1:length(S_2);
3054 if (S_1(i)=='R'& S_2(j)=='D');
3055     h(i,j)=.15*10^(-9)+Rr+Rd+4*Rt1;
3056 else
3057 if (S_1(i)=='R'& S_2(j)=='E');
3058        h(i,j)=.15*10^(-9)+Rr+Re+4*Rt1;
3059         else
3060 if (S_1(i)=='D'& S_2(j)=='R');
3061 h(i,j)=.15*10^(-9)+Rd+Rr+4*Rt1;
3062 else
3063 if (S_1(i)=='D'& S_2(j)=='H');
3064 h(i,j)=.15*10^(-9)+Rd+Rh+4*Rt1;
3065   else
```

```
3066 if (S_1(i)=='D'& S_2(j)=='R');
3067 h(i,j)=.15*10^(-9)+Rd+Rr+4*Rt1;
3068 else
3069     if (S_1(i)=='D'& S_2(j)=='H');
3070     h(i,j)=.15*10^(-9)+Rd+Rh+4*Rt1;
3071   else
3072 if (S_1(i)=='D'& S_2(j)=='K');
3073 h(i,j)=.15*10^(-9)+Rd+Rk+4*Rt1;
3074  else
3075 if (S_1(i)=='E')& (S_2(j)=='R');
3076 h(i,j)=.15*10^(-9)+Re+Rr+4*Rt1;
3077         else
3078 if (S_1(i)=='E'& S_2(j)=='H');
3079 h(i,j)=.15*10^(-9)+Re+Rh+4*Rt1;
3080       else
3081 if (S_1(i)=='E'& S_2(j)=='K');
3082 h(i,j)=.15*10^(-9)+Re+Rk+4*Rt1;
3083   else
3084 if (S_1(i)=='H'& S_2(j)=='D')
3085 h(i,j)=.15*10^(-9)+Rh+Rd+4*Rt1;
3086 else
3087 if (S_1(i)=='H'& S_2(j)=='E')
3088 h(i,j)=.15*10^(-9)+Rh+Re+4*Rt1;
3089 else
3090 if (S_1(i)=='R'& S_2(j)=='R')
3091     h(i,j)=.4*10^(-9)+Rr+Rr+4*Rt1;
3092    else
3093   if (S_1(i)=='R'& S_2(j)=='H')
3094    h(i,j)=.4*10^(-9)+Rr+Rh+4*Rt1;
3095   else
3096 if (S_1(i)=='R'& S_2(j)=='H')
3097     h(i,j)=.4*10^(-9)+Rr+Rh+4*Rt1;
3098 else
3099   if (S_1(i)=='R'& S_2(j)=='K')
3100     h(i,j)=.4*10^(-9)+Rr+Rk+4*Rt1;
3101  else
3102 if (S_1(i)=='D'& S_2(j)=='E');
3103         h(i,j)=.4*10^(-9)+Rd+Re+4*Rt1;
3104 else
3105    if (S_1(i)=='D'& S_2(j)=='D');
3106    h(i,j)=.4*10^(-9)+Rd+Rd+4*Rt1;
3107   else
3108 if (S_1(i)=='H'& S_2(j)=='R')
3109     h(i,j)=.4*10^(-9)+Rh+Rr+4*Rt1;
3110 else
3111   if (S_1(i)=='H'& S_2(j)=='H')
3112      h(i,j)=.4*10^(-9)+Rh+Rh+4*Rt1;
3113  else
3114    if (S_1(i)=='H'& S_2(j)=='K')
3115      h(i,j)=.4*10^(-9)+Rh+Rk+4*Rt1;
3116    else
3117 if (S_1(i)=='K'& S_2(j)=='R')
```

```
3118          h(i,j)=.4*10^(-9)+Rk+Rr+4*Rt1;
3119 else
3120  if (S_1(i)=='K'& S_2(j)=='H')
3121        h(i,j)=.4*10^(-9)+Rk+Rh+4*Rt1;
3122  else
3123   if (S_1(i)=='K'& S_2(j)=='K')
3124      h(i,j)=.4*10^(-9)+Rk+Rk+4*Rt1;
3125  else
3126 if (S_1(i)=='N'& S_2(j)=='Q')
3127          h(i,j)=.25*10^(-9)+Rn+Rq+4*Rt1;
3128 else
3129  if (S_1(i)=='N'& S_2(j)=='S')
3130     h(i,j)=.25*10^(-9)+Rn+Rs+4*Rt1;
3131   else
3132   if (S_1(i)=='N'& S_2(j)=='Y')
3133      h(i,j)=.25*10^(-9)+Rn+Ry+4*Rt1;
3134  else
3135 if (S_1(i)=='Q'& S_2(j)=='S')| (S_1(i)=='Q')& (S_2(j)=='Y');
3136          h(i,j)=.25*10^(-9)+Rq+Rs+4*Rt1;
3137 else
3138  if   (S_1(i)=='Q')& (S_2(j)=='Y');
3139       h(i,j)=.25*10^(-9)+Rq+Ry+4*Rt1;
3140 else
3141 if (S_1(i)=='S'& S_2(j)=='Y');
3142          h(i,j)=.25*10^(-9)+Rs+Ry+4*Rt1;
3143 else
3144 if (S_1(i)=='I'& S_2(j)=='V')|(S_1(i)=='I'& S_2(j)=='L')|...
3145 (S_1(i)=='I'& S_2(j)=='F')|(S_1(i)=='I'& S_2(j)=='W')|...
3146 (S_1(i)=='I'& S_2(j)=='Y')|(S_1(i)=='I'& S_2(j)=='M')|...
3147 (S_1(i)=='I'& S_2(j)=='A')|(S_1(i)=='V'& S_2(j)=='V')|...
3148 (S_1(i)=='V'& S_2(j)=='L')|(S_1(i)=='V'& S_2(j)=='F')|...
3149 (S_1(i)=='V'& S_2(j)=='W')|(S_1(i)=='V'& S_2(j)=='M')|...
3150 (S_1(i)=='V'& S_2(j)=='A')|(S_1(i)=='L'& S_2(j)=='F')|...
3151 (S_1(i)=='L'& S_2(j)=='W')|(S_1(i)=='L'& S_2(j)=='Y')|...
3152 (S_1(i)=='L'& S_2(j)=='M')|(S_1(i)=='L'& S_2(j)=='A')|...
3153 (S_1(i)=='F'& S_2(j)=='W')|(S_1(i)=='F'& S_2(j)=='F')|...
3154 (S_1(i)=='F'& S_2(j)=='Y')|(S_1(i)=='F'& S_2(j)=='M')|...
3155 (S_1(i)=='F'& S_2(j)=='A')|(S_1(i)=='W'& S_2(j)=='W')|...
3156 (S_1(i)=='W'& S_2(j)=='Y')|(S_1(i)=='W'& S_2(j)=='M')|...
3157 (S_1(i)=='W'& S_2(j)=='A')|(S_1(i)=='Y'& S_2(j)=='Y')|...
3158 (S_1(i)=='Y'& S_2(j)=='M')|(S_1(i)=='Y'& S_2(j)=='A')|...
3159 (S_1(i)=='M'& S_2(j)=='M')|(S_1(i)=='M'& S_2(j)=='A')|...
3160 (S_1(i)=='A'& S_2(j)=='A')|(S_1(i)=='I'& S_2(j)=='I')|...
3161 (S_1(i)=='V'& S_2(j)=='V')|(S_1(i)=='L'& S_2(j)=='L')|...
3162 (S_1(i)=='F'& S_2(j)=='F')|(S_1(i)=='W'& S_2(j)=='W')|...
3163 (S_1(i)=='P'& S_2(j)=='I')|(S_1(i)=='P'& S_2(j)=='V')|...
3164 (S_1(i)=='P'& S_2(j)=='L')|(S_1(i)=='P'& S_2(j)=='F')|...
3165 (S_1(i)=='P'& S_2(j)=='W')|(S_1(i)=='P'& S_2(j)=='Y')|...
3166  (S_1(i)=='P'& S_2(j)=='M')|(S_1(i)=='P'& S_2(j)=='A');
3167      h(i,j)=.36*10^(-9)+(0.736*10^(-9))*2;
3168  else
3169   if (S_2(j)=='I'& S_1(i)=='V')|(S_2(j)=='I'& S_1(i)=='L')|...
```

```
3170 (S_2(j)=='I'& S_1(i)=='F')|(S_2(j)=='I'& S_1(i)=='W')|...
3171 (S_2(j)=='I'& S_1(i)=='Y')|(S_2(j)=='I'& S_1(i)=='M')|...
3172 (S_2(j)=='I'& S_1(i)=='A')|(S_2(j)=='V'& S_1(i)=='V')|...
3173 (S_2(j)=='V'& S_1(i)=='L')|(S_2(j)=='V'& S_1(i)=='F')|...
3174 (S_2(j)=='V'& S_1(i)=='W')|(S_2(j)=='V'& S_1(i)=='M')|...
3175 (S_2(j)=='V'& S_1(i)=='A')|(S_2(j)=='L'& S_1(i)=='F')|...
3176 (S_2(j)=='L'& S_1(i)=='W')|(S_2(j)=='L'& S_1(i)=='Y')|...
3177 (S_2(j)=='L'& S_1(i)=='M')|(S_2(j)=='L'& S_1(i)=='A')|...
3178 (S_2(j)=='F'& S_1(i)=='W')|(S_2(j)=='F'& S_1(i)=='F')|...
3179 (S_2(j)=='F'& S_1(i)=='Y')|(S_2(j)=='F'& S_1(i)=='M')|...
3180 (S_2(j)=='F'& S_1(i)=='A')|(S_2(j)=='W'& S_1(i)=='W')|...
3181 (S_2(j)=='W'& S_1(i)=='Y')|(S_2(j)=='W'& S_1(i)=='M')|...
3182 (S_2(j)=='W'& S_1(i)=='A')|(S_2(j)=='Y'& S_1(i)=='Y')|...
3183 (S_2(j)=='Y'& S_1(i)=='M')|(S_2(j)=='Y'& S_1(i)=='A')|...
3184 (S_2(j)=='M'& S_1(i)=='M')|(S_2(j)=='M'& S_1(i)=='A')|...
3185 (S_2(j)=='A'& S_1(i)=='A')|(S_2(j)=='I'& S_1(i)=='I')|...
3186 (S_2(j)=='V'& S_1(i)=='V')|(S_2(j)=='L'& S_1(i)=='L')|...
3187 (S_2(j)=='F'& S_1(i)=='F')|(S_2(j)=='W'& S_1(i)=='W')|...
3188 (S_2(j)=='P'& S_1(i)=='I')|(S_2(j)=='P'& S_1(i)=='V')|...
3189 (S_2(j)=='P'& S_1(i)=='L')|(S_2(j)=='P'& S_1(i)=='F')|...
3190 (S_2(j)=='P'& S_1(i)=='W')|(S_2(j)=='P'& S_1(i)=='Y')|...
3191 (S_2(j)=='P'& S_1(i)=='M')|(S_2(j)=='P'& S_1(i)=='A');
3192         h(i,j)=.36*10^(-9)+(0.736*10^(-9))*2;
3193 else
3194         h(i,j)=(0.71286*10^(-9))*2+0.3*10^(-9)+4*Rt1;
3195 end
3196 end
3197 end
3198 end
3199 end
3200 end
3201 end
3202 end
3203 end
3204 end
3205 end
3206 end
3207 end
3208 end
3209 end
3210 end
3211 end
3212 end
3213 end
3214 end
3215 end
3216 end
3217 end
3218 end
3219 end
3220 end
3221 end
```

```
3222 end
3223 end
3224 end
3225 end
3226 end
3227 end
3228 end
3229
3230 function[A]=electrostatic(Q1,Q2, R1,R2,h,M,N,N1,epsilon)
3231 for i=1:N
3232     for j=1:M
3233         if R1(i)>R2(j)
3234             gamma(i,j)=R1(i)/R2(j);
3235         else
3236             if  R1(i)<R2(j)
3237                 gamma(i,j)=R2(j)/R1(i);
3238               else if R1(i)==R2(j);
3239      gamma(i,j)=R2(j)/R1(i);
3240          end
3241              end
3242         end
3243         if h(i,j)>(R1(i)+R2(j))
3244             r(i,j)=h(i,j)/(R1(i)+R2(j));
3245         else if  h(i,j)<=(R1(i)+R2(j))
3246             r(i,j)=(R1(i)+R2(j))/h(i,j);
3247         end
3248         end
3249     y(i,j)=(((r(i,j)^2*(1+gamma(i,j))^2)-...
3250     (1+(gamma(i,j))^2))/(2*gamma(i,j)));
3251     beta(i,j)=acosh(y(i,j));
3252     z(i,j)=exp(-beta(i,j));
3253     S12=0;
3254     S22=0;
3255     S11=0;
3256     for k=1:N1
3257         gamma1(i,j)=R2(j)/R1(i);
3258  S_1(k)=(z(i,j)^k)/(((1-z(i,j)^(2*k)))*((gamma(i,j)+...
3259    y(i,j))-(y(i,j)^2-1)^(1/2)*...
3260   (1+z(i,j)^(2*k))/(1-z(i,j)^(2*k))));
3261         S11=S11+S_1(k);
3262   S_2(k)=(z(i,j)^(2*k))/(1-(z(i,j)^(2*k)));
3263       S12=S12+S_2(k);
3264   S_3(k)=(z(i,j)^k)/(((1-z(i,j)^(2*k)))*...
3265         ((1-gamma(i,j)*y(i,j))-...
3266         gamma(i,j)*(y(i,j)^2-1)^(1/2)*...
3267         (1+z(i,j)^(2*k))/(1-z(i,j)^(2*k))));
3268         S22=S22+S_3(k);
3269     end
3270     epsilon0=8.85418781762*10^(-12);
3271 c11(i,j)=(2*gamma(i,j)*((y(i,j)^2-1)^(1/2))).*S11;
3272 c22(i,j)=(2*gamma(i,j)*((y(i,j)^2-1)^(1/2))).*S22;
3273 c12(i,j)=-((2*gamma(i,j)*((y(i,j)^2-1))^(1/2))/(r(i,j)*...
```

```
3274         (1+gamma(i,j)))).*S12;
3275         delta(i,j)=((c11(i,j)*c22(i,j)-c12(i,j)^2));
3276         k=1/(4*pi*epsilon0);
3277         k1=1/(4*pi*epsilon0*epsilon);
3278             alpha(i,j)=Q2(j)/Q1(i);
3279         if R1(i)>R2(j)
3280             gamma(i,j)=R1(i)/R2(j);
3281       W1(i,j)=((1/k1)*R2(j)*gamma(i,j))*...
3282       ((1+gamma(i,j))/(2*alpha(i,j)))*...
3283       ((alpha(i,j)^2*c11(i,j)-2*alpha(i,j)*...
3284       c12(i,j)+c22(i,j))/delta(i,j));
3285             else if (R1(i)<R2(j))
3286                 gamma(i,j)=R2(j)/R1(i);
3287      W1(i,j)=((1/k1)*R1(i)*gamma(i,j))*...
3288      ((1+gamma(i,j))/(2*alpha(i,j)))*...
3289      ((alpha(i,j)^2*c11(i,j)-2*alpha(i,j)*...
3290      c12(i,j)+c22(i,j))/delta(i,j));
3291          else if R1(i)==R2(j);
3292      W1(i,j)=((1/k1)*R1(i)*gamma(i,j))*...
3293      ((1+gamma(i,j))/(2*alpha(i,j)))*...
3294      ((alpha(i,j)^2*c11(i,j)-2*alpha(i,j)*...
3295      c12(i,j)+c22(i,j))/delta(i,j));
3296                 end
3297                 end
3298         end
3299         W2(i,j)=(k*(Q1(i)*Q2(j)))/(R1(i)+R2(j));
3300         A1(i,j)=W1(i,j);
3301         A2(i,j)=W2(i,j);
3302         A(i,j)=A1(i,j)/A2(i,j);
3303         end
3304     end
3305     return
3306 
3307     function[cond2]=condmy(A)
3308     [U,S,V]=SVD_2(A);
3309     lambda_max=max(diag(S));
3310     lambda_min=min(diag(S));
3311     cond_1=(((lambda_max)/(lambda_min)));
3312     cond2=(log(cond_1))/(log(10));
3313     return
3314 
3315     function [Uout,Sout,Vout] = SVD_2(A)
3316           m = size(A,1);
3317           n = size(A,2);
3318           U = eye(m);
3319           V = eye(n);
3320           e = eps*fro(A);
3321           while (sum(abs(A(~eye(m,n)))) > e)
3322           for i = 1:n
3323           for j = i+1:n
3324               [J1,J2] = jacobi(A,m,n,i,j);
3325               A = mtimes(J1,mtimes(A,J2));
```

```
3326            U = mtimes(U,J1');
3327            V = mtimes(J2',V);
3328        end
3329        for j = n+1:m
3330            J1 = jacobi2(A,m,n,i,j);
3331            A = mtimes(J1,A);
3332            U = mtimes(U,J1');
3333        end
3334        end
3335        end
3336        S = A;
3337        if (nargout < 3)
3338           Uout = diag(S);
3339        else
3340   Uout = U; Sout = times(S,eye(m,n)); Vout = V;
3341        end
3342        end
3343      function [J1,J2] = jacobi(A,m,n,i,j)
3344         B = [A(i,i), A(i,j); A(j,i), A(j,j)];
3345         [U,S,V] = tinySVD(B); %
3346         J1 = eye(m);
3347         J1(i,i) = U(1,1);
3348         J1(j,j) = U(2,2);
3349         J1(i,j) = U(2,1);
3350         J1(j,i) = U(1,2);
3351         J2 = eye(n);
3352         J2(i,i) = V(1,1);
3353         J2(j,j) = V(2,2);
3354         J2(i,j) = V(2,1);
3355         J2(j,i) = V(1,2);
3356      end
3357      function J1 = jacobi2(A,m,n,i,j)
3358         B = [A(i,i), 0; A(j,i), 0];
3359         [U,S,V] = tinySVD(B);
3360         J1 = eye(m);
3361         J1(i,i) = U(1,1);
3362         J1(j,j) = U(2,2);
3363         J1(i,j) = U(2,1);
3364         J1(j,i) = U(1,2);
3365      end
3366      function [Uout,Sout,Vout] = tinySVD(A)
3367   t = rdivide((minus(A(1,2),A(2,1))),(plus(A(1,1),A(2,2))));
3368        c = rdivide(1,sqrt(1+t^2));
3369        s = times(t,c);
3370        R = [c,-s;s,c];
3371        M = mtimes(R,A);
3372        [U,S,V] = tinySymmetricSVD(M);
3373        U = mtimes(R',U);
3374        if (nargout < 3)
3375            Uout = diag(S);
3376        else
3377            Uout = U; Sout = S; Vout = V;
```

```
3378        end
3379        end
3380      function [Uout,Sout,Vout] = tinySymmetricSVD(A)
3381        if (A(2,1) == 0)
3382            S = A;
3383            U = eye(2);
3384            V = U;
3385        else
3386            w = A(1,1);
3387            y = A(2,1);
3388            z = A(2,2);
3389            ro = rdivide(minus(z,w),times(2,y));
3390  t2=rdivide(sign(ro),plus(abs(ro),sqrt(plus(times(ro,ro),1))));
3391            t = t2;
3392            c = rdivide(1,sqrt(plus(1,times(t,t))));
3393            s = times(t,c);
3394            U = [c, -s; s, c];
3395            V = [c,  s;-s, c];
3396            S = mtimes(U,mtimes(A,V));
3397            U = U';
3398            V = V';
3399        end
3400        [U,S,V] = fixSVD(U,S,V);
3401        if (nargout < 3)
3402            Uout = diag(S);
3403        else
3404            Uout = U; Sout = S; Vout = V;
3405        end
3406        end
3407      function [U,S,V] = fixSVD(U,S,V)
3408        Z = [sign(S(1,1)),0; 0,sign(S(2,2))]; %
3409        U = mtimes(U,Z);
3410        S = mtimes(Z,S);
3411        if (S(1,1) < S(2,2))
3412            P = [0,1;1,0];
3413            U = mtimes(U,P);
3414            S = mtimes(P,mtimes(S,P));
3415            V = mtimes(P,V);
3416        end
3417        end
3418      function f = fro(M)
3419        f = sqrt(sum(sum(times(M,M))));
3420      end
3421      function s = sign(x)
3422          if (x > 0)
3423              s = 1;
3424          else
3425              s = -1;
3426          end
3427          end
```

References

1. A.L. Fink, Protein aggregation: folding aggregates, inclusion bodies and amyloid. Fold Des. **3**(1), R1–R23 (1998)
2. E. Dijk, A. Hoogeveen, S. Abeln, The hydrophobic temperature dependence of amino acids directly calculated from protein structures. PLOS Comput. Biol. **11**(5), 1–17 (2015)
3. Ya.G. Barkan, *Organic Chemistry* (Vysshaya Shkola, Moscow, 1973)
4. N.A. Chebotareva, S.G. Roman, B.I. Kurganov, Dissociative mechanism for irreversible thermal denaturation of oligomeric proteins. Biophys. Rev. **8**(4), 397–407 (2016)
5. V.I. Rezyapkin, V.N. Bourdi, *Foundations of Biochemistry* (Grodno State University, Grodno, 2012)
6. K.P. Petrov, *Methods of Biochemistry of Plant Products* (Kiev, 1978)
7. H. Neurath, K.C. Bailey, *The Proteins: Chemistry, Biological Activity, and Methods* (Academic Press, New York, 1953)
8. F. Haurowitz, *Chemistry and Biology of Proteins* (Academic Press, New York, 1950)
9. V. Karantza, A.D. Baxevanis, E. Freire, E.N. Moudrianakis, Thermodynamic studies of the core histones: ionic strength and pH dependence of H2A–H2B dimer stability. Biochemistry **34**(17), 5988–5996 (1995)
10. Andreas D. Baxevanis, Jamie E. Godfrey, Evangelos N. Moudrianakis, Associative behavior of the histone $(H3 - H4)_2$ tetramer: dependence on ionic environment. Biochemistry **30**(36), 8817–8823 (1991)
11. Thomas H. Eickbush, Evangelos N. Moudrianakis, The histone core complex: an octamer assembled by two sets of protein-protein interactions. Biochemistry **17**(23), 4955–4964 (1978)
12. V. Bhat, M.B. Olenick, B.J. Schuchardt, D.C. Mikles, B.J. Deegan, C.B. McDonald, K.L. Seldeen, D. Kurouski, M.H. Faridi, M.M. Shareef, V. Gupta, I.K. Lednev, A. Farooq, Heat-induced fibrillation of Bcl-xl apoptotic repressor. Biophys. Chem. **179**, 12–25 (2013)
13. V. Bhat, D. Kurouski, M.B. Olenick, C.B. McDonald, D.C. Mikles, B.J. Deegan, K.L. Seldeen, I.K. Lednev, A. Farooq, Acidic pH promotes oligomerization and membrane insertion of the Bcl-xl apoptotic repressor. Arch. Biochem. Biophys. **528**(1), 32–44 (2012)
14. V. Bhat, C.B. McDonald, D.C. Mikles, B.J. Deegan, K.L. Seldeen, M.L. Bates, A. Farooq, Ligand binding and membrane insertion compete with oligomerization. J. Mol. Biol. **416**(1), 57–77 (2012)
15. K.G. Kulikov, T.V. Koshlan, Mathematical simulation of interactions of protein molecules and prediction of their reactivity. Tech. Phys. **61**(10), 1572–1579 (2016)
16. T.V. Koshlan, K.G. Kulikov, Mathematical modeling of the temperature effect on the character of linking between monomeric proteins in aqueous solutions. Tech. Phys. **62**(11), 1736–1743 (2017)
17. T.V. Koshlan, K.G. Kulikov, Mathematical modeling the formation of a histone octamer. Tech. Phys. **62**(5), 684–690 (2017)
18. T.V. Chalikian, M. Totrov, R. Abagyan, K.J. Breslauer, The hydration of globular proteins as derived from volume and compressibility measurements: cross correlating thermodynamic and structural data. J. Mol. Biol. **260**(4), 588–603 (1996)
19. *High Pressure Bioscience: Basic Concepts, Applications and Frontiers*. ed. by K. Akasaka, H. Matsuki (Springer, Berlin, 2015)
20. C.G. Malmberg, A.A. Maryott, Dielectric constant of water from $0°$ to $100°$. J. Res. Natl. Bur. Stand. **56**(1), 1–8 (1956)
21. http://www.uniprot.org/

Chapter 4
Mathematical Modelling of the Effect of a Monovalent Salt Solution on the Interaction of Protein Molecules

Abstract This chapter is devoted to the development of a mathematical model that will allow us to describe the behavior of biological complexes in vitro on the example of the formation of two histone dimers H2A–H2B and H3–H4 from the corresponding monomeric proteins H2A, H2B, H3, and H4 in solutions with different concentrations of monovalent salt. The calculations were performed taking into account the screening of the electrostatic charge of charged amino acids at different concentrations of monovalent salt using the Guy–Chapman theory. It should be noted that the screening of non-polar, polar, aromatic amino acids in solutions with different ionic strength was not taken into account in this chapter.

4.1 Introduction

In this chapter, a physical model is constructed that simulates the initial stage of the formation of a histone octamer, namely, the formation of H2A–H2B and H3–H4 dimers, taking into account the different concentrations of the monovalent salt solution. Since in the present chapter we consider the pairwise interaction of amino acid residues, which are represented as uniformly charged spheres, the article [1] should be mentioned, in which various pairs of amino acid residues were investigated for their ability to interact on the basis of physicochemical properties: size, charge, hydrophobic interactions. The results obtained in the study characterize the ability of an amino acid to bind to another amino acid. However, it should be noted that in the above work, no criterion is given for quantifying the electrostatic interaction forces between protein units leading to the assembly or dissociation of the histone octamer at different concentrations of monovalent salt.

Thus, the approach developed in this chapter will make it possible to quantify the electrostatic interaction between histone proteins in solutions with different concentrations of monovalent salt, taking into account the screening effect of charged amino acid residues, and also the criteria for the stability of various protein

T. Koshlan and K. Kulikov, *Mathematical Modeling of Protein Complexes*, Biological and Medical Physics, Biomedical Engineering,
https://doi.org/10.1007/978-3-319-98304-2_4

compounds depending on the concentration of the salt solution. It should be noted that the stability of most biological complexes with given amino acid sequences must be checked experimentally method.

The chapter consists of several parts.The first part describes the structure of the histone core previously conducted experimental work on the effect of salt solutions on the nature of the interaction of the biological complexes. In the second part, the electrostatic problem of interaction of protein molecules is considered taking into account the screening effect in a salt solution of different concentrations. In the third part, a new algorithm is proposed for determining the formation of a biochemical complex from two compounds by analyzing the potential energy matrix of the pairwise electrostatic interaction of biological reagents. In the fourth part, numerical calculations of the formation of protein complexes in solutions with different concentrations and analysis of the data obtained are presented.

4.2 General Principles for the Formation of Dimers H2A–H2B, H3–H4 and the Behavior of These Compounds in Solutions with Different Concentrations of Monovalent Salt

To analyze the interaction of protein molecules, taking into account the effect of the concentration of the monovalent salt of the solution, we chose four histones H2A, H2B, H3, and H4, forming a histone octamer of DNA. Let us turn to a brief examination of the structure of DNA and histones.

In [2] the central histone octamer contains two copies of each of the core histone proteins, H2A, H2B, H3, and H4 as established 3.1 Å crystal structure of the histone octamer.

The core histones are assembled into four histone-fold heterodimers (two each of H2A/H2B and H3/H4). Each of the core histones contains the histone fold domain, composed of three α-helices connected by two loops, which allow heterodimeric interactions between core histones known as the handshake motif, more in detail with the structure of the histone domain can read in articles [3–5].

Let's move on to the behavior of selected protein complexes in solutions with different concentrations of monovalent salt.

In [6] reported that at low ionic strength, H3–H4 and $(H3{-}H4)_2$ crosslinked products are favored and, as the ionic strength is raised, increased aggregation is observed in the form of higher molecular weight products until, at 2 M NaCl, the products of the cross-linking reaction are too large to enter the gel. Association H2A–H2B with the $(H3{-}H4)_2$ tetramer indicates that the H2A–H2B dimer could prevent small-scale aggregation of H3 and H4 by serving as a molecular "cap", i.e., by binding to the sites available on either side of the $(H3{-}H4)_2$ tetramer, the H2A–H2B dimer may be blocking tetramer ≪sticky≫ regions responsible for self-aggregation.

In [7] authors examine the stability of the isolated H2A–H2B heterodimer using urea denaturation in the presence of a variety of salts. The results presented here show that the salt stabilization of the H2A–H2B dimer involves a combination of enhancing the hydrophobic effect (via the Hofmeister effect or preferential hydration) and screening of electrostatic repulsion. The most highly charged regions of the dimer are the N-terminal tails, sites of posttranslational modifications such as acetylation and phosphorylation. These modifications, which alter the charge density of the tails, are involved in regulation of nucleosome dynamics.

In [8] the thermodynamic properties $(H3-H4)_2$ can not be studied directly though, since its thermal denaturation is completely irreversible even at the lowest salt concentrations.

Below we have considered the problem of modelling the processes of electrostatic interaction of the formation of dimers from the complete amino acid sequences of selected histone proteins.

Note that the biological objects interact in a solution that may have different ionic strengths, i.e., that contains a variety of dissolved ions. In these biological systems, the interactions between the ions are of great importance, which strongly depend on the ionic strength of the solution. This value is a measure of the intensity of the electric field created by ions in solution. To consider the effect of the magnitude of the ionic strength in solution on the stability of the studied biological compounds, we used the Gouy–Chapman theory to calculate the screening potential of a charged amino acid sequence of the protein at various concentrations of monovalent salt in solution with biological objects.

4.3 Shielding Effect in a Salt Solution

To take into account the mechanism of formation of the compensating layer of ions in the solution (shielding effect), which is formed due to forces of electrostatic attraction to distributed surface charge, we used the Gouy–Chapman theory [9–11]. In this theory the ions of the electrolyte are described by point charges of both signs in a water medium with a certain dielectric permittivity. If the energy of the ions in the field of attraction to the surface charge is of the order of kT (where k is the Boltzmanns constant, T is the absolute temperature), then the thermal motion must make such layer diffusive. Thus, the spatial distribution of counterions (ions that have the opposite charge) is determined by the fact that they are in a state of thermal motion and, at the same time, are attracted to the surface charge, which results in the formation of a diffuse layer of a certain length. Note that the length at low concentrations of the electrolyte can be significant. The electric field in the double layer must monotonically decrease with distance from a charged surface because its charge is shielded by the charge of counterions located between the given remote point and the charged surface. At the outer boundary of the electric double layer the electric field has to vanish. Thus, the only variable on which the function of the potential decay depends is the distance from the charged surface. Note that the radius of the particle is collinear with the vector of the distance from the charged surface. In accordance with this model, the functions of the electric potential and corresponding average charge distribution are computed in the neighborhood of a charged surface. The calculation of the electric double layer for the charged surface of the sphere was performed for the five charged amino acids, i.e., aspartic acid (D), glutamic acid (E), lysine (K), arginine (R), and histidine (H).

Let us write the Poisson equation for a flat surface

$$\frac{d^2\varphi(x)}{dx^2} = -\frac{\rho}{\varepsilon\varepsilon_0}, \tag{4.1}$$

where ρ is the charge density defined at a distance x from the surface and $\varphi(x)$ is the potential, ε is dielectric permittivity of the medium, and ε_0 is the electric constant. According to [9–11], we write the total charge density per unit volume for a particular ion:

$$\rho = \sum_{i=1}^{N} n_i z_i e = \sum_{i=1}^{N} n_i^0 z_i e \exp\left[\frac{-z_i e\varphi(x)}{kT}\right], \tag{4.2}$$

where n_i^0 is the concentration of the ions in solution, e is the charge of the electron, and z is the ion charge.

Combining (4.1) and (4.2), we get the Poisson–Boltzmann equation

$$\frac{d^2\varphi(x)}{dx^2} = -\frac{e}{\varepsilon\varepsilon_0}\sum_{i=1}^{N} n_i^0 z_i \exp\left[\frac{-z_i e\varphi(x)}{kT}\right]. \tag{4.3}$$

The equation must be supplemented by the boundary conditions [11]:

$$\varphi(0) = \varphi_0, \quad \varphi|_{x\to\infty} = 0. \tag{4.4}$$

Multiplying (4.3) by $\frac{d\varphi}{dx}$ on the left and right, we get

$$\frac{1}{2}\frac{d}{dx}\left(\frac{d\varphi}{dx}\right)^2 = -\frac{e}{\varepsilon\varepsilon_0}\sum_{i=1}^{N} n_i^0 z_i \exp\left[\frac{-z_i e\varphi(x)}{kT}\right]\left(\frac{d\varphi}{dx}\right),$$

$$\frac{1}{2}\frac{d}{dx}\left(\frac{d\varphi}{dx}\right)^2 = \frac{d}{dx}\frac{kT}{\varepsilon\varepsilon_0}\sum_{i=1}^{N} n_i^0 \exp\left[\frac{-z_i e\varphi(x)}{kT}\right]. \tag{4.5}$$

After integrating (4.5) and taking into account the conversion of the derivative of the potential far from the surface to zero (whereby the integration constant is determined), we have

$$\frac{1}{2}\left(\frac{d\varphi}{dx}\right)^2 = \frac{kT}{\varepsilon\varepsilon_0}\sum_{i=1}^{N} n_i^0 \exp\left[\left[\frac{-z_i e\varphi(x)}{kT}\right] - 1\right], \tag{4.6}$$

$$\frac{d\varphi}{dx} = \pm \left[\frac{2kT}{\varepsilon\varepsilon_0} \sum_{i=1}^{N} n_i^0 \exp \left[\left[\frac{-z_i e\varphi(x)}{kT} \right] - 1 \right] \right]^{1/2}. \tag{4.7}$$

Note that the obtained (4.7) can be integrated in the case of arbitrary potentials of the surface, but only for a symmetric electrolyte as follows:

$$z_+ = -z_- = z, \;\; n_+^0 = n_-^0 = n^0.$$

We convert the expression included in the righthand side of (4.7)

$$\sum_{i=1}^{N} n_i^0 \exp \left[\left[\frac{-z_i e\varphi(x)}{kT} \right] - 1 \right]$$

as follows:

$$\sum_{i=1}^{N} n_i^0 \exp \left[\left[\frac{-z_i e\varphi(x)}{kT} \right] - 1 \right] = 2n^0 \left[\mathrm{ch} \left[\frac{ze\varphi(x)}{kT} \right] - 1 \right] = 4n^0 \mathrm{sh}^2 \left[\frac{ze\varphi(x)}{2kT} \right]. \tag{4.8}$$

Then, taking into account expressions (4.8), (4.7) takes the form

$$\frac{d\varphi}{dx} = -\sqrt{\frac{8n^0 kT}{\varepsilon\varepsilon_0}} \mathrm{sh} \left[\frac{ze\varphi(x)}{2kT} \right]. \tag{4.9}$$

After integrating (4.9) taking into account the boundary conditions (4.4), we get

$$\varphi(x) = \frac{4kT}{ze} \mathrm{Arth}(\mathrm{th}(ze\varphi_0/4kT)\exp(-x\kappa)), \tag{4.10}$$

where κ is the characteristic length of the Debye radius. It is defined as follows [9]:

$$\kappa^{-1} = \left[\frac{\varepsilon\varepsilon_0 kT}{\sum_{i=1}^{N} n_i^0 z^2 e^2} \right]^{1/2}, \tag{4.11}$$

Note that, at this distance, the field of a charged particle is screened due to the accumulation of the charge of opposite sign around it. We assume that when a charged amino acid residue of the protein is placed in a solution with a given ionic strength, the shielding of a charge of its sphere occurs, i.e., its potential is reduced and its effective radius is increased due to the characteristic length of the Debye radius. Thus, from (4.10) we obtain the values of the potential of the sphere on the boundary of shielding, and from the expression (4.11) we obtain the Debye radius. Using these data, we find the new value of the charge for each sphere placed in a saline solution.

Table 4.1 Debye length at various concentrations of a monovalent salt at 20 °C configuration

$N^{\underline{0}}$	Ionic concentration, M	Debye length
1	0.1	0.479
2	0.2	0.336
3	0.4	0.234
4	0.6	0.188
5	0.9	0.157
6	1	0.141

4.3.1 Debye Length

The Debye length (κ^{-1}), is a measure of the electric double layer thickness, and is a property of the electrolyte solution. It should be noted that this parameter contains information about the dielectric permittivity of the solvent, as well as the valence, z, and bulk concentration, n_i^0, of the ions. However, no information regarding the properties of the charged surface is present in the Debye length. Although it is normally referred to as the thickness of the electric double layer, the actual thickness of a double layer extends well beyond κ^{-1}. Typically, the Debye length represents a characteristic distance from the charged surface to a point where the electric potential decays to approximately 33% of the surface potential [12].

Now, n_i^0, the ionic number concentration, is given by

$$n_i^0 = \left[\mathrm{M}\frac{\mathrm{mol}}{\mathrm{L}}\right] \times \left[1000\frac{\mathrm{L}}{\mathrm{m}^3}\right] \times \left[\mathrm{N_a}\frac{1}{\mathrm{mol}}\right]$$

or $n_i^0 = 1000\mathrm{N}_a\mathrm{M}$, with the Avogadro number $\mathrm{N}_a = 6.022 \times 10^{23}\ \mathrm{mol}^{-1}$ and M being the molar concentration (mol/L) of the electrolyte. The values of the Debye length, κ^{-1}, for different electrolyte concentrations for the case of $z = 1$ are shown in the following table. In this case, the ionic strength is equal to the molar concentration of the electrolyte. It is clear from the tabulated results that κ^{-1} decreases as the electrolyte concentration increases. At high molarity, the electric double-layer thickness becomes very small. In a non-electrolyte system, however, the double-layer thickness can be thought of as extending to infinity (i.e., a is equal to the molar concentration of the electrolyte. It is clear from the tabulated results that κ^{-1} decreases as the electrolyte concentration increases. The Table 4.1 shows the debye length at various concentrations of a monovalent salt at 20 °C.

As the concentration of the salt increases, the Debye length decreases. We calculated the charge at different distances 0.2, 0.25, 0.5, 0.75nm from the sphere boundary at different concentrations of saline solution 0.1 M, 0.2 M, 0.4 M, 0.6 M, 0.8 M, 1.0 M. The results are shown in Fig. 4.1. As can be seen in the graph, with an increase in the salt solution concentration, the coulomb charge decreases due to a screening of the sphere charge by salt counter-ions.

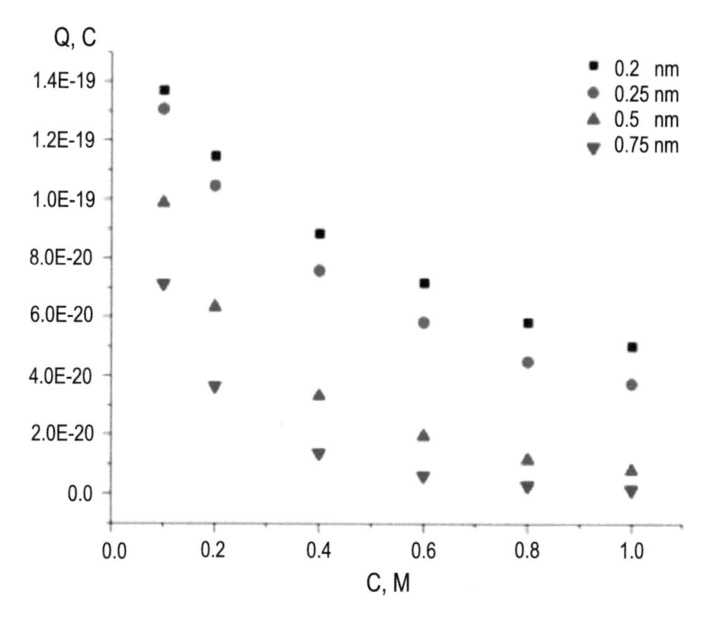

Fig. 4.1 Graph of the Coulomb charge decay with increasing salt solution concentration 0.1 M, 0.2 M, 0.4 M, 0.6 M, 0.8 M, 1.0 M at different distances from the sphere boundary 0.2, 0.25, 0.5, 0.75 nm

4.4 Description of the Physical Model

In this section, we turn to the description of the developed physical model for accounting for the effect of the concentration of a monovalent salt solution on the nature of selected histone proteins dimmers binding. In this physical model, we took into account the screening of the charged amino acid residues arginine (R), aspartic acid (D), phenylalanine (F), histidine (H), lysine (K) in an aqueous solution at various concentrations of a monovalent salt from 0.1 M to 0.8 M at a temperature of 20 °C. It was suggested that when the concentration of the monovalent salt of the aqueous solution increases, the interaction between charged amino acid residues decreases due to a screening of their charge by salt counter-ions in the aqueous solution.

At the same time, we suppose that:

1. the charge of charged amino acids is screened at a certain distance delta$_i$, ($i = \overline{1, 5}$) from the boundary of the amino acid residue, which we represent as a conducting sphere. This distance increases as the salt concentration in the solution increases and charges amino acids decrease as the salt concentration in the solution increases (see Table 4.2).

Table 4.2 The magnitude of the delta and the charges of the charged amino acid residues depending on the concentration of the salt solution

№	Ionic concentration, M	delta, nm	$Q_D \times 10^{-19}$	$Q_E \times 10^{-19}$	$Q_R \times 10^{-19}$	$Q_H \times 10^{-19}$	$Q_K \times 10^{-19}$
1	0.2	0.003902	1.596	1.5954	1.59489	1.5955	1.59544
2	0.4	0.00702	1.583	1.589	1.580	1.582	1.581
3	0.6	0.00741	1.567	1.565	1.564	1.565	1.565
4	0.8	0.00780	1.553	1.551	1.569	1.551	1.551
5	1.0	0.001482	1.491	1.488	1.485	1.488	1.488

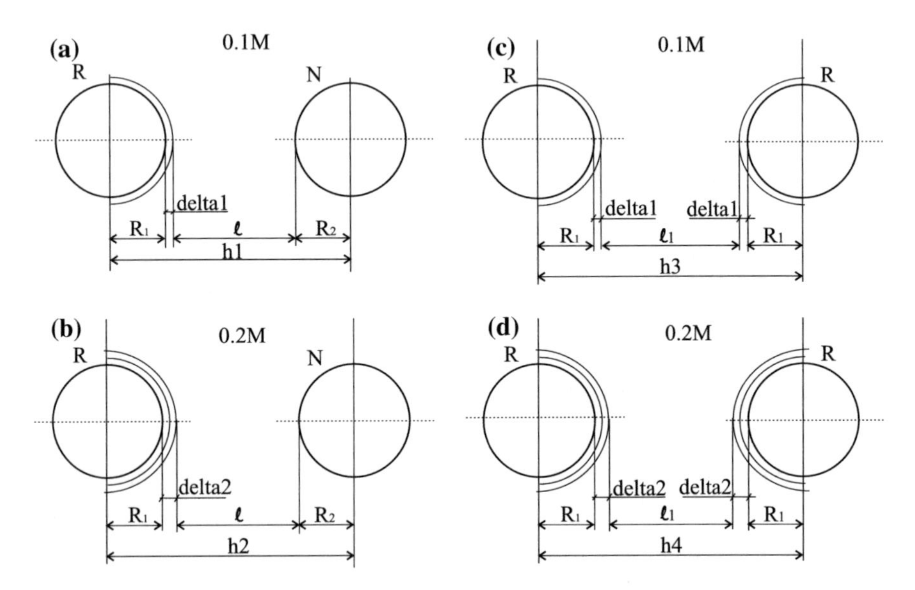

Fig. 4.2 Schema of increasing the distance between charged amino acids, as well as between charged amino acids and all other amino acids

2. Taking into account the increase in the concentration of the monovalent salt of the solution, the distance between the centers of the charged amino acid residues also increases, as well as between charged amino acids and all other amino acid residues, as shown in Fig. 4.2.

Thus, with an increase in the salt concentration in an aqueous solution at a constant temperature of 20 °C, the value of delta increases, as shown in Fig. 4.2. Between the charged amino acids, the distance increases from each side of each amino acid. The schema for calculating the distance between the amino acids R and N at a monovalent salt concentration 0.1 M is shown in Fig. 4.2a.

The schema for calculating distance between the amino acids R and N at a monovalent salt concentration of 0.2 M is shown in Fig. 4.2b.Thus, $h_1 < h_2$.

Table 4.3 Values of the dielectric constant of the medium from the concentration of the monovalent salt at temperatures 20 °C configuration

№	Ionic concentration, M	Dielectric constant
1	0.1	79.800
2	0.2	78.602
3	0.3	77.409
4	0.4	76.223
5	0.5	75.045
6	0.6	73.878
7	0.7	72.724
8	0.8	71.584
9	0.9	70.460
10	1.0	69.354

A schema for calculating distance between two charged amino acid residues R and R at a monovalent salt concentration of 0.1 M is given in Fig. 4.2c.

In Fig. 4.2d, a schema for calculating distance between the charged amino acid residues R and R is shown with an increase in the concentration of the monovalent salt in the aqueous solution to 0.2 M. Thus $h_3 < h_4$.

With this, the values of l and l_1 remain constant. Since each concentration of the monovalent salt of the aqueous solution corresponds to its electrical permeability value, the value for concentrations from 0.1 M to1 M, we have combined these data in Table 4.3.

To analyze the biochemical processes we use the notion of condition number matrix of the potential energy of the pair electrostatic interaction between peptides. In this physical formulation of the problem, it will characterize the degree of stability of the configuration of the biological complex. In order to choose a more stable biochemical compound between proteins, we select the matrix of potential energy of electrostatic interaction with the **smallest** value of the condition number (see Chap. 2).

4.5 Results of Numerical Simulation. Conclusion

In this section, we present the numerical simulation results of the effect of the concentration of a monovalent salt solution on the nature of binding of protein complexes. Let us consider the behavior of the histone monomeric proteins H2A, H2B, H3 and H4 when they are bound to the H2A–H2B and H3–H4 dimers and the concentration of the monovalent salt in the aqueous solution increases.

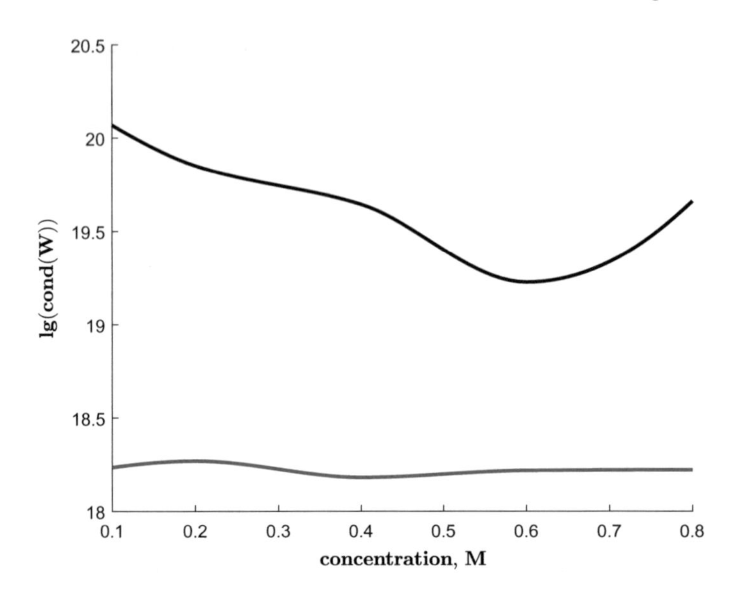

Fig. 4.3 influence of the concentration of a monovalent salt of an aqueous solution for the formation of histone dimer H2A−H2B and dimer H3−H4

Figure 4.3 shows the results of the influence of the concentration of a monovalent salt solution for the formation of histone dimer H2A−H2B and dimer H3−H4. Dimer H2A−H2B is described by the black curve and dimer H3−H4 is described by the red curve.

The first value of lg(cond(W)) for the interaction of H2A and H2B proteins at a concentration of 0.1 M was 20.068. All subsequent values of lg(cond(W)) do not rise above the first value. The curve obtained for the H3−H4 dimer is in the range of much lower values of lg(cond(W)). Note that the first value of lg(cond(W)) for the interaction of H3 and H4 monomers at 0.1 M is 18.233. Thus, from Fig. 4.3 it follows that when the dimer H2A−H2B is formed, an increase in the concentration of the monovalent salt leads to an increase the stabilization of the complex. At the same time, the analysis of the H3−H4 plot demonstrates a lower range of values of the lg(cond(W)) value compared to the values of lg(cond(W)) obtained by the interaction of H2A and H2B. The presence of a lower range of lg(cond(W)) values is interpreted as a possible propensity to aggregate of this biological complex, in this case, the H3−H4 dimer.

Thus, the mathematical modelling performed in this work on biological objects using histone proteins H2A, H2B, H3, and H4 has demonstrated the ability to predict the stability of the biological complex in the case of in vitro solutions with different ion strengths. An analysis of the calculations performed showed that different concentrations of the monovalent salt of the solutions correspond to the formation of more stable biological complexes. The introduced criterion lg(cond(W)) allows predicting a decrease or increase in the binding strength of histone proteins in the formation of histone dimers, taking into account the charge screening of charged amino acid residues of proteins.

The effect of the concentration of a monovalent salt of an aqueous solution was studied on the same histone proteins at 20 °C. A comparison was made of the results obtained for the histone dimers H2A–H2B and H3–H4 with an increase in the concentration of the monovalent salt at a temperature of 20 °C and the behavior of the histone dimers in the salt-free aqueous solution at 20 °C. It should be noted that in the previous experimental article [13] the unfolding transition temperature of the 28kDa H2A–H2B dimer increases as a function of the ionic strength of the solvent.

It should be noted that we cannot at the moment interpret a specific concentration of monovalent salt on the achievement of which aggregation of protein complexes and their precipitation occurs. This requires additional experiments to study the behavior of biological complexes in solutions with different salt concentrations. In this chapter, screening of charged amino acid residues was taken into account, but we did not take into account how the other amino acid residues interact, in particular, how the charge of amino acid residues changes with increasing salt concentration, what changes occur with the volume of aa. and a possible change in other physical parameters of a.a. To obtain these additional data, further theoretical and experimental studies are required.

4.6 Matlab Script for Mathematical Modelling of the Effect of a Monovalent Salt Solution on the Interaction of Protein Molecules

Input parameters:

1. S_1, S_{20} are amino acid sequences of biological complexes ($S_1 \geq S_{20}$)
2. concentration of a monovalent solution
3. epsilon is the dielectric constant of the medium

Output parameters:

lg(cond(W) is the common logarithm of the condition number of the matrix W, where its elements are composed of the electrostatic potential energy which is created based on the interaction between pair of amino acid residues of biological complexes.

Compute:

lg(cond(W) is the common logarithm of the condition number of the matrix W, which will allow a prediction the reactivity of the studied biological complexes.

```
1  clc
2  clear all
3  format long e
4  %H2A
5  S_1=['M'  'S'  'G'  'G'  'K' 'G'  'G'  'K'  'A' ...
6  'G'  'S'  'A' 'A'  'K'  'A'  'S'  'Q'  'S'  'R' ...
7  'S' 'A'  'K'  'A'  'G'  'L' 'T'  'F'  'P'  'V' ...
8  'G' 'R'  'V'  'H'  'R'  'L'  'L'  'R'  'R' 'G' ...
9  'N' 'Y'  'A'  'Q'  'R'  'I'  'G'  'S'  'G'  'A' ...
10 'P' 'V'  'Y'  'L'  'T'  'A'  'V'  'L'  'E'  'Y' ...
11 'L' 'A'  'A'  'E'  'I'  'L'  'E'  'L'  'A'  'G' ...
12 'N' 'A'  'A'  'R'  'D'  'N'  'K'  'K'  'T'  'R' ...
13 'I' 'I'  'P'  'R'  'H'  'L'  'Q'  'L'  'A'  'I' ...
14 'R'  'N'  'D'  'D'  'E'  'L'  'N'  'K'  'L'  'L' ...
15 'G' 'N'  'V'  'T'  'I'  'A'  'Q'  'G'  'G'  'V' ...
16 'L' 'P'  'N'  'I'  'H'  'Q'  'N'  'L'  'L'  'P' ...
17 'K' 'K'  'S'  'A'  'K'  'A'  'T'  'K'  'A'  'S' ...
18 'Q' 'E'  'L' ]
19 %H2B
20 S_20=['M'  'S'  'A'  'K'  'A'  'E' 'K'  'K'  'P'...
21 'A' 'S'  'K'  'A'  'P'  'A'  'E'  'K'  'K'  'P' ...
22 'A' 'A'  'K'  'K'  'T'  'S'  'T'  'S'  'T'  'D' ...
23 'G' 'K'  'K'  'R'  'S'  'K'  'A'  'R'  'K'  'E' ...
24 'T' 'Y'  'S'   'S'  'Y'  'I'  'Y'  'K'  'V'  'L' ...
25 'K'  'Q'  'T'  'H'  'P'  'D'  'T'  'G'  'I'  'S' ...
26 'Q' 'K'  'S'  'M'  'S'  'I'  'L'  'N'  'S'  'F' ...
27 'V'  'N'  'D'  'I'  'F'  'E'  'R'  'I'  'A'  'T' ...
28 'E'  'A'  'S'  'K'  'L'  'A'  'A'  'Y'  'N'  'K'...
29 'K' 'S'  'T'  'I'  'S'  'A'  'R'  'E'  'I'  'Q' ...
30 'T'  'A'  'V'  'R'  'L'  'I'  'L'  'P'  'G'  'E' ...
31 'L' 'A'  'K'  'H'  'A'  'V'  'S'  'E'  'G'  'T'  ...
32 'R'  'A'  'V'  'T'  'K'  'Y'  'S'  'S'  'S'  'T' ...
33 'Q'  'A'  ]
34 t=20;
35 delta1=0.0003902;
36 a0=10;
37 a1=18;
38 a2=19;
39 a3=20;
40 a4=38;
41 a5=50;
42 nn=0.1;
43 epsilon1=79.8;
44 epsilon=epsilon1;
45 delta=a0*delta1;
46 N1=length(S_1);
47 [S_1,S_2,Q1,Q2,R1,R2,h,M,N]=...
48 potential_salt(nn,delta,epsilon1,S_1,S_20);
49 [A1]=electrostatic(Q1,Q2, R1,R2,h,M,N,N1,epsilon);
50 [R_1]=condmy(A1)
51 nn=0.2;
```

```
52 epsilon1=78.6;
53 epsilon=epsilon1;
54 delta=a1*delta1;
55 [S_1,S_2,Q1,Q2,R1,R2,h,M,N]=...
56 potential_salt(nn,delta,epsilon1,S_1,S_20);
57 [A2]=electrostatic(Q1,Q2, R1,R2,h,M,N,N1,epsilon);
58 [R_2]=condmy(A2)
59 nn=0.4;
60 epsilon1=76.22;
61 epsilon=epsilon1;
62 delta=a2*delta1;
63 [S_1,S_2,Q1,Q2,R1,R2,h,M,N]=...
64 potential_salt(nn,delta,epsilon1,S_1,S_20);
65 [A3]=electrostatic(Q1,Q2, R1,R2,h,M,N,N1,epsilon);
66 [R_3]=condmy(A3)
67 nn=0.6;
68 epsilon1=73.87;
69 epsilon=epsilon1;
70 delta=a3*delta1;
71 [S_1,S_2,Q1,Q2,R1,R2,h,M,N]=...
72 potential_salt(nn,delta,epsilon1,S_1,S_20);
73 [A4]=electrostatic(Q1,Q2, R1,R2,h,M,N,N1,epsilon);
74 [R4]=condmy(A4)
75 nn=0.8;
76 epsilon1=71.58;
77 epsilon=epsilon1;
78 delta=a4*delta1;
79 [S_1,S_2,Q1,Q2,R1,R2,h,M,N]=...
80 potential_salt(nn,delta,epsilon1,S_1,S_20);
81 [A5]=electrostatic(Q1,Q2, R1,R2,h,M,N,N1,epsilon);
82 [R5]=condmy(A5)
83 %------------------------------------------------
84 %H3
85 S_1=['M' 'A' 'R' 'T' 'K' 'Q' 'T' ...
86 'A' 'R' 'K' 'S' 'T' 'G' 'G' 'K' 'A'...
87 'P' 'R' 'K' 'Q' 'L' 'A' 'S' 'K' 'A'...
88 'A' 'R' 'K' 'S' 'A' 'P' 'S' 'T' 'G' ...
89 'G' 'V' 'K' 'K' 'P' 'H' 'R' 'Y' 'K'...
90 'P' 'G' 'T' 'V' 'A' 'L' 'R' 'E' 'I' ...
91 'R' 'R' 'F' 'Q' 'K' 'S' 'T' 'E' 'L' ...
92 'L' 'I' 'R' 'K' 'L' 'P' 'F' 'Q' 'R' ...
93 'L' 'V' 'R' 'E' 'I' 'A' 'Q' 'D'...
94 'F' 'K' 'T' 'D' 'L' 'R' 'F' ...
95 'Q' 'S' 'S' 'A' 'I' 'G' 'A' ...
96 'L' 'Q' 'E' 'S' 'V' 'E' 'A' 'Y'...
97 'L' 'V' 'S' 'L' 'F' 'E' 'D' 'T' ...
98 'N' 'L' 'A' 'A' 'I' 'H' 'A' 'K' ...
99 'R' 'V' 'T' 'I' 'Q' 'K' 'K' 'D' 'I'...
100 'K' 'L' 'A' 'R' 'R' 'L' 'R' 'G' 'E' ...
101 'R' 'S' ]
102 %H4
103 S_20=['M' 'S' 'G' 'R' 'G' 'K'
```

```
                                                   ...
104 'G'  'G'  'K' 'G' 'L' 'G'  'K'   'G' ...
105 'G'  'A'  'K'  'R'  'H' 'R' 'K'  'I' ...
106 'L' 'R'  'D'  'N'  'I'  'Q'  'G' 'I' ...
107 'T'  'K'  'P'  'A'  'I'  'R' 'R' 'L' ...
108 'A' 'R' 'R'  'G'  'G'  'V'  'K'  'R' ...
109 'I'  'S'  'G' 'L' 'I'  'Y'  'E'  'E' ...
110 'V'  'R'  'A'  'V'  'L' 'K' 'S'  'F' ...
111 'L'  'E'  'S'  'V'  'I'  'R'  'D' ...
112 'S' 'V'  'T'  'Y'  'T'  'E' 'H' ...
113 'A'  'K'  'R'  'K' 'T'  'V'  'T' 'S' ...
114 'L'  'D'  'V'  'V'  'Y' 'A' 'L'  'K' ...
115 'R'  'Q'  'G'  'R'  'T'  'L'  'Y' 'G' ...
116 'F'  'G'  'G'  ]
117 nn=0.1;
118 epsilon1=79.8;
119 epsilon=epsilon1;
120 delta=a0*delta1;
121 [S_1,S_2,Q1,Q2,R1,R2,h,M,N]=...
122 potential_salt(nn,delta,epsilon1,S_1,S_20);
123 [A7]=electrostatic(Q1,Q2, R1,R2,h,M,N,N1,epsilon);
124 [R7]=condmy(A7)
125 nn=0.2;
126 epsilon1=78.6;
127 epsilon=epsilon1;
128 delta=a1*delta1;
129 [S_1,S_2,Q1,Q2,R1,R2,h,M,N]=...
130 potential_salt(nn,delta,epsilon1,S_1,S_20);
131 [A8]=electrostatic(Q1,Q2, R1,R2,h,M,N,N1,epsilon);
132 [R8]=condmy(A8)
133 nn=0.4;
134 epsilon1=76.22;
135 epsilon=epsilon1;
136 delta=a2*delta1;
137 [S_1,S_2,Q1,Q2,R1,R2,h,M,N]=...
138 potential_salt(nn,delta,epsilon1,S_1,S_20);
139 [A9]=electrostatic(Q1,Q2, R1,R2,h,M,N,N1,epsilon);
140 [R9]=condmy(A9)
141 nn=0.6;
142 epsilon1=73.87;
143 epsilon=epsilon1;
144 delta=a3*delta1;
145 [S_1,S_2,Q1,Q2,R1,R2,h,M,N]=...
146 potential_salt(nn,delta,epsilon1,S_1,S_20);
147 [A10]=electrostatic(Q1,Q2, R1,R2,h,M,N,N1,epsilon);
148 [R10]=condmy(A10)
149 nn=0.8;
150 epsilon1=71.58;
151 epsilon=epsilon1;
152 delta=a4*delta1;
153 [S_1,S_2,Q1,Q2,R1,R2,h,M,N]=...
154 potential_salt(nn,delta,epsilon1,S_1,S_20);
155 [A11]=electrostatic(Q1,Q2, R1,R2,h,M,N,N1,epsilon);
```

```
156 [R11]=condmy(A11)
157 %--------------------------------------------------
158 R_11=[R_1  R_2  R_3  R4  R5];
159 R_12=[R7  R8  R9  R10  R11];
160 nn=  [0.1 0.2 0.4 0.6 0.8];
161 h = .01 ;
162 x = nn;
163 h1=.8;
164 xi = 0.1:h:h1;
165 y1 = R_11;
166 y2 = R_12;
167 xi = 0.1:h:0.8;
168 yi1 = interp1(x,y1, xi,'cubic');
169 yi2 = interp1(x,y2, xi,'cubic');
170 N5=14;
171 set(0,'DefaultTextInterpreter', 'latex');
172 hold on
173 plot(xi, yi2, '-r', 'LineWidth' ,2.5)
174 plot(xi, yi1, '-k', 'LineWidth' ,2.5)
175 set(0,'DefaultTextFontSize',N5,...
176 'DefaultTextFontName','Arial Cyr');
177 xlabel('concentration, M');
178 ylabel('lg(cond(W))');
179 %------------------------------------------------------------
180 function [QD,FD]=Debai_D(nn,delta,epsilon1)
181 t=20;
182 rD=0.665*1E-9;
183 k=1.38*10^(-23);
184 T=t+273;
185 Na=6.022e+23;
186 e=1.6*10^(-19);
187 Zna=1;
188 Zcl=-1;
189 z=2;
190 e0=8.8*10^(-12);
191 epsilon=epsilon1*e0;
192 fs=e/(4*pi* epsilon*rD);
193 lambdaD=(epsilon.*k.*T)./(2.*nn*1000*Na.*e.^2.*z.^2);
194 kD1=sqrt(lambdaD);
195 x5=delta*1E-9   ;
196 FD=(4*k*T)./(Zna.*e).*...
197 atanh((tanh(Zna.*e.*fs./(4.*k.*T).*exp(-1./kD1*x5))));
198 QD= 4*pi.*epsilon.*FD.*(rD+x5) ;
199
200 function [QE,FE]=Debai_E(nn,delta,epsilon1)
201 t=20;
202 rE=0.735*1E-9;
203 k=1.38*10^(-23);
204 T=t+273;
205 Na=6.022e+23;
206 e=1.6*10^(-19);
207 Zna=1;
```

```
208 Zcl=-1;
209 z=2;
210 e0=8.8*10^(-12);
211 epsilon=epsilon1*e0;
212 fs=e/(4*pi* epsilon*rE);
213 lambdaD=(epsilon.*k.*T)./(2.*nn*1000*Na.*e.^2.*z.^2);
214 kD1=sqrt(lambdaD);
215 x5=delta*1E-9  ;
216 FE=(4*k*T)./(Zna.*e).*...
217 atanh((tanh(Zna.*e.*fs./(4.*k.*T).*exp(-1./kD1*x5))));
218 QE= 4*pi.*epsilon.*FE.*(rE+x5);
219 
220 function [QH,FH]=Debai_H(nn,delta,epsilon1)
221 t=20;
222 rH=0.732*1E-9;
223 k=1.38*10^(-23);
224 T=t+273;
225 Na=6.022e+23;
226 e=1.6*10^(-19);
227 Zna=1;
228 Zcl=-1;
229 z=2;
230 e0=8.8*10^(-12);
231 epsilon=epsilon1*e0;
232 fs=e/(4*pi* epsilon*rH);
233 lambdaD=(epsilon.*k.*T)./(2.*nn*1000*Na.*e.^2.*z.^2);
234 kD1=sqrt(lambdaD);
235 x5=delta*1E-9
236 FH=(4*k*T)./(Zna.*e).*...
237 atanh((tanh(Zna.*e.*fs./(4.*k.*T).*exp(-1./kD1*x5))));
238 QH= 4*pi.*epsilon.*FH.*(rH+x5) ;
239 
240 function [QK,FK]=Debai_K(nn,delta,epsilon1)
241 t=20;
242 rK=0.737*1E-9;
243 k=1.38*10^(-23);
244 T=t+273;
245 Na=6.022e+23;
246 e=1.6*10^(-19);
247 Zna=1;
248 Zcl=-1;
249 z=2;
250 e0=8.8*10^(-12);
251 epsilon=epsilon1*e0;
252 fs=e/(4*pi* epsilon*rK);
253 lambdaD=(epsilon.*k.*T)./(2.*nn*1000*Na.*e.^2.*z.^2);
254 kD1=sqrt(lambdaD);
255 x5=delta*1E-9;
256 FK=(4*k*T)./(Zna.*e).*...
257 atanh((tanh(Zna.*e.*fs./(4.*k.*T).*exp(-1./kD1*x5))));
258 QK= 4*pi.*epsilon.*FK.*(rK+x5);
259 
```

```
260 function [QR,FR]=Debai_R(nn,delta,epsilon1)
261 t=20;
262 rR=0.809*1E-9;
263 k=1.38*10^(-23);
264 T=t+273;
265 Na=6.022e+23;
266 e=1.6*10^(-19);
267 Zna=1;
268 Zcl=-1;
269 z=2;
270 e0=8.8*10^(-12);
271 epsilon=epsilon1*e0;
272 fs=e/(4*pi* epsilon*rR);
273 lambdaD=(epsilon.*k.*T)./(2.*nn*1000*Na.*e.^2.*z.^2);
274 kD1=sqrt(lambdaD);
275 x5=delta*1E-9 ;
276 FR=(4*k*T)./(Zna.*e).*...
277 atanh((tanh(Zna.*e.*fs./(4.*k.*T).*exp(-1./kD1*x5))));
278 QR= 4*pi.*epsilon.*FR.*(rR+x5) ;
279
280
281 function [S_1,S_2,Q1,Q2,R1,R2,h,M,N]=...
282 potential_salt(nn,delta,epsilon1,S_1,S_20);
283 N=length(S_1);
284 M=length(S_20);
285 S_2=S_20;
286 Q1=[];
287 Q2=[];
288 R1=[];
289 R2=[];
290 for i=1:length(S_1);
291  for j=1:length(S_2);
292  [QD,FD]=Debai_D(nn,delta,epsilon1);
293  [QE,FE]=Debai_E(nn,delta,epsilon1);
294  [QR,FR]=Debai_R(nn,delta,epsilon1);
295  [QH,FH]=Debai_H(nn,delta,epsilon1);
296  [QK,FK]=Debai_K(nn,delta,epsilon1);
297 if (S_1(i)=='D' & S_2(j)=='E')| (S_1(i)=='E' & S_2(j)=='D');
298 Q1(i)= -QD;
299 Q2(j)= -QE;
300 else
301 if (S_1(i)=='D' & S_2(j)=='D');
302 Q1(i)= -QD;
303 Q2(j)= -QD;
304 else
305 if (S_1(i)=='D' & S_2(j) =='C')|(S_1(i)=='C' & S_2(j) =='D');
306 Q1(i)= 0.05e-19;
307 Q2(j)= 0.05e-19;
308 else
309 if (S_1(i)=='D' &S_2(j)=='N')|(S_1(i)=='N' &S_2(j)=='D')|...
310 (S_1(i)=='D' &S_2(j)=='F')|(S_1(i)=='D' &S_2(j)=='Y')|...
311 (S_1(i)=='D' &S_2(j)=='Q')|(S_1(i)=='D' &S_2(j)=='S')|...
```

```
312 (S_1(i)=='F' &S_2(j)=='D')|(S_1(i)=='Y' &S_2(j)=='D')|..
313 (S_1(i)=='Q' &S_2(j)=='D')|(S_1(i)=='S' &S_2(j)=='D');
314 Q1(i)=  0.57e-19;
315 Q2(j)= 0.57e-19;
316 else
317 if ((S_1(i)=='D' & S_2(j)=='M')|(S_1(i)=='D' & S_2(j)=='T')|..
318 (S_1(i)=='D' & S_2(j)=='I')|(S_1(i)=='D' & S_2(j)=='G')|...
319 (S_1(i)=='D' & S_2(j)=='V')|(S_1(i)=='D' & S_2(j)=='W')|...
320 (S_1(i)=='D' & S_2(j)=='L')|(S_1(i)=='D' & S_2(j)=='A')|...
321 (S_1(i)=='M' & S_2(j)=='D')|(S_1(i)=='T' & S_2(j)=='D')|...
322 (S_1(i)=='I' & S_2(j)=='D')|(S_1(i)=='G' & S_2(j)=='D')|...
323 (S_1(i)=='V' & S_2(j)=='D')|(S_1(i)=='W' & S_2(j)=='D')|...
324 (S_1(i)=='L' & S_2(j)=='D')|(S_1(i)=='A' & S_2(j)=='D'));
325 Q1(i)= 0.64e-19;
326 Q2(j)= 0.64e-19;
327 else
328 if ((S_1(i)=='D'& S_2(j)=='P')|(S_1(i)=='P'& S_2(j)=='D'));
329 Q1(i)= 0.78e-19;
330 Q2(j)= 0.78e-19;
331 else
332 if ((S_1(i)=='D' & S_2(j)=='H')|(S_1(i)=='H'& S_2(j)=='D'));
333 Q1(i)= -QD;
334 Q2(j)= QH;
335 else
336 if ((S_1(i)=='D'& S_2(j)=='K')|(S_1(i)=='K'& S_2(j)=='D'));
337 Q1(i)= -QD;
338 Q2(j)= QK;
339 else
340 if ((S_1(i)=='D' & S_2(j)=='R')|(S_1(i)=='R'& S_2(j)=='D'));
341 Q1(i)= -QD;
342 Q2(j)= QR;
343 else
344 if ((S_1(i)=='E'&S_2(j)=='E'));
345 Q1(i)= -QE;
346 Q2(j)= -QE;
347 else
348 if((S_1(i)=='E'  & S_2(j)=='C')|(S_1(i)=='E' & S_2(j)=='F')|..
349 (S_1(i)=='E' & S_2(j)=='N')|(S_1(i)=='C'  & S_2(j)=='E')|...
350 (S_1(i)=='F' & S_2(j)=='E')|(S_1(i)=='N' & S_2(j)=='E'));
351 Q1(i)= 0.55e-19;
352 Q2(j)= 0.55e-19;
353 else
354 if((S_1(i)=='E' & S_2(j)=='Q')|(S_1(i)=='E'& S_2(j)=='Y')|...
355  (S_1(i)=='E' & S_2(j)=='S')|(S_1(i)=='E' & S_2(j)=='M')|...
356  (S_1(i)=='E' & S_2(j)=='T')|(S_1(i)=='E' & S_2(j)=='I')|...
357  (S_1(i)=='E' & S_2(j)=='G')|(S_1(i)=='E' & S_2(j)=='V')|...
358  (S_1(i)=='E' & S_2(j)=='W')|(S_1(i)=='E' & S_2(j)=='L')|...
359  (S_1(i)=='E' & S_2(j)=='A')|(S_1(i)=='Q' & S_2(j)=='E')|...
360  (S_1(i)=='Y' & S_2(j)=='E')| (S_1(i)=='S' & S_2(j)=='E')|...
361  (S_1(i)=='M' & S_2(j)=='E')|(S_1(i)=='T' & S_2(j)=='E')|...
362  (S_1(i)=='I' & S_2(j)=='E')| (S_1(i)=='G' & S_2(j)=='E')|
```

```
...
363  (S_1(i)=='V' & S_2(j)=='E')|(S_1(i)=='W' & S_2(j)=='E')|...
364  (S_1(i)=='L' & S_2(j)=='E')|(S_1(i)=='A' & S_2(j)=='E'));
365 Q1(i)= 0.64e-19;
366 Q2(j)= 0.64e-19;
367 else
368 if ((S_1(i)=='E' & S_2(j)=='P' )|(S_1(i)=='P' & S_2(j)=='E'));
369 Q1(i)= 0.78e-19;
370 Q2(j)= 0.78e-19;
371 else
372 if ((S_1(i)=='E' & S_2(j)=='H')|(S_1(i)=='H' &S_2(j)=='E'));
373 Q1(i)= -QE;
374 Q2(j)= QH;
375 else
376 if (S_1(i)=='E'& S_2(j)=='K')| (S_1(i)=='K'& S_2(j)=='E');
377 Q1(i)=-QE;
378 Q2(j)= QK;
379 else
380 if (S_1(i)=='E' & S_2(j)=='R')|(S_1(i)=='R' & S_2(j)=='E');
381 Q1(i)= -QE;
382 Q2(j)= QR;
383 else
384 if (S_1(i)=='C' & S_2(j)=='C')|(S_1(i)=='C' & S_2(j)=='F')|...
385 (S_1(i)=='C' & S_2(j)=='Q')|(S_1(i)=='C'& S_2(j)=='Y')|...
386 (S_1(i)=='C' & S_2(j)=='S')|(S_1(i)=='C' & S_2(j)=='M')|...
387 (S_1(i)=='C' & S_2(j)=='T')|(S_1(i)=='C' & S_2(j)=='I')|...
388 (S_1(i)=='C' & S_2(j)=='G')|(S_1(i)=='C' & S_2(j)=='V')|...
389 (S_1(i)=='C' & S_2(j)=='W')|(S_1(i)=='C' & S_2(j)=='L')|...
390 (S_1(i)=='C' & S_2(j)=='L')|(S_1(i)=='C' & S_2(j)=='A')|...
391 (S_1(i)=='F' & S_2(j)=='C')|(S_1(i)=='Q' & S_2(j)=='C')|...
392 (S_1(i)=='Y'& S_2(j)=='C')|(S_1(i)=='S' & S_2(j)=='C')|...
393 (S_1(i)=='M' & S_2(j)=='C')|(S_1(i)=='T' & S_2(j)=='C')|...
394 (S_1(i)=='I' & S_2(j)=='C')|(S_1(i)=='G' & S_2(j)=='C')|...
395 (S_1(i)=='V' & S_2(j)=='C')|(S_1(i)=='W' & S_2(j)=='C')|...
396 (S_1(i)=='L' & S_2(j)=='C')|( S_1(i)=='A' & S_2(j)=='C');
397 Q1(i)=0.74e-19;
398 Q2(j)=0.74e-19;
399 else
400 if (S_1(i)=='C' & S_2(j)=='H')| (S_1(i)=='H' & S_2(j)=='C');
401 Q1(i)= 0.99e-19;
402 Q2(j)= 0.99e-19;
403 else
404 if (S_1(i)=='C' & S_2(j)=='K')|(S_1(i)=='K' & S_2(j)=='C');
405 Q1(i)= 1.34e-19;
406 Q2(j)= 1.34e-19;
407 else
408 if (S_1(i)=='C' & S_2(j)=='R')|(S_1(i)=='R' & S_2(j)=='C');
409 Q1(i)= 1.59e-19;
410 Q2(j)= 1.59e-19;
411 else
412 if (S_1(i)=='N' & S_2(j)=='N')|(S_1(i)=='N' & S_2(j)=='F')|...
413 (S_1(i)=='N' & S_2(j)=='Q')|(S_1(i)=='N' & S_2(j)=='Y')|...
```

```
414 (S_1(i)=='N' & S_2(j)=='S')|(S_1(i)=='N'& S_2(j)=='M')|...
415 (S_1(i)=='F' & S_2(j)=='N')|(S_1(i)=='Q' & S_2(j)=='N')|...
416 (S_1(i)=='Y' & S_2(j)=='N')|(S_1(i)=='S' & S_2(j)=='N')|...
417 (S_1(i)=='M'& S_2(j)=='N');
418 Q1(i)=0.74e-19;
419 Q2(j)=0.74e-19;
420 else
421 if (S_1(i)=='N' & S_2(j)=='H')|(S_1(i)=='H' & S_2(j)=='N')
422 Q1(i)= 0.99e-19;
423 Q2(j)=0.99e-19;
424 else
425 if(S_1(i)=='N' & S_2(j)=='K')|(S_1(i)=='K' & S_2(j)=='N');
426 Q1(i)= 1.05e-19;
427 Q2(j)= 1.05e-19;
428 else
429 if (S_1(i)=='N' & S_2(j)=='R')|(S_1(i)=='R' & S_2(j)=='N');
430 Q1(i)= 1.1e-19;
431 Q2(j)= 1.1e-19;
432 else
433 if ((S_1(i)=='F' & S_2(j)=='F')|(S_1(i)=='F' & S_2(j)=='Q'));
434 Q1(i)=0.74e-19;
435 Q2(j)=0.74e-19;
436 else
437 if ((S_1(i)=='F' & S_2(j)=='Y')|(S_1(i)=='F' & S_2(j)=='S')|..
438 (S_1(i)=='F' & S_2(j)=='M')|(S_1(i)=='Q' & S_2(j)=='F')|...
439 (S_1(i)=='Y' & S_2(j)=='F'));
440 Q1(i)=0.74e-19;
441 Q2(j)=0.74e-19;
442 else
443 if (S_1(i)=='S' & S_2(j)=='F')|(S_1(i)=='M' & S_2(j)=='F');
444 Q1(i)=0.74e-19;
445 Q2(j)=0.74e-19;
446 else
447 if (S_1(i)=='F' & S_2(j)=='H')|(S_1(i)=='H' & S_2(j)=='F');
448 Q1(i)= 0.99e-19;
449 Q2(j)= 0.99e-19;
450 else
451 if (S_1(i)=='F' & S_2(j)=='K')|(S_1(i)=='K' & S_2(j)=='F');
452 Q1(i)= 1.05e-19;
453 Q2(j)= 1.05e-19;
454 else
455 if (S_1(i)=='F' & S_2(j)=='R')|(S_1(i)=='R' & S_2(j)=='F');
456 Q1(i)= 1.1e-19;
457 Q2(j)= 1.1e-19;
458 else
459 if (S_1(i)=='Q' & S_2(j)=='H')|(S_1(i)=='H' & S_2(j)=='Q');
460 Q1(i)= 0.99e-19;
461 Q2(j)= 0.99e-19;
462 else
463 if (S_1(i)=='Q' & S_2(j)=='K')|(S_1(i)=='K' & S_2(j)=='Q');
464 Q1(i)= 1.05e-19;
465 Q2(j)= 1.05e-19;
```

```
466 else
467 if (S_1(i)=='Q' & S_2(j)=='R')|(S_1(i)=='R' & S_2(j)=='Q');
468 Q1(i)= 1.1e-19;
469 Q2(j)= 1.1e-19;
470 else
471 if (S_1(i)=='Q' & S_2(j)=='H')|(S_1(i)=='H' & S_2(j)=='Q');
472 Q1(i)= 0.99e-19;
473 Q2(j)= 0.99e-19;
474 else
475 if (S_1(i)=='Y' & S_2(j)=='K')|(S_1(i)=='K' & S_2(j)=='Y')
476 Q1(i)= 1.05e-19;
477 Q2(j)= 1.05e-19;
478 else
479 if (S_1(i)=='Y' & S_2(j)=='R')|(S_1(i)=='R' & S_2(j)=='Y');
480 Q1(i)= 1.1e-19;
481 Q2(j)= 1.1e-19;
482 else
483 if (S_1(i)=='S' & S_2(j)=='H')|(S_1(i)=='H' & S_2(j)=='S');
484 Q1(i)= 0.99e-19;
485 Q2(j)= 0.99e-19;
486 else
487 if (S_1(i)=='S' & S_2(j)=='K')|(S_1(i)=='K' & S_2(j)=='S');
488 Q1(i)= 1e-19;
489 Q2(j)= 1e-19;
490 else
491 if (S_1(i)=='S' & S_2(j)=='R')|(S_1(i)=='R' & S_2(j)=='S');
492 Q1(i)= 1.1e-19;
493 Q2(j)= 1.1e-19;
494 else
495 if (S_1(i)=='M' & S_2(j)=='H')|(S_1(i)=='H' & S_2(j)=='M');
496 Q1(i)= 0.99e-19;
497 Q2(j)= 0.99e-19;
498 else
499 if (S_1(i)=='M' & S_2(j)=='K')|(S_1(i)=='K' & S_2(j)=='M');
500 Q1(i)= 1e-19;
501 Q2(j)= 1e-19;
502 else
503 if (S_1(i)=='M' & S_2(j)=='R')|(S_1(i)=='R' & S_2(j)=='M');
504 Q1(i)= 1.1e-19;
505 Q2(j)= 1.1e-19;
506 else
507 if (S_1(i)=='T' & S_2(j)=='H')|(S_1(i)=='H' & S_2(j)=='T');
508 Q1(i)= 0.99e-19;
509 Q2(j)= 0.99e-19;
510 else
511 if (S_1(i)=='T' & S_2(j)=='K')|(S_1(i)=='K' & S_2(j)=='T');
512 Q1(i)= 1e-19;
513 Q2(j)= 1e-19;
514 else
515 if (S_1(i)=='T' & S_2(j)=='R')|(S_1(i)=='R' & S_2(j)=='T');
516 Q1(i)= 1.05e-19;
517 Q2(j)= 1.05e-19;
```

```
518 else
519 if (S_1(i)=='I' & S_2(j)=='H')|(S_1(i)=='H' & S_2(j)=='I');
520 Q1(i)= 0.99e-19;
521 Q2(j)= 0.99e-19;
522 else
523 if (S_1(i)=='I' & S_2(j)=='K')|(S_1(i)=='K' & S_2(j)=='I');
524 Q1(i)= 1e-19;
525 Q2(j)= 1e-19;
526 else
527 if (S_1(i)=='I' & S_2(j)=='R')|(S_1(i)=='R' & S_2(j)=='I');
528 Q1(i)= 1.05e-19;
529 Q2(j)= 1.05e-19;
530 else
531 if (S_1(i)=='G' & S_2(j)=='H')|(S_1(i)=='H' & S_2(j)=='G');
532 Q1(i)= 0.99e-19;
533 Q2(j)= 0.99e-19;
534 else
535 if (S_1(i)=='G' & S_2(j)=='K')|(S_1(i)=='K' & S_2(j)=='G');
536 Q1(i)= 1e-19;
537 Q2(j)= 1e-19;
538 else
539 if (S_1(i)=='G' & S_2(j)=='R')|(S_1(i)=='R' & S_2(j)=='G');
540 Q1(i)= 1.05e-19;
541 Q2(j)= 1.05e-19;
542 else
543 if (S_1(i)=='V' & S_2(j)=='H')|(S_1(i)=='H' & S_2(j)=='V');
544 Q1(i)= 0.99e-19;
545 Q2(j)= 0.99e-19;
546 else
547 if (S_1(i)=='V' & S_2(j)=='K')|(S_1(i)=='K' & S_2(j)=='V');
548 Q1(i)= 1e-19;
549 Q2(j)= 1e-19;
550 else
551 if (S_1(i)=='V' & S_2(j)=='R')|(S_1(i)=='R' & S_2(j)=='V');
552 Q1(i)= 1.05e-19;
553 Q2(j)= 1.05e-19;
554 else
555 if (S_1(i)=='W' & S_2(j)=='H')|(S_1(i)=='H' & S_2(j)=='W');
556 Q1(i)= 0.99e-19;
557 Q2(j)= 0.99e-19;
558 else
559 if (S_1(i)=='W' & S_2(j)=='K')|(S_1(i)=='K' & S_2(j)=='W');
560 Q1(i)= 1e-19;
561 Q2(j)= 1e-19;
562 else
563 if (S_1(i)=='W' & S_2(j)=='R')|(S_1(i)=='R' & S_2(j)=='W');
564 Q1(i)= 1.05e-19;
565 Q2(j)= 1.05e-19;
566 else
567 if (S_1(i)=='L' & S_2(j)=='H')|(S_1(i)=='H' & S_2(j)=='L');
568 Q1(i)= 0.99e-19;
569 Q2(j)= 0.99e-19;
```

```
570 else
571 if (S_1(i)=='L' & S_2(j)=='K')|(S_1(i)=='K' & S_2(j)=='L');
572 Q1(i)= 1e-19;
573 Q2(j)= 1e-19;
574 else
575 if (S_1(i)=='L' & S_2(j)=='R')|(S_1(i)=='R' & S_2(j)=='L');
576 Q1(i)= 1.05e-19;
577 Q2(j)= 1.05e-19;
578 else
579 if (S_1(i)=='A' & S_2(j)=='H')|(S_1(i)=='H' & S_2(j)=='A');
580 Q1(i)= 0.99e-19;
581 Q2(j)= 0.99e-19;
582 else
583 if (S_1(i)=='A' & S_2(j)=='K')|(S_1(i)=='K' & S_2(j)=='A');
584 Q1(i)= 1e-19;
585 Q2(j)= 1e-19;
586 else
587 if (S_1(i)=='A' & S_2(j)=='R')|(S_1(i)=='R' & S_2(j)=='A');
588 Q1(i)= 1.05e-19;
589 Q2(j)= 1.05e-19;
590 else
591 if (S_1(i)=='P' & S_2(j)=='H')|(S_1(i)=='H' & S_2(j)=='P');
592 Q1(i)= 0.99e-19;
593 Q2(j)= 0.99e-19;
594 else
595 if (S_1(i)=='P' & S_2(j)=='K')|(S_1(i)=='K' & S_2(j)=='P');
596 Q1(i)= 0.82e-19;
597 Q2(j)= 0.82e-19;
598 else
599 if (S_1(i)=='P' & S_2(j)=='R')|(S_1(i)=='R' & S_2(j)=='P');
600 Q1(i)= 0.96e-19;
601 Q2(j)= 0.96e-19;
602 else
603 if (S_1(i)=='H' & S_2(j)=='H');
604 Q1(i)= QH;
605 Q2(j)= QH;
606 else
607 if (S_1(i)=='H' & S_2(j)=='K')|(S_1(i)=='K' & S_2(j)=='H');
608 Q1(i)= QH;
609 Q2(j)= QK;
610 else
611 if (S_1(i)=='H' & S_2(j)=='R')|(S_1(i)=='R' & S_2(j)=='H');
612 Q1(i)= QH;
613 Q2(j)= QR;
614 else
615 if (S_1(i)=='K' & S_2(j)=='K');
616 Q1(i)= QK;
617 Q2(j)= QK;
618 else
619 if (S_1(i)=='K' & S_2(j)=='R')|(S_1(i)=='R' & S_2(j)=='K');
620 Q1(i)= QK;
621 Q2(j)= QR;
```

```
622 else
623 if (S_1(i)=='R' & S_2(j)=='R');
624 Q1(i)= QR;
625 Q2(j)= QR;
626 else
627 Q1(i)= 0.824e-19;
628 Q2(j)= -0.824e-19;
629 end
630 end
631 end
632 end
633 end
634 end
635 end
636 end
637 end
638 end
639 end
640 end
641 end
642 end
643 end
644 end
645 end
646 end
647 end
648 end
649 end
650 end
651 end
652 end
653 end
654 end
655 end
656 end
657 end
658 end
659 end
660 end
661 end
662 end
663 end
664 end
665 end
666 end
667 end
668 end
669 end
670 end
671 end
672 end
673 end
```

```
674 end
675 end
676 end
677 end
678 end
679 end
680 end
681 end
682 end
683 end
684 end
685 end
686 end
687 end
688 end
689 end
690 end
691 end
692 end
693 end
694 end
695 end
696 end
697 end
698 end
699 end
700 end
701 end
702 end
703  R1=[];
704  R2=[];
705 for i=1:length(S_1);
706       if (S_1(i)=='A');
707             R1(i)=0.6E-9;
708         else
709         if (S_1(i)=='R');
710             R1(i)=0.805E-9+delta*1E-9;
711             %R2(j)=Rr+Rt1;
712         else
713 if (S_1(i)=='N');
714             R1(i)=0.682E-9;
715 else
716 if (S_1(i)=='D');
717     R1(i)=0.665E-9+delta*1E-9;
718 else
719       if (S_1(i)=='C');
720             R1(i)=0.629E-9;
721         else
722 if (S_1(i)=='Q');
723            R1(i)=0.725E-9;
724 else
725         if (S_1(i)=='E');
```

```
726                 R1(i)=0.714E-9+delta*1E-9;
727             else
728               if (S_1(i)=='G');
729                 R1(i)=0.537E-9;
730             else
731   if (S_1(i)=='H');
732         R1(i)=0.732E-9+delta*1E-9;
733   else
734       if (S_1(i)=='I');
735                 R1(i)=0.732E-9+delta*1E-9;
736             else
737       if (S_1(i)=='L');
738                 R1(i)=0.734E-9;
739             else
740       if (S_1(i)=='K')
741                R1(i)=0.738E-9+delta*1E-9;
742             else
743       if (S_1(i)=='M')
744                 R1(i)=0.741E-9;
745             else
746       if (S_1(i)=='F')
747                 R1(i)=0.781E-9;
748       else
749           if (S_1(i)=='P');
750                 R1(i)=0.672E-9;
751             else
752              if (S_1(i)=='S');
753                 R1(i)=0.615E-9;
754             else
755              if (S_1(i)=='T');
756                 R1(i)=0.659E-9;
757             else
758              if (S_1(i)=='W');
759                 R1(i)=0.826E-9;
760             else
761              if (S_1(i)=='Y');
762                 R1(i)=0.781E-9;
763              else
764                 if (S_1(i)=='V');
765                 R1(i)=0.694E-9;
766  end
767  end
768  end
769  end
770  end
771  end
772  end
773  end
774  end
775  end
776  end
777  end
```

```
778 end
779 end
780 end
781 end
782 end
783 end
784 end
785 end
786 end
787     for j=1:length(S_2);
788      if (S_2(j)=='A');
789             R2(j)=0.6E-9;
790         else
791         if (S_2(j)=='R');
792             R2(j)=0.809E-9+delta*1E-9;
793         else
794 if (S_2(j)=='N');
795            R2(j)=0.682E-9;
796 else
797 if (S_2(j)=='D');
798     R2(j)=0.665E-9+delta*1E-9;
799 else
800       if (S_2(j)=='C');
801             R2(j)=0.629E-9;
802         else
803 if (S_2(j)=='Q');
804          R2(j)=0.725E-9;
805 else
806         if (S_2(j)=='E');
807             R2(j)=0.714E-9+delta*1E-9;
808         else
809           if (S_2(j)=='G');
810             R2(j)=0.537E-9;
811         else
812  if (S_2(j)=='H');
813       R2(j)=0.732E-9+delta*1E-9;
814  else
815      if (S_2(j)=='I');
816             R2(j)=0.735E-9;
817         else
818      if (S_2(j)=='L');
819             R2(j)=0.734E-9;
820         else
821      if (S_2(j)=='K')
822            R2(j)=0.738E-9+delta*1E-9;
823         else
824      if (S_2(j)=='M')
825             R2(j)=0.741E-9;
826         else
827      if (S_2(j)=='F')
828            R2(j)=0.781E-9;
829      else
```

```
830          if (S_2(j)=='P');
831              R2(j)=0.672E-9;
832           else
833            if (S_2(j)=='S');
834              R2(j)=0.615E-9;
835           else
836            if (S_2(j)=='T');
837               R2(j)=0.659E-9;
838           else
839            if (S_2(j)=='W');
840               R2(j)=0.826E-9;
841           else
842            if (S_2(j)=='Y');
843               R2(j)=0.781E-9;
844            else
845               if (S_2(j)=='V');
846               R2(j)=0.694E-9;
847               else
848                R2(j)=0;
849  end
850  end
851  end
852  end
853  end
854  end
855  end
856  end
857  end
858  end
859  end
860  end
861  end
862  end
863  end
864  end
865  end
866  end
867  end
868  end
869  end
870   Rr=0.809E-9+delta*1E-9;
871   Rd=0.665E-9+delta*1E-9;
872   Re=0.714E-9+delta*1E-9;
873   Rh=0.732E-9+delta*1E-9;
874   Rk=0.737E-9+delta*1E-9;
875   Ra=0.6E-9;
876   Rn=0.682E-9;
877   Rc=0.629E-9;
878   Rq=0.725E-9;
879   Rg=0.725E-9;
880   Ri=0.735E-9;
881   Rl=0.734E-9;
```

```
882  Rm=0.741E-9;
883  Rf=0.781E-9;
884  Rp=0.672E-9;
885  Rs=0.615E-9;
886  Rt=0.659E-9;
887  Rw=0.826E-9;
888  Ry=0.781E-9;
889  Rv=0.694E-9;
890  for i=1:length(S_1);
891  for j=1:length(S_2);
892     if (S_1(i)=='R'& S_2(j)=='D');
893          h(i,j)=.15*10^(-9)+Rr+Rd;
894  else
895     if (S_1(i)=='R'& S_2(j)=='E');
896         h(i,j)=.15*10^(-9)+Rr+Re;
897          else
898  if (S_1(i)=='D'& S_2(j)=='R');
899          h(i,j)=.15*10^(-9)+Rd+Rr;
900  else
901
902     if (S_1(i)=='D'& S_2(j)=='H');
903         h(i,j)=.15*10^(-9)+Rd+Rh;
904   else
905   if (S_1(i)=='D'& S_2(j)=='R');
906          h(i,j)=.15*10^(-9)+Rd+Rr;
907  else
908
909     if (S_1(i)=='D'& S_2(j)=='H');
910         h(i,j)=.15*10^(-9)+Rd+Rh;
911   else
912   if (S_1(i)=='D'& S_2(j)=='K');
913     h(i,j)=.15*10^(-9)+Rd+Rk;
914   else
915           if (S_1(i)=='E')& (S_2(j)=='R');
916          h(i,j)=.15*10^(-9)+Re+Rr;
917          else
918            if (S_1(i)=='E'& S_2(j)=='H');
919          h(i,j)=.15*10^(-9)+Re+Rh;
920        else
921            if (S_1(i)=='E'& S_2(j)=='K');
922        h(i,j)=.15*10^(-9)+Re+Rk;
923   else
924  if (S_1(i)=='H'& S_2(j)=='D')
925
926          h(i,j)=.15*10^(-9)+Rh+Rd;
927  else
928   if (S_1(i)=='H'& S_2(j)=='E')
929
930     h(i,j)=.15*10^(-9)+Rh+Re;
931   else
932  if (S_1(i)=='R'& S_2(j)=='R')
933          h(i,j)=.4*10^(-9)+Rr+Rr;
```

```
934      else
935     if (S_1(i)=='R'& S_2(j)=='H')
936      h(i,j)=.4*10^(-9)+Rr+Rh;
937     else
938   if (S_1(i)=='R'& S_2(j)=='H')
939       h(i,j)=.4*10^(-9)+Rr+Rh;
940   else
941     if (S_1(i)=='R'& S_2(j)=='K')
942       h(i,j)=.4*10^(-9)+Rr+Rk;
943    else
944   if (S_1(i)=='D'& S_2(j)=='E');
945           h(i,j)=.4*10^(-9)+Rd+Re;
946   else
947      if (S_1(i)=='D'& S_2(j)=='D');
948
949        h(i,j)=.4*10^(-9)+Rd+Rd;
950
951     else
952   if (S_1(i)=='H'& S_2(j)=='R')
953           h(i,j)=.4*10^(-9)+Rh+Rr;
954   else
955     if (S_1(i)=='H'& S_2(j)=='H')
956        h(i,j)=.4*10^(-9)+Rh+Rh;
957    else
958
959      if (S_1(i)=='H'& S_2(j)=='K')
960        h(i,j)=.4*10^(-9)+Rh+Rk;
961
962      else
963   if (S_1(i)=='K'& S_2(j)=='R')
964           h(i,j)=.4*10^(-9)+Rk+Rr;
965   else
966    if (S_1(i)=='K'& S_2(j)=='H')
967         h(i,j)=.4*10^(-9)+Rk+Rh;
968    else
969     if (S_1(i)=='K'& S_2(j)=='K')
970       h(i,j)=.4*10^(-9)+Rk+Rk;
971    else
972    if (S_1(i)=='N'& S_2(j)=='Q')
973           h(i,j)=.25*10^(-9)+Rn+Rq;
974   else
975    if (S_1(i)=='N'& S_2(j)=='S')
976      h(i,j)=.25*10^(-9)+Rn+Rs;
977     else
978     if (S_1(i)=='N'& S_2(j)=='Y')
979       h(i,j)=.25*10^(-9)+Rn+Ry;
980    else
981   if (S_1(i)=='Q'& S_2(j)=='S')|...
982   (S_1(i)=='Q')&(S_2(j)=='Y');
983           h(i,j)=.25*10^(-9)+Rq+Rs;
984   else
985    if  (S_1(i)=='Q')& (S_2(j)=='Y');
```

```
986          h(i,j)=.25*10^(-9)+Rq+Ry;
987  else
988
989  if (S_1(i)=='S'& S_2(j)=='Y');
990          h(i,j)=.25*10^(-9)+Rs+Ry;
991
992  else
993      if (S_1(i)=='X')|(S_2(j)=='X');
994           h(i,j)=10*10^(-2);
995           else
996          h(i,j)=1.76*10^(-9);
997  end
998  end
999  end
1000 end
1001 end
1002 end
1003 end
1004 end
1005 end
1006 end
1007 end
1008 end
1009 end
1010 end
1011 end
1012 end
1013 end
1014 end
1015 end
1016 end
1017 end
1018 end
1019 end
1020 end
1021 end
1022 end
1023 end
1024 end
1025 end
1026 end
1027 end
1028 end
1029 end
1030
1031 function[A]=electrostatic(Q1,Q2, R1,R2,h,M,N,N1,epsilon)
1032 for i=1:N
1033     for j=1:M
1034         if R1(i)>R2(j)
1035             gamma(i,j)=R1(i)/R2(j);
1036         else
1037             if  R1(i)<R2(j)
```

```
1038                  gamma(i,j)=R2(j)/R1(i);
1039                else if R1(i)==R2(j);
1040       gamma(i,j)=R2(j)/R1(i);
1041           end
1042              end
1043          end
1044          if h(i,j)>(R1(i)+R2(j))
1045              r(i,j)=h(i,j)/(R1(i)+R2(j));
1046          else if  h(i,j)<=(R1(i)+R2(j))
1047              r(i,j)=(R1(i)+R2(j))/h(i,j);
1048          end
1049          end
1050      y(i,j)=(((r(i,j)^2*(1+gamma(i,j))^2)-...
1051      (1+(gamma(i,j))^2))/(2*gamma(i,j)));
1052      beta(i,j)=acosh(y(i,j));
1053      z(i,j)=exp(-beta(i,j));
1054      S12=0;
1055      S22=0;
1056      S11=0;
1057      for k=1:N1
1058          gamma1(i,j)=R2(j)/R1(i);
1059          S_1(k)=(z(i,j)^k)/(((1-z(i,j)^(2*k)))*...
1060  ((gamma(i,j)+y(i,j))-(y(i,j)^2-1)^(1/2)*...
1061  (1+z(i,j)^(2*k))/(1-z(i,j)^(2*k))));
1062          S11=S11+S_1(k);
1063          S_2(k)=(z(i,j)^(2*k))/(1-(z(i,j)^(2*k)));
1064          S12=S12+S_2(k);
1065          S_3(k)=(z(i,j)^k)/(((1-z(i,j)^(2*k)))*...
1066          ((1-gamma(i,j)*y(i,j))-gamma(i,j)*...
1067  (y(i,j)^2-1)^(1/2)*(1+z(i,j)^(2*k))/(1-z(i,j)^(2*k))));
1068          S22=S22+S_3(k);
1069      end
1070      epsilon0=8.85418781762*10^(-12);
1071  c11(i,j)=(2*gamma(i,j)*...
1072  ((y(i,j)^2-1)^(1/2))).*S11;
1073  c22(i,j)=(2*gamma(i,j)*...
1074  ((y(i,j)^2-1)^(1/2))).*S22;
1075  c12(i,j)=-((2*gamma(i,j)*...
1076  ((y(i,j)^2-1))^(1/2))/(r(i,j)*(1+gamma(i,j)))).*S12;
1077  delta(i,j)=((c11(i,j)*c22(i,j)-c12(i,j)^2));
1078      k=1/(4*pi*epsilon0);
1079      k1=1/(4*pi*epsilon0*epsilon);
1080          alpha(i,j)=Q2(j)/Q1(i);
1081      if R1(i)>R2(j)
1082          gamma(i,j)=R1(i)/R2(j);
1083    W1(i,j)=((1/k1)*R2(j)*gamma(i,j))*...
1084     ((1+gamma(i,j))/(2*alpha(i,j)))*...
1085  ((alpha(i,j)^2*c11(i,j)-2*alpha(i,j)*...
1086  c12(i,j)+c22(i,j))/delta(i,j));
1087          else if (R1(i)<R2(j))
1088              gamma(i,j)=R2(j)/R1(i);
1089  W1(i,j)=((1/k1)*R1(i)*gamma(i,j))*
```

```
                                                            ...
1090 ((1+gamma(i,j))/(2*alpha(i,j)))*...
1091 ((alpha(i,j)^2*c11(i,j)-2*alpha(i,j)*...
1092 c12(i,j)+c22(i,j))/delta(i,j));
1093      else if R1(i)==R2(j);
1094 W1(i,j)=((1/k1)*R1(i)*gamma(i,j))*...
1095 ((1+gamma(i,j))/(2*alpha(i,j)))*...
1096 ((alpha(i,j)^2*c11(i,j)-2*alpha(i,j)*...
1097 c12(i,j)+c22(i,j))/delta(i,j));
1098              end
1099              end
1100     end
1101     W2(i,j)=(k*(Q1(i)*Q2(j)))/(R1(i)+R2(j));
1102     A1(i,j)=W1(i,j);
1103     A2(i,j)=W2(i,j);
1104      A(i,j)=A1(i,j)/A2(i,j);
1105
1106     end
1107 end
1108 return
1109
1110 function[cond2]=condmy(A)
1111 [U,S,V]=SVD_2(A);
1112 lambda_max=max(diag(S));
1113 lambda_min=min(diag(S));
1114 cond_1=(((lambda_max)/(lambda_min)));
1115 cond2=(log(cond_1))/(log(10));
1116 return
1117
1118 function [Uout,Sout,Vout] = SVD_2(A)
1119       m = size(A,1);
1120       n = size(A,2);
1121       U = eye(m);
1122       V = eye(n);
1123       e = eps*fro(A);
1124       while (sum(abs(A(~eye(m,n)))) > e)
1125       for i = 1:n
1126       for j = i+1:n
1127           [J1,J2] = jacobi(A,m,n,i,j);
1128           A = mtimes(J1,mtimes(A,J2));
1129           U = mtimes(U,J1');
1130           V = mtimes(J2',V);
1131       end
1132       for j = n+1:m
1133           J1 = jacobi2(A,m,n,i,j);
1134           A = mtimes(J1,A);
1135           U = mtimes(U,J1');
1136       end
1137       end
1138       end
1139       S = A;
1140       if (nargout < 3)
```

```
1141          Uout = diag(S);
1142        else
1143           Uout = U; Sout = times(S,eye(m,n)); Vout = V;
1144        end
1145        end
1146      function [J1,J2] = jacobi(A,m,n,i,j)
1147         B = [A(i,i), A(i,j); A(j,i), A(j,j)];
1148         [U,S,V] = tinySVD(B); %
1149         J1 = eye(m);
1150         J1(i,i) = U(1,1);
1151         J1(j,j) = U(2,2);
1152         J1(i,j) = U(2,1);
1153         J1(j,i) = U(1,2);
1154         J2 = eye(n);
1155         J2(i,i) = V(1,1);
1156         J2(j,j) = V(2,2);
1157         J2(i,j) = V(2,1);
1158         J2(j,i) = V(1,2);
1159      end
1160      function J1 = jacobi2(A,m,n,i,j)
1161         B = [A(i,i), 0; A(j,i), 0];
1162         [U,S,V] = tinySVD(B);
1163         J1 = eye(m);
1164         J1(i,i) = U(1,1);
1165         J1(j,j) = U(2,2);
1166         J1(i,j) = U(2,1);
1167         J1(j,i) = U(1,2);
1168      end
1169      function [Uout,Sout,Vout] = tinySVD(A)
1170   t = rdivide((minus(A(1,2),A(2,1))),(plus(A(1,1),A(2,2))));
1171        c = rdivide(1,sqrt(1+t^2));
1172        s = times(t,c);
1173        R = [c,-s;s,c];
1174        M = mtimes(R,A);
1175        [U,S,V] = tinySymmetricSVD(M);
1176        U = mtimes(R',U);
1177        if (nargout < 3)
1178           Uout = diag(S);
1179        else
1180           Uout = U; Sout = S; Vout = V;
1181        end
1182        end
1183      function [Uout,Sout,Vout] = tinySymmetricSVD(A)
1184        if (A(2,1) == 0)
1185           S = A;
1186           U = eye(2);
1187           V = U;
1188        else
1189           w = A(1,1);
1190           y = A(2,1);
1191           z = A(2,2);
1192           ro = rdivide(minus(z,w),times(2,y));
```

```
1193 t2 = rdivide(sign(ro),plus(abs(ro),sqrt(plus(times(ro,ro),1))));
1194          t = t2;
1195          c = rdivide(1,sqrt(plus(1,times(t,t))));
1196          s = times(t,c);
1197          U = [c, -s; s, c];
1198          V = [c,  s;-s, c];
1199          S = mtimes(U,mtimes(A,V));
1200          U = U';
1201          V = V';
1202       end
1203       [U,S,V] = fixSVD(U,S,V);
1204       if (nargout < 3)
1205           Uout = diag(S);
1206       else
1207           Uout = U; Sout = S; Vout = V;
1208       end
1209       end
1210     function [U,S,V] = fixSVD(U,S,V)
1211       Z = [sign(S(1,1)),0; 0,sign(S(2,2))]; %
1212       U = mtimes(U,Z);
1213       S = mtimes(Z,S);
1214       if (S(1,1) < S(2,2))
1215           P = [0,1;1,0];
1216           U = mtimes(U,P);
1217           S = mtimes(P,mtimes(S,P));
1218           V = mtimes(P,V);
1219       end
1220       end
1221     function f = fro(M)
1222       f = sqrt(sum(sum(times(M,M))));
1223     end
1224     function s = sign(x)
1225        if (x > 0)
1226            s = 1;
1227        else
1228            s = -1;
1229        end
1230        end
```

References

1. J.C. Biro, Amino acid size, charge, hydropathy indices and matrices for protein structure analysis. Theor. Biol. Med. Model **3**(15) (2006)
2. G. Arents, R.W. Burlingame, B.C. Wang, W.E. Love, E.N. Moudrianakis, The nucleosomal core histone octamer at 3.1 A resolution: a tripartite protein assembly and a left-handed superhelix. Proc. Natl. Acad. Sci. U. S. A. **88**(22), 10148–10152 (1991)
3. L. Marino-Ramirez, M.G. Kann, B.A. Shoemaker, D. Landsman, Histone structure and nucleosome stability. Expert Rev. Proteomics **2**(5), 719–729 (2005)
4. G. Arents, E.N. Moudrianakis, The histone fold: a ubiquitous architectural motif utilized in DNA compaction and protein dimerization. Proc. Natl. Acad. Sci. U.S.A. **92**(24), 11170–11174 (1995)

5. A.D. Baxevanis, G. Arents, E.N. Moudrianakis, D. Landsman, A variety of DNA-binding and multimeric proteins contain the histone fold motif. Nucleic Acids Res. **23**(14), 2685–2691 (1995)
6. A.D. Baxevanis, J.E. Godfrey, E.N. Moudrianakis, Associative behavior of the histone $(H3–H4)_2$ tetramer: dependence on ionic environment. Biochem. **30**(36), 8817–8823 (1991)
7. L.M. Gloss, B.J. Placek, The effect of salts on the stability of the H2A–H2B histone dimer. Biochem. **41**(50), 14951–14959 (2002)
8. V. Karantza, A.D. Baxevanis, E. Freire, E.N. Moudrianakis, Thermodynamic studies of the core histones: pH and ionic strength effects on the stability of the (H3–H4)/(H3–H4)2system. Biochem. **34**(17), 5988–5996 (1995)
9. S.S. Dukhin, *Dielectric Phenomena and Double Layer in Disperse Systems and Polyelectrolytes* (Naukova Dumka, Kiev, 1972)
10. K.B.J. Oldham, Electroanal. Chem. **613**, 131–138 (2008)
11. J.H. Masliyah, S. Bhattacharjee, *Electrokinetic and Colloid Transport Phenomena* (Wiley, Hoboken, 2006)
12. H. Masliyah, S. Bhattacharjee, *Electrokinetic and Colloid Transport Phenomena* (Wiley, New York, 2006)
13. Vassiliki Karantza, Andreas D. Baxevanis, Ernesto Freire, Evangelos N. Moudrianakis, Thermodynamic studies of the core histones: ionic strength and pH dependence of H2A–H2B dimer stability. Biochem. **34**(17), 5988–5996 (1995)

Chapter 5
Mathematical Modeling Identification of Active Sites Interaction of Protein Molecules

Abstract In this chapter, two algorithms are developed: Algorithm 1 and Algorithm 2. Algorithm 1 was developed in order to search for the interaction of a polypeptide chain of a full-length protein with short active region. Algorithm 2 was developed to determine the most active sites of interaction between full-length proteins when dimers are formed in the direction from the N-terminus to C-terminus. Numerical calculations were made using proteins Mdm2, Nap1, P53.

5.1 Introduction

For modern proteomics, research and prediction of protein interactions are very important tasks, since they determine the function of proteins at levels from the cell to the whole organism. For proteins whose structure is known, the search for intermolecular interactions according to known data on the conformation of their tertiary structure reduces to the problem of searching for geometric complementarity of the sections of two interacting molecular surfaces and modelling their contacts, the so-called molecular docking [1]. The task of molecular docking is the task of a conformational search algorithm, which reduces to a search for the conformational space of the formed biological complex due to the variation of the torsion angles of protein molecules.

Modern conformational search algorithms in most cases find conformations that are generally close to the experimentally found structures in a relatively short time. However, there are factors that also have a significant impact on the success of the docking, which are often not taken into account in standard algorithms. One such factor is the conformational mobility of the target protein. The mobility range can be different - beginning with a small ≪adjustment≫ of the side chains and ending with scale domain movements [2]. These movements play an important role. At first glance, the most logical solution to this problem is to take into account the mobility

T. Koshlan and K. Kulikov, *Mathematical Modeling of Protein Complexes*,
Biological and Medical Physics, Biomedical Engineering,
https://doi.org/10.1007/978-3-319-98304-2_5

of the protein in a docking program. Unfortunately, modern computational tools do not allow such modelling to be performed in an acceptable time frame since a protein molecule is very large, and allowing for mobility over all degrees of freedom can lead to a so-called ≪combinatorial explosion≫ (an astronomical increase in the number of possible variants). Only in some programs is there a limited mobility of protein binding sites (usually at the level of a small adaptation of conformations of the side chains of the active center residues). Another approach to this problem consists in docking the same protein in several different conformations and then selecting the best solutions from each docking run. The third approach is to find a universal structure of the target protein in which docking would produce fairly good results for different classes of ligands. In this case, the number of ≪missed≫ (but correct) solutions decreases, but the number of incorrect options [3] also increases significantly. It should also be noted that most programs for the theoretical docking of proteins work according to the following principle: one protein is fixed in space, and the second is rotated around it in a variety of ways.

At the same time, for each rotation configuration, estimates are made for the evaluation function. The evaluation function is based on surface complementarity (the mutual correspondence of complementary structures (macromolecules, radicals), determined by their chemical properties), electrostatic interactions, van der Waals repulsion, and so on. The problem with this approach is that calculations throughout the configuration space require a lot of time, rarely leading to a single solution [1, 2], which in turn does not allow us to speak of the uniqueness of the target protein and ligand interaction variant. So in the work [2], while modelling by the methods of molecular dynamics, from 200 to 10 000 possible combinations of the formation of a protein complex with a ligand were found. Such a large number of modifications, along with the lack of a criterion for selecting the most probable variants of the bound structures of biological complexes (which would allow a radical reduction in their number) makes it very difficult to interpret the theoretical results obtained for practical use, namely, the finding of catalytic centers and a qualitative assessment of the dissociation constant of interacting substances.

In contrast to the above computer simulation algorithms, mathematical algorithms have been developed in this chapter that allow determining the detection of proteins active regions and detecting the stability of different regions of protein complexes (linear docking) by analyzing the potential energy matrix of pairwise electrostatic interaction between different sites of the biological complex, such as the homodimer of the histone chaperone Nap1–Nap1, the heterodimer of the p53–Mdm2 proteins, and the homodimer Mdm2–Mdm2, which are responsible for the entry of a whole protein molecule into biochemical reactions.

The chapter consists of several parts.

The first part describes the structure and function of proteins Mdm2, P53, Nap1. The second part describes developed algorithms. The third part presents numerical calculations and their analysis The amino acid sequences of the studied proteins P53, Nap1, Mdm2 were taken in [4] with the numbers: P04637, P25293, Q00987, respectively.

5.2 The Structure and Function of the Protein P53

The P53 protein was discovered in 1979 and received its name on the molecular weight (53KDa) [5–7].

Protein P53 is transcription factor regulating the cell cycle, and it suppresses the formation of malignant tumors [8, 9].

The P53 protein in the activated state regulates the transcription of a large number genes, and also interacts with a large number of other proteins, thereby affecting many intracellular processes [10, 11].

One of the functions of protein P53 is the control of the state of cellular DNA [10, 12].

P53 is activated when it receives deviation signals from normal cellular prosesses, and it recognaser damage in the gemetic apparatus. This leads to either an acceleration, or it stop the cell cycle and with strong stress stimulus-to apoptosis [13].

The P53 protein undergoes phosphorylation during cellular stress [14] and its level of concentration in the cell increases [15].

This activates the protein genes, which are involved in cellular apoptosis, such as the protein Mdm2, which is involved in negative regulation of P53 protein [6].

The MDM2 protein binds to the transactivation domain of the P53 protein, which is located on the flexible N-terminus of the P53 protein, which forwards to the regulation of the amount of P53 protein in the cell and leads to its subsequent degradation [16, 17].

As is known, the functions of proteins depend on their three-dimensional structure. The human protein P53 has 393 amino acid residues. Protein domains are independent folding units, which usually have sizes from 40 to 200 amino acid residues [16].

Thus, the P53 protein contains several domains. In the structure of this protein are three main domains: the transactivation domain (1–70); a sequence which specifically binds to $DNA_{(94-293)}$; tetramerization domain (324–355) [16]. Figure 5.1 shows the structure of protein P53. The P53 protein domains are connected by linker regions. Proline-rich domain (71–93) binds to the transactivation domain of the domain, which is responsible for binding to DNA. DNA binding domain is target for a large number of mutations P53 [16].

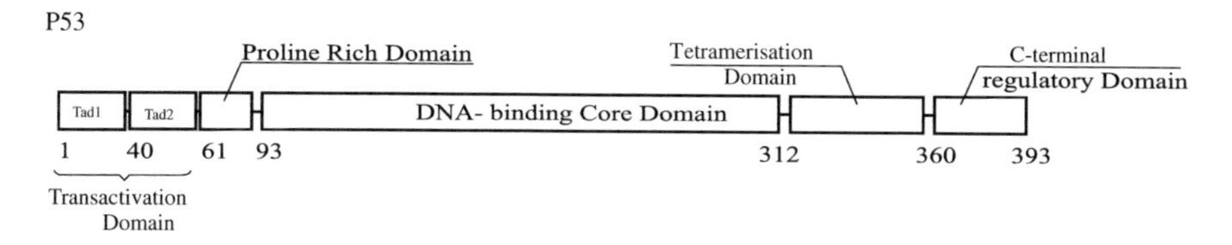

Fig. 5.1 The structure of the protein p53

5.3 The Structure and Functions of the Protein Mdm2

Human Murine double minute 2 (Mdm2) is a 491-amino acid (a.a.) -long phosphoprotein [18, 19]. Mdm2 is an oncogene with both P53-dependent and P53-independent oncogenic activities, and often has increased expression levels in a variety of human cancers [20].

Inhibition functions of protein P53, when it binds with the Mdm2 protein is carried out different ways. These ways help block the transactivation domain P53 and simultaneously promote export P53 from the nucleus to proteasome degradation systems [21].

A detailed study of the structure formed by proteins Mdm2 and P53 showed that the amino terminal domain of Mdm2 forms a deep hydrophobic cleft into which the transactivation domain of P53 binds, thereby concealing itself from interaction with the transcriptional machinery [22, 23].

The direct interaction between the two proteins has been localized to a relatively small hydrophobic pocket domain at the N-terminus of Mdm2 and 15 a.a. amphipathic peptide at the N-terminus of P53. The P53 binding domain of human Mdm2 which can be identified within residues 18–101 and interact with residues 15–29 of P53. Various signals, for example the destruction of cellular DNA leads to an abnormal interaction of Mdm2 and P53, which is the cause of activation P53-dependent cell responses [16, 22]. In the protein Mdm2 several areas were identified see Fig. 5.2 at N-tail of the main region is binding to P53.

In the central part of the protein there are many acidic regions. Moreover, the C-tail contains a zinc-binding domain. This part of the protein interacts with a variety of regulatory factors as well as multiple ribosomal or nucleolar proteins.

The C-terminus also contains a RING domain that has been shown to be responsible for the E3 ubiquitin ligase activity, as well as the binding of the Mdmx and Mdm2 [24].

In this chapter, we have performed simulations of the protein Mdm2 interaction and protein P53 and the interaction between the same proteins of Mdm2 with the formation of a homodimer Mdm2–Mdm2.

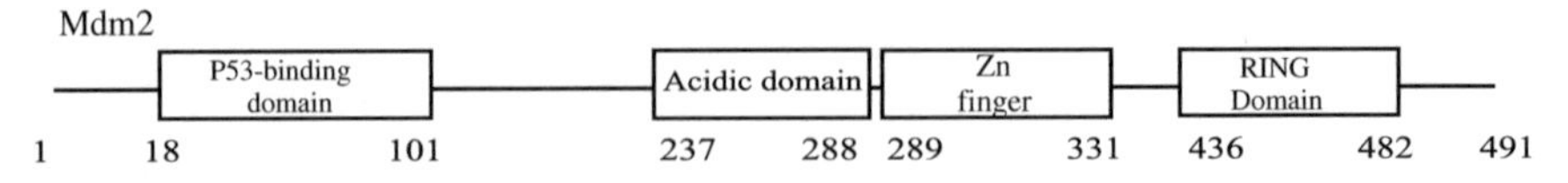

Fig. 5.2 The structure of the protein Mdm2 [16]

5.4 The Structure and Functions of the Protein Nap1

Nucleosome assembly protein 1 (Nap1) is an integral component in the establishment, maintenance, and dynamics of eukaryotic chromatin. It shuttles histones into the nucleus, assembles nucleosomes, and promotes chromatin fluidity [25].

The article [26, 27] presents various functions of the histone chaperone Nap1 protein, mainly its role in nucleosome assembly and disassembly, and the interactions of Nap1 with different chromatin remodelling factors; information is given on various binding sites of Nap1 with other proteins.

Let us consider in more detail the structure of the protein of the histone chaperone Nap1. In [25] was found that of a total of 417 a.a. well structured central sites residues, whereas the N- and C-terminal regions were largely disordered. The central region is defined in [25] is core region (74–365).

The structure of the protein Nap1 can be divided into several sections: unstructured N- and C-tails, Domain I and Domain II. Consider in more detail Domain 1 which is responsible for the dimerization of Nap1 [25].

During the dimerization prosess, an interaction occurs between the long $\alpha 2$-helices of two proteins in opposite directions. The dimer is further stabilized by the $\alpha 2$–$\alpha 3$ loop, the $\alpha 3$-helix, and the $\alpha 3$–$\alpha 4$ loop that wrap around the base of the $\alpha 2$-helix of the dimerization partner. In Fig. 2.2 a schematic representation of the Domain I and Domain II structures is presented. Domain I includes $\alpha 1$-, $\alpha 2$-, $\alpha 3$-helices, which are in most degree responsible for dimerization of the protein Nap1. Domain II spans residues 181–370 of Nap1. Figure 5.4 shows a scheme of the formation of the Nap1–Nap1 homodimer by two proteins of the histone chaperone. In blue color which is denoted by Domain 1 of the first histone chaperone Nap1 with indication of $\alpha 2$ and $\alpha 3$-helices, which take part in homodimerization. Orange color represents Domain 1 of the second protein of the histone chaperone Nap1 amd also shows the $\alpha 2$ and $\alpha 3$-helices. In monochrome, the rest of the proteins that are not actively involved in the formation of the Nap1–Nap1 homodimer are shown.

The dimer interface is characterized mainly by hydrophobic interactions over the entire length of the involved amino acid residues in $\alpha 2$-helices. A wide region of dimerization, covering all dimer diagonally is indicated in color Fig. 5.4.

5.5 Description of the Algorithms

5.5.1 *Algorithm 1*

This algorithm has been developed to search for protein sites responsible for protein interactions.

During the development of this algorithm we have made the following assumptions:

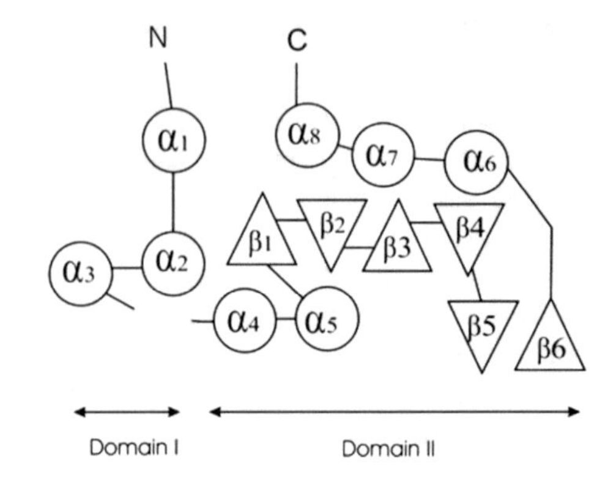

Fig. 5.3 Scheme of domains I and II of protein Nap1 [25]

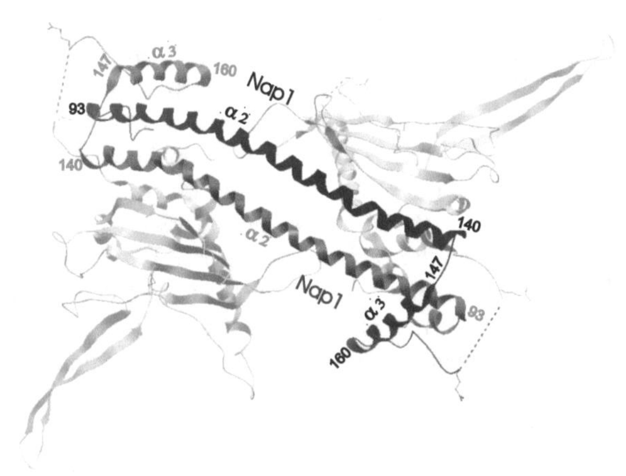

Fig. 5.4 The scheme for the formation of the Nap1–Nap1 homodimer [28]

– we know the short amino acid sequence of one protein, which takes an active part in binding to another protein, with formation large numbers nearly located interacting amino acid residues, for example, the formation of homodimers Mdm2−Mdm2 and Nap1–Nap1,

– we do not know the active site of the whole protein responsible for binding to the short polypeptide sequence.

Thus, using Algorithm 1, we find the active site on the polypeptide sequence of the whole protein. This algorithm (see Fig. 5.5) presents two vectors:

– the one-dimensional array 1 ≪DTEVNKL≫ and one-dimensional array 2 ≪ADHVNRTYOIK≫, which are amino acid sequences of the proteins P_1 and P_2, respectively,

– the one-dimensional array of the P_1 protein has a smaller number of amino acid residues in its polypeptide sequence than the one-dimensional array of the protein P_2. As each step occurs, a shot section of the amino acid sequence of protein P_2 forms, which is equal to the length of the shoter one dimernsional array of protein P_1.

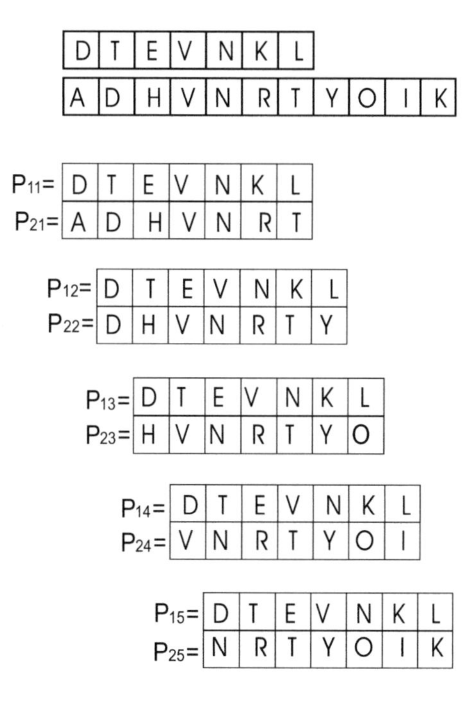

Fig. 5.5 The scheme Algorithm 1

The more short amino acid sequence of one-dimensional array P_1 moves along the more long amino acid sequence one-dimensional array of the protein P_2 with some step, in our example, the step is equal one amino acid.

In each step occurs the formation of a short section of the amino acid sequence of protein P_2, equal to the length of the shorter one-dimensional array of protein P_1.

Each new segment of the one-dimensional array P_2 corresponds to the length of the polypeptide chainone-dimensional array P_1. Five pairs of one-dimansional arrays P_1 and P_2 were successfully formed when the one-dimensional array 1 shifted by one amino acid residue along the one-dimensional array P_2.

$P_{11} = \text{DTEVNKL}\ P_{21} = \text{ADHVNRT}$
$P_{12} = \text{DTEVNKL}\ P_{22} = \text{DHVNRTY}$
$P_{13} = \text{DTEVNKL}\ P_{23} = \text{HVNRTYO}$
$P_{14} = \text{DTEVNKL}\ P_{24} = \text{VNRTYOI}$
$P_{15} = \text{DTEVNKL}\ P_{25} = \text{NRTYOIK}$

Note that the vector 1 remains unchanged in all formed pairs one-dimensional arrays, i.e. $P_{11} = P_{12} = P_{13} = P_{14} = P_{15}$.

After finding all the participating pairs of vectors, we build a matrix of potential energy electrostatic interaction between their amino acid residues. These matrices will have a square form.

Further, from each of these matrices we calculate the value lg(cond(W)) and construct a graph of the dependence of lg(cond(W)) on the order number of the

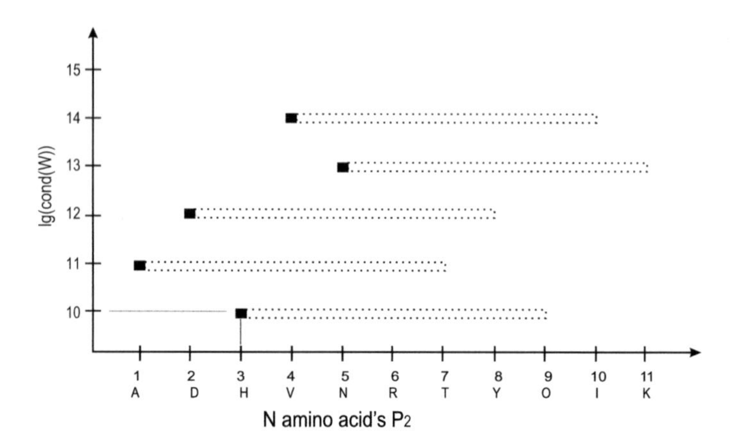

Fig. 5.6 Scheme dependence of lg(cond(W)) from Amino acid sequence

amino acid residue of one-dimensional array P_2, where cond(W_k) is the condition number.

Thus, each resulting value of lg(cond(W)) will correspond to a strictly defined segment of one-dimensional array P_2. Note that on the graph it is postponed opposite the first amino acid residue of a segments P_{21}, P_{22}, P_{23}, P_{24} or P_{25} one-dimensional array P_2.

Figure 5.6 shows graph dependence of values lg(cond(W)) on the sequence number of the polypeptide chains of participating one-dimensional arrays P_1 and P_2. In this example, as we see from graph the smallest value of lg(cond(W)) corresponds to the interaction of the vectors P_{13} is ≪DTEVNKL≫ and P_{23} is ≪HVNRTYO≫. The dotted line indicates the amino acid sequence of one-dimensional array P_2, which participates in the formation of a biological complex with a one-dimensional array P_1. The value of the value lg(cond(W)) is placed opposite the first amino acid of the remainder of the segment of the one-dimensional array P_2. For data processing we will choose several of the smallest values of lg(cond(W)) (see Chap. 2).

We suppose that for the most stable complex of interacting sites has the largest number of nearby points with a minimum value of lg(cond(W)). We call this area a cluster.

5.5.2 *Algorithm 2*

We developed a second algorithm for detecting interacting regions of protein molecules. The scheme search interacting sections is shown in Fig. 5.7. In this algorithm we take whole amino acid sequences of the two proteins P_1 and P_2. For selecting interacting sites, we shift the frame of a specific size along two one-dimensional arrays of proteins P_1 and P_2.

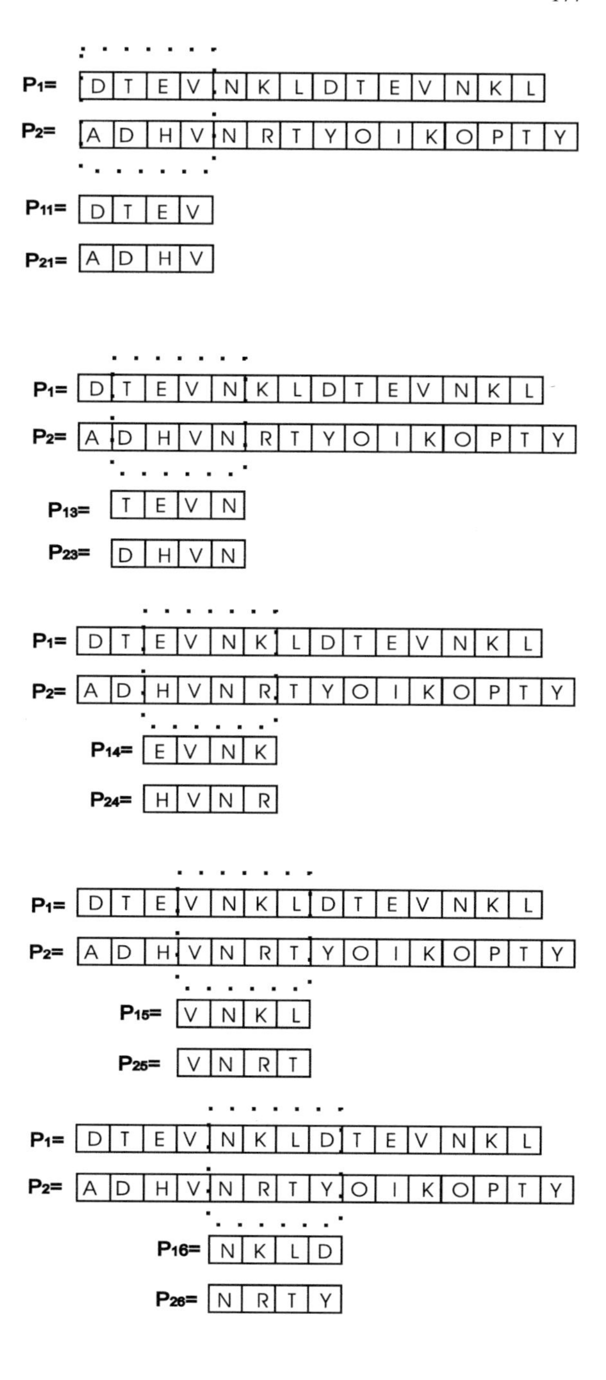

Fig. 5.7 The scheme Algorithm 2

According to the method, we test the interaction sites for the following pairs of proteins: Nap1–Nap1, Mdm2–Mdm2.

Note that Algorithm 2 can be used to analyze the interaction of two proteins, which have identical sites of interaction during the formation of a dimer.

After finding all the participating pairs of one-dimensional arrays, we build a matrix of potential energy electrostatic interactions. These matrices will have a square form. Further, for all matrices we calculate the values lg(cond(W)) and construct a graph of the dependence of lg(cond(W)) on the order number of the amino acid residue of the participating one-dimensional arrays.

The sequence number will be the same for the two considered one-dimensional arrays. In this case, the amino acid residues corresponding to the ordinal number can be different, if the interactions of different proteins are investigated.

5.6 Numerical Simulation of the Formation of Heterodimers and Homodimers According to Algorithm 1

In this section, the interactions of the short amino acid sequencesMdm2$_{(436-482)}$ and Nap1$_{(81-150)}$, which take an active part in the formation of Nap1–Nap1 and Mdm2-Mdm2 homodimers, have been modeled. Preliminary information on the activity of these sites of Mdm2$_{(436-482)}$ and Nap 1$_{(81-150)}$ was obtained from previous experimental studies [25, 29].

Numerical simulation was performed according to the developed Algorithm 1. Thus, the purpose of this section is to test the developed Algorithm 1 in determining the most active interaction regions of a full-length protein with a short polypeptide sequence. Numerical calculations were performed with $\varepsilon = 1$ (air) and $\varepsilon = 80$ (water). At the same time, the authors tried to choose a common scale for presenting the obtained graphic data for all the calculations performed in order to facilitate understanding and to allow the reader to visually compare the results of the obtained data. In addition to the graphical representation, we also gave 10 minimum values of lg(cond(W)) with a list of the corresponding interacting amino acid sequences at $\varepsilon = 1$ and $\varepsilon = 80$, and the data were tabulated. We assume that the more precisely the active region of one protein is given when interacting with the whole amino acid sequence of the second protein, the more qualitative the result of the interaction of the two proteins is. It is assumed that a large amount of a.a. (a significant length of polypeptide chain) from each protein corresponds to the formation of the biological complex.

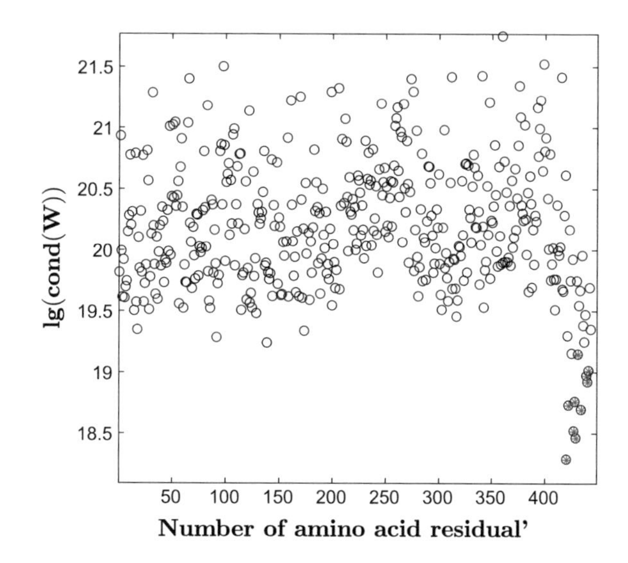

Fig. 5.8 Results of numerical simulation of the interactions of $Mdm2_{(436-482)}$ with Mdm2, $\varepsilon = 1$. The red dots denote the 10 minimum values of lg(cond(W))

5.6.1 *Numerical Calculation of the Interaction of* **$Mdm2_{(436-482)}$** *Mdm2*

A search was carried out for the polypeptide chain Mdm2, which is most inclined to form a complex with the Mdm2 protein Ring domain. To achieve this goal, we took the Mdm2 protein domain polypeptide region (436–482):

$$[EPCVICQGRPKNGCIVHGKTGHLMACFTCAKKLKKRNKPCPVCRQPI]$$

and numerically calculated its interaction with the Mdm2 protein according to Algorithm 1 developed earlier. In this case, the short sequence of the protein $Mdm2_{(436-482)}$ shifted along the long sequence of the protein Mdm2 at intervals of 1 a.a. As a result, for each pair of the obtained one-dimensional arrays, a matrix of potential energy of electrostatic interaction was formed, and the value of lg(cond(W)) was calculated. The value of lg(cond(w)) was plotted opposite the first a.a. section of the Mdm2 protein upon interaction with $Mdm2_{(436-482)}$. The results are shown in Figs. 5.8 and 5.9. In this section, we present one graph that contains all the values of lg(cond(W)) obtained for the interaction of $Mdm2_{(436-482)}$ with Mdm2. We will use scaled graphs of the smallest values of lg(cond (W)), since we will analyze these values.

As can be seen from the Figs. 5.8 and 5.9, the set of minimum values form a cluster from the C-terminus of the Mdm2 protein. The ten minimum values of lg(cond(W)) for the interaction of $Mdm2_{(436-482)}$ with Mdm2, as well as the corresponding amino acid sequences at $\varepsilon = 1$ and $\varepsilon = 80$ are summarized in Tables 5.1 and 5.2.

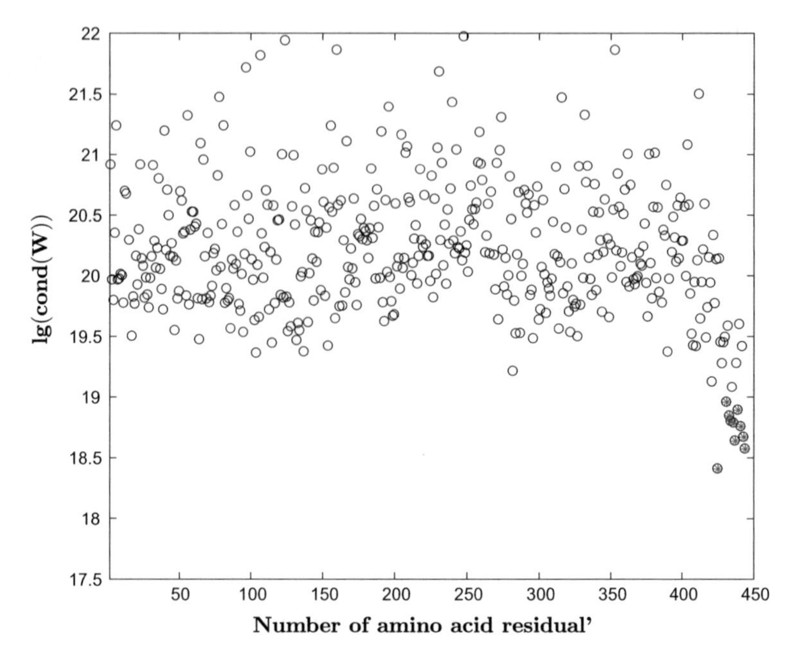

Fig. 5.9 Results of numerical simulation of the interactions of $\text{Mdm2}_{(436-482)}$ with Mdm2, $\varepsilon = 80$. The red dots denote the 10 minimum values of lg(cond(W))

It follows from Tables 5.1 and 5.2 that the areas with the minimum values of lg(cond(W)) are almost identical at $\varepsilon = 1$ and $\varepsilon = 80$ and are in the previously identified range of the polypeptide chain of the Mdm2 protein responsible for binding to $\text{Mdm2}_{(436-482)}$ from the C-Terminus. Thus, it can be concluded that the sequence $\text{Mdm2}_{(436-482)}$ plays an active role in the dimerization of the Mdm2 protein and is most likely to form stable biological complexes in the C-terminus region of the Mdm2 protein. This result is in good agreement with [29–31].

Note that we gave 10 minimum values of lg(cond(W)), characterizing the interactions of $\text{Mdm2}_{(436-482)}$ with Mdm2, but we did not exclude the presence of other regions of interaction of $\text{Mdm2}_{(436-482)}$ with Mdm2 that did not fall within the given range of the 10 minimum values of lg(cond(W)) from Table 5.1 and 5.2 for $\varepsilon = 1$ and $\varepsilon = 80$. We also did not exclude the existence of other possible sites for binding $\text{Mdm2}_{(436-482)}$ to Mdm2.

Table 5.1 The ten minimum values of lg(cond(W)) and the corresponding amino acid sequences of the detected regions of the Mdm2 protein when interacting with $Mdm2_{(436-482)}$, $\varepsilon = 1$

Sequence number	Amino acid sequence $Mdm2_{(436-482)}$	lg(cond(W))
422	KEESVESSLPLNAIEPCVICQGRPKNGCIVHGKTGHLMACFTCAKKL	18.285
430	PLNAIEPCVICQGRPKNGCIVHGKTGHLMACFTCAKKLKKRNKPCPV	18.460
428	SLPLNAIEPCVICQGRPKNGCIVHGKTGHLMACFTCAKKLKKRNKPC	18.516
435	EPCVICQGRPKNGCIVHGKTGHLMACFTCAKKLKKRNKPCPVCRQPI	18.691
423	ESVESSLPLNAIEPCVICQGRPKNGCIVHGKTGHLMACFTCAKKLKK	18.727
429	LPLNAIEPCVICQGRPKNGCIVHGKTGHLMACFTCAKKLKKRNKPCP	18.757
441	QGRPKNGCIVHGKTGHLMACFTCAKKLKKRNKPCPVCRQPIQMIVLT	18.917
440	CQGRPKNGCIVHGKTGHLMACFTCAKKLKKRNKPCPVCRQPIQMIVL	18.969
442	GRPKNGCIVHGKTGHLMACFTCAKKLKKRNKPCPVCRQPIQMIVLTY	19.012
432	NAIEPCVICQGRPKNGCIVHGKTGHLMACFTCAKKLKKRNKPCPVCR	19.146

lg(cond(W)) is common logarithm of condition number

Table 5.2 The ten minimum values of lg(cond(W)) and the corresponding amino acid sequences of the detected regions of the Mdm2 protein when interacting with $Mdm2_{(436-482)}$, $\varepsilon = 80$.

Sequence number	Amino acid sequence $Mdm2_{(436-482)}$	lg(cond(W))
425	VESSLPLNAIEPCVICQGRPKNGCIVHGKTGHLMACFTCAKKLKKRN	18.408
444	PKNGCIVHGKTGHLMACFTCAKKLKKRNKPCPVCRQPIQMIVLTYFP	18.572
437	CVICQGRPKNGCIVHGKTGHLMACFTCAKKLKKRNKPCPVCRQPIQM	18.638
443	RPKNGCIVHGKTGHLMACFTCAKKLKKRNKPCPVCRQPIQMIVLTYF	18.669
441	QGRPKNGCIVHGKTGHLMACFTCAKKLKKRNKPCPVCRQPIQMIVLT	18.756
436	PCVICQGRPKNGCIVHGKTGHLMACFTCAKKLKKRNKPCPVCRQPIQ	18.786
434	IEPCVICQGRPKNGCIVHGKTGHLMACFTCAKKLKKRNKPCPVCRQP	18.801
433	AIEPCVICQGRPKNGCIVHGKTGHLMACFTCAKKLKKRNKPCPVCRQ	18.844
439	ICQGRPKNGCIVHGKTGHLMACFTCAKKLKKRNKPCPVCRQPIQMIV	18.892
431	LNAIEPCVICQGRPKNGCIVHGKTGHLMACFTCAKKLKKRNKPCPVC	18.957

lg(cond(W)) is common logarithm of condition number

5.6.2 Numerical Calculations of the Interaction $\mathbf{Nap1_{(81-150)}-Nap1}$

In this section, we consider the results of the numerical modelling of the interaction of the protein region with the whole amino acid sequence of the histone chaperone Nap1 protein.

We selected a region of the protein $Nap1_{(81-150)}$ which takes an active part in the dimerization of the Nap1 protein and made a numerical calculation of the interaction of this region with the polypeptide sequence of the whole protein Nap1.

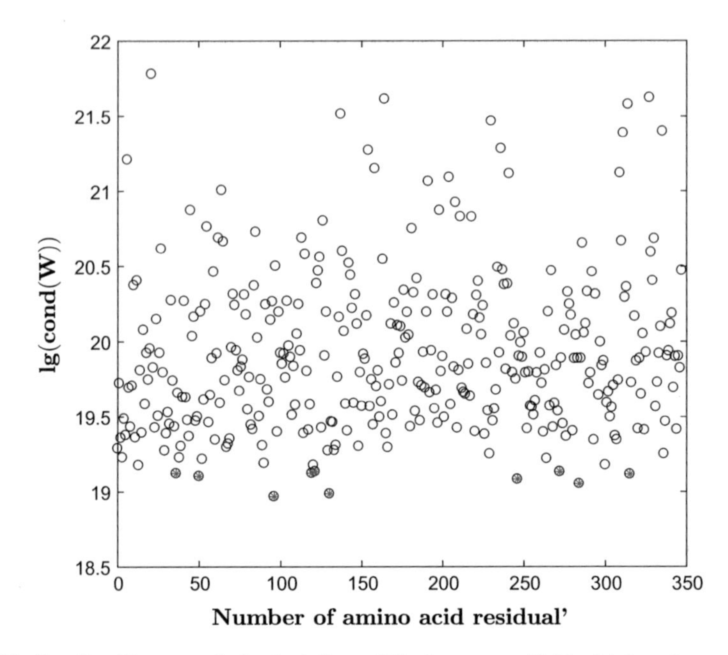

Fig. 5.10 Results of the numerical calculations of $\text{Nap1}_{(81-150)}$ with Nap1 interactions according to Algorithm 1, $\varepsilon = 1$. The red dots denote the 10 minimum values of lg(cond(W))

The amino acid sequence $\text{Nap1}_{(81-150)}$ was moved along the polypeptide sequence of the protein Nap1 at intervals of 1 a.a. according to Algorithm 1. Arrays 70 a.a. long were formed, one of which was represented as the amino acid sequence of the active site of the protein ($\text{Nap1}_{(81-150)}$), and the second array was characterized by serially changing amino acid residues of the protein Nap1. For each pair of the obtained arrays, a matrix of potential energy of pairwise electrostatic interaction was formed and the value of lg(cond(W)) was calculated.

We assume that the more precisely the active site of interaction of one protein is initially plotted, the more accurate the results will be obtained when the active site is located on another protein as they are bound to the biological complex. Numerical calculation of Nap1 interaction with $\text{Nap1}_{(81-150)}$ was carried out according to Algorithm 1 for $\varepsilon = 1$ (air) and $\varepsilon = 80$ (water). The results are shown in Figs. 5.10 and 5.11.

As can be seen from the Fig. 5.11 for $\varepsilon = 80$ a cluster of the smallest values of lg(cond(W)) is observed, with the smallest value obtained at 74 a.a. Note that this 74 a.a. is the first amino acid residue of domain 1 responsible for the formation of the Nap1–Nap1 homodimer. The results obtained during the interaction of $\text{Nap1}_{(81-150)}$ with Nap1, $\varepsilon = 1$, shown in Fig. 5.10 do not demonstrate the existence of cluster

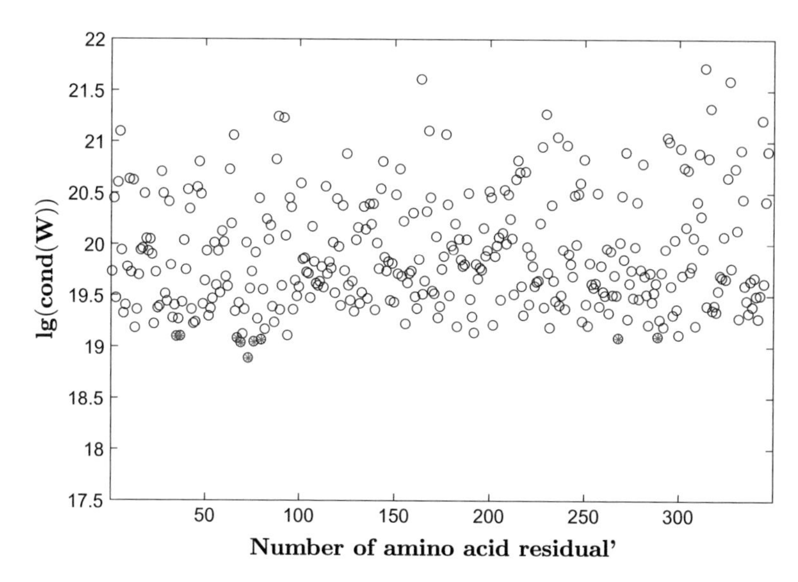

Fig. 5.11 Results of the numerical calculations of Nap1$_{(81-150)}$ with Nap1 interactions according to Algorithm 1, $\varepsilon = 80$. The red dots denote the 10 minimum values of lg(cond(W))

formation from the minimum values. The minimum value for $\varepsilon = 1$ is obtained at 97 a.a., which is also a satisfactory result because 97 a.a. is located in the region of the beginning of domain 1, while other areas of possible interaction of Nap1$_{(81-150)}$ with Nap1 in the domain 2 region are observed.

The ten minimum values of lg(cond(W)) for Nap1$_{(81-150)}$ interaction with Nap1 at $\varepsilon = 1$ and $\varepsilon = 80$ are presented in Tables 5.3 and 5.4.

The results of Table 5.4 show the formation of a cluster from the minimum values of lg(cond(W)) in the region of domain 1 responsible for the formation of the Nap1–Nap1 homodimer at $\varepsilon = 80$. The results of Table 5.3 for $\varepsilon = 1$ do not show the formation of such a cluster from the minimum values of lg(cond (W)). Thus, calculations performed at $\varepsilon = 80$ made it possible to obtain a better result and subsequent analysis than the numerical results obtained at $\varepsilon = 1$.

As can be seen from the Fig. 5.11 and Table 5.3, the polypeptide sequence of the histone chaperone protein Nap1 has its main binding site with the short sequence Nap1$_{(81-150)}$ in the Domain 1 region. The smallest value of lg(cond(W)) was obtained at 74 a.a. and amounted to 18.884. This amino acid residue (74 a.a.) in domain 1 is the first responsible a.a. for the formation of the Nap1–Nap1 homodimer and the result is in good agreement with previous work [25].

Table 5.3 The ten minimum values of lg(cond(W)) and the corresponding amino acid sequences of the detected regions of the protein Nap1 in interaction with $\mathrm{Nap1}_{(81-150)}$, $\varepsilon = 1$

$N^{\underline{0}}$	Amino acid sequence $\mathrm{Nap1}_{(81-150)}$	lg(cond(W))
97	LSLKTLQSELFEVEKEFQVEMFELENKFLQKYKPIWEQRSRIISGQEQPKP EQIAKGQEIVESLNETELL	18.968
131	IWEQRSRIISGQEQPKPEQIAKGQEIVESLNETELLVDEEEKAQNDSEEEQVK GIPSFWLTALENLPIVC	18.986
285	VDLEMRKQRNKTTKQVRTIEKITPIESFFNFFDPPKIQNEDQDEELEE DLEERLALDYSIGEQLKDKLIP	19.053
247	ILCKTYFYQKELGYSGDFIYDHAEGCEISWKDNAHNVTVDLEMRKQRNKTTK QVRTIEKITPIESFFNFF	19.084
51	IGTINEEDILANQPLLLQSIQDRLGSLVGQDSGYVGGLPKNVKEKLLSLKTL QSELFEVEKEFQVEMFEL	19.105
325	FDPPKIQNEDQDEELEEDLEERLALDYSIGEQLKDKLIPRAVDWF TGAALEFEFEEDEEEADEDEDEEED	19.117
37	GNPVRAQAQEQDDKIGTINEEDILANQPLLLQSIQDRLGSLVGQDSGYVGGLP KNVKEKLLSLKTLQSEL	19.123
120	LENKFLQKYKPIWEQRSRIISGQEQPKPEQIAKGQEIVESLNETELLVDEEE KAQNDSEEEQVKGIPSFW	19.125
273	EISWKDNAHNVTVDLEMRKQRNKTTKQVRTIEKITPIESFFNFFDPPKIQNED QDEELEEDLEERLALDY	19.133
122	NKFLQKYKPIWEQRSRIISGQEQPKPEQIAKGQEIVESLNETELLVDE EEKAQNDSEEEQVKGIPSFWL	19.135

lg(cond(W)) is common logarithm of condition number
$N^{\underline{0}}$ is number of amino acid residual

Table 5.4 The ten minimum values of lg(cond(W)) and the corresponding amino acid sequences of the detected regions of the protein Nap1 in interaction with $Nap1_{(81-150)}$, $\varepsilon = 80$

$N^{\underline{0}}$	Amino acid sequence $Nap1_{(81-150)}$	lg(cond(W))
74	LGSLVGQDSGYVGGLPKNVKEKLLSLKTLQSELFEVEK EFQVEMFELENKFLQKYKPIWEQRSRIISGQE	18.884
70	IQDRLGSLVGQDSGYVGGLPKNVKEKLLSLKTLQSE LFEVEKEFQVEMFELENKFLQKYKPIWEQRSRII	19.035
77	LVGQDSGYVGGLPKNVKEKLLSLKTLQSELFEVEKEFQVEMFEL ENKFLQKYKPIWEQRSRIISGQEQPK	19.043
81	DSGYVGGLPKNVKEKLLSLKTLQSELFEVEKEFQVEMFELENKFL QKYKPIWEQRSRIISGQEQPKPEQI	19.063
68	QSIQDRLGSLVGQDSGYVGGLPKNVKEKLLSLKTLQSELFEVEKEFQV EMFELENKFLQKYKPIWEQRSR	19.078
269	AEGCEISWKDNAHNVTVDLEMRKQRNKTTKQVRTIEKITPIESF FNFFDPPKIQNEDQDEELEEDLEERL	19.084
36	NGNPVRAQAQEQDDKIGTINEEDILANQPLLLQSIQDRLGS LVGQDSGYVGGLPKNVKEKLLSLKTLQSE	19.094
290	RKQRNKTTKQVRTIEKITPIESFFNFFDPPKIQNEDQDEELEEDLEER LALDYSIGEQLKDKLIPRAVDW	19.096
1	MSDPIRTKPKSSMQIDNAPTPHNTPASVLNPSYLKNGNPVRAQA QEQDDKIGTINEEDILANQPLLLQSI	19.096
38	NPVRAQAQEQDDKIGTINEEDILANQPLLLQSIQDRLGSLVGQDSGYVG GLPKNVKEKLLSLKTLQSELF	19.098

lg(cond(W)) is common logarithm of condition number
$N^{\underline{0}}$ is number of amino acid residual

5.7 Numerical Simulation of the Formation of Protein Dimers According to Algorithm 2

In this section, numerical calculations of the interaction of amino acid sequences were performed to determine the active regions of proteins Mdm2 and Nap1 with the formation of the homodimers Nap1–Nap1 and Mdm2–Mdm2.

We used whole amino acid sequences of identical proteins Nap1 and Mdm2 and we shifted a frame of a given size depending on the type of protein along two one-dimensional arrays describing the identical amino acid sequences Nap1 and Nap1, Mdm2 and Mdm2 at intervals of 1 a.a. By the method of successively increasing the size of the frameshift from 20 a.a. up to 70 a.a. we obtained a set of data that, with varying degrees of approximation, allows us to obtain a qualitative result of the interaction of the selected proteins. In this case, we assume that the closer the size of the frameshift to the size of the interacting regions of proteins is given, the more qualitative the result will be of the interaction of the two proteins, provided that amino acid sequences with close order number of the participating amino acid residues of proteins play an important role in the interaction. It is assumed that a large amount of a.a. (a significant length of the polypeptide chain) from each protein corresponds to the formation of the biological complex.

5.7.1 Numerical Calculation of the Interaction of Two Polypeptide Chains of the Mdm2 Protein

In this section, we analyzed the numerical calculation of the homodimer formation Mdm2–Mdm2 according to Algorithm 2. We used the amino acid sequences of the proteins Mdm2 and Mdm2 and formed the corresponding one-dimensional arrays, by shifting the frame along two polypeptide chains at intervals of 1 a.a., in order to identify the most active regions of the interaction of the studied proteins. The ten minimum values of lg(cond(W)) for each calculation obtained are tabulated in the corresponding graph.

We now turn to the analysis of numerical calculations of the interaction of two Mmd2 and Mdm2 proteins at a frameshift length equal to 20 a.a. The results of the numerical interaction are shown in Fig. 5.12 for $\varepsilon = 1$ and in Fig. 5.13 for $\varepsilon = 80$. Ten minimum values of lg(cond(W)) are presented in Tables 5.5 and 5.6.

We now turn to the analysis of numerical calculations of the interaction of two Mdm2 and Mdm2 proteins at a frameshift length equal to 30 a.a. The results of the numerical interaction are are shown in Fig. 5.14 for $\varepsilon = 1$ and in Fig. 5.15 for $\varepsilon = 80$. Ten minimum values of lg(cond(W)) are presented in Tables 5.7 and 5.8.

We now turn to the analysis of numerical calculations of the interaction of two Mdm2 and Mdm2 proteins at a frameshift length equal to 40 a.a. The results of the numerical interaction are are shown in Fig. 5.16 for $\varepsilon = 1$ and in Fig. 5.17 for $\varepsilon = 80$. Ten minimum values of lg(cond(W)) are presented in Tables 5.9 and 5.10.

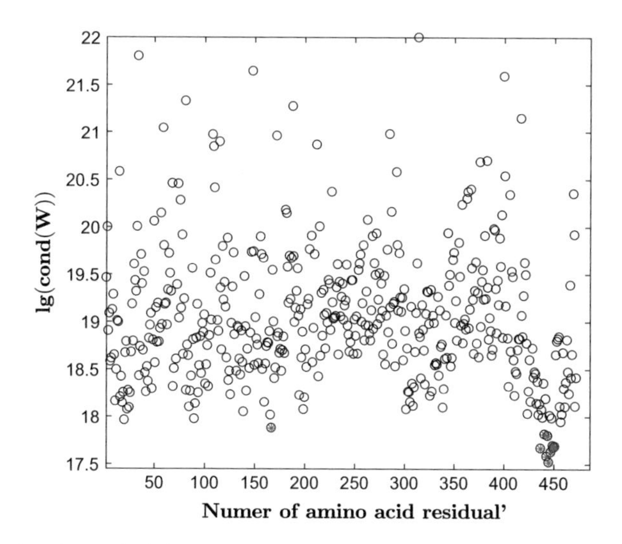

Fig. 5.12 Results of numerical simulation of the interaction of Mdm2 with Mdm2, frame size equal to 20 a.a., $\varepsilon = 1$. The red dots denote the 10 minimum values of lg(cond(W))

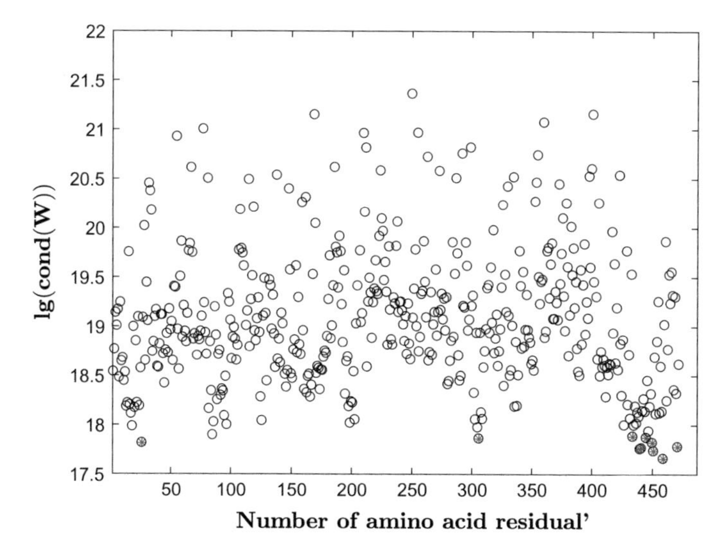

Fig. 5.13 Results of numerical simulation of the interaction of Mdm2 with Mdm2, frame size equal to 20 a.a. for $\varepsilon = 80$. The red dots denote the 10 minimum values of lg(cond(W))

We now turn to the analysis of numerical calculations of the interaction of two Mdm2 and Mdm2 proteins at a frameshift length of 50 a.a. The results of the numerical interaction are are shown in Fig. 5.18 for $\varepsilon = 1$ and in Fig. 5.19 for $\varepsilon = 80$. Ten minimum values of lg(cond(W)) are presented in Tables 5.11 and 5.12.

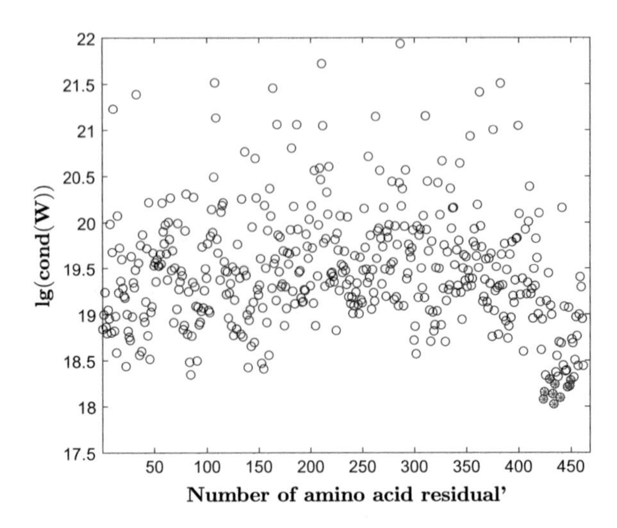

Fig. 5.14 Results of numerical simulation of the interaction of Mdm2 with Mdm2, frame size equal to 30 a.a., $\varepsilon = 1$. The red dots denote the 10 minimum values of lg(cond(W))

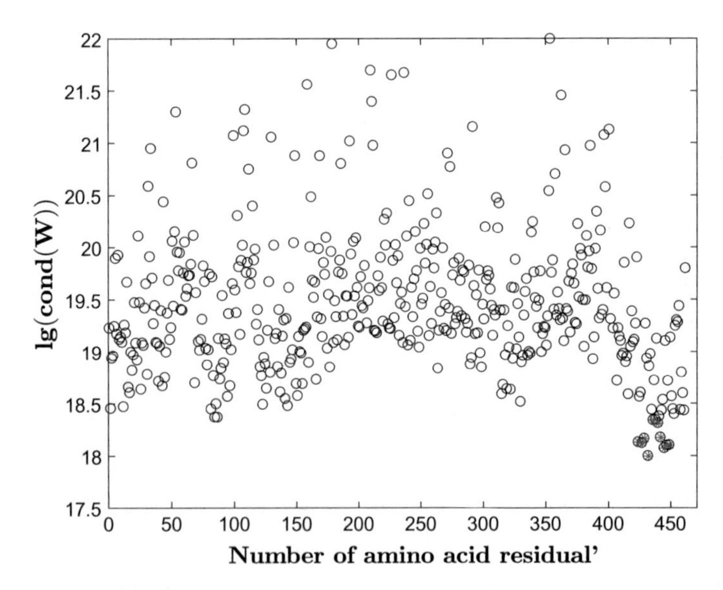

Fig. 5.15 Results of numerical simulation of the interaction of Mdm2 with Mdm2, frame size equal to 30 a.a., $\varepsilon = 80$. The red dots denote the 10 minimum values of lg(cond(W))

As can be seen from Tables 5.1 and 5.8, the greatest number of minimum values lies at the C-terminus of the Mdm2 protein. It also follows from the graphs Figs. 5.12, 5.13, 5.14, 5.15, 5.16, 5.17, 5.18 and 5.19 that, as a result of numerical calculation of the interaction of two polypeptide chains of the Mdm2 protein, according to Algorithm 2, a cluster of the smallest values in the C-terminus regions of two Mdm2

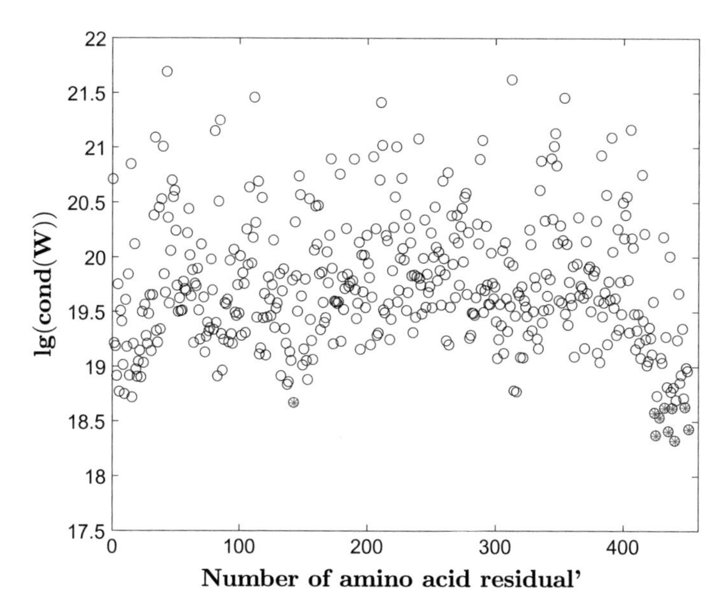

Fig. 5.16 Results of numerical simulation of the interaction of Mdm2 with Mdm2, frame size equal to 40, $\varepsilon = 1$. The red dots denote the 10 minimum values of lg(cond(W))

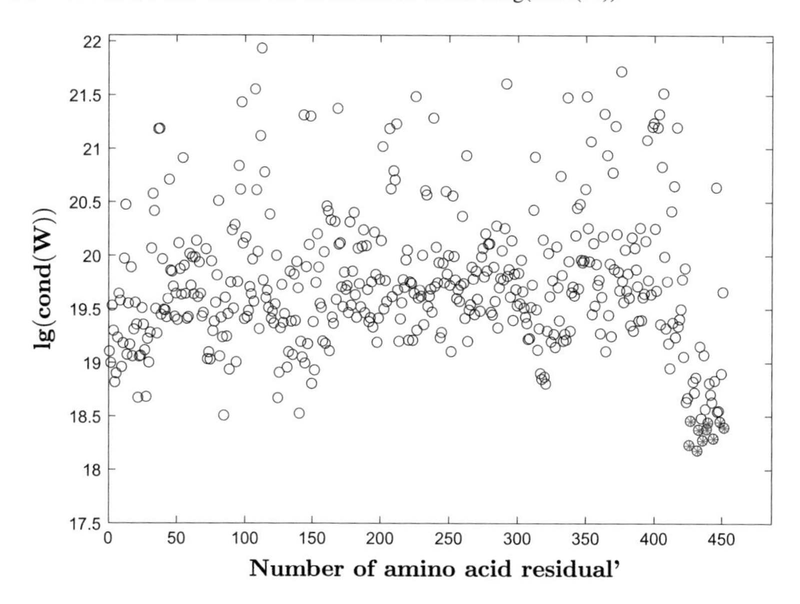

Fig. 5.17 Results of numerical simulation of the interaction of Mdm2 with Mdm2, frame size equal to 40 a.a., $\varepsilon = 80$. The red dots denote the 10 minimum values of lg(cond(W))

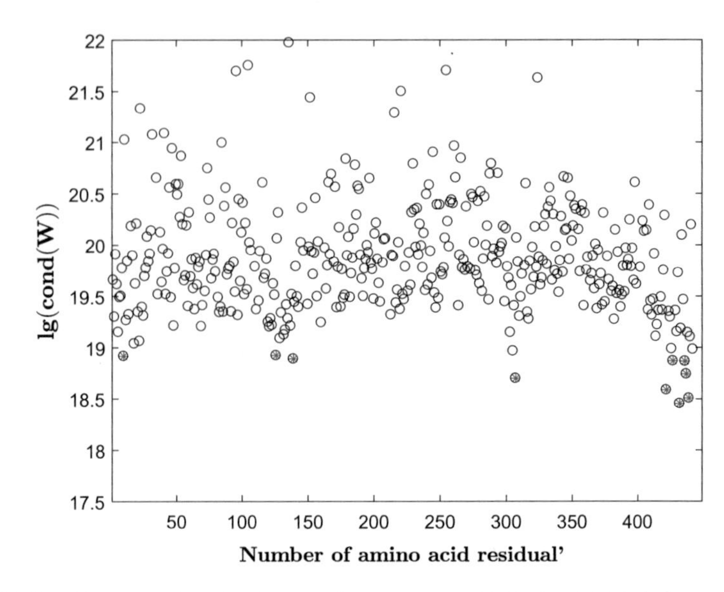

Fig. 5.18 Results of numerical simulation of the interaction of Mdm2 with Mdm2, frame size equal to 50, $\varepsilon = 1$. The red dots denote the 10 minimum values of lg(cond(W))

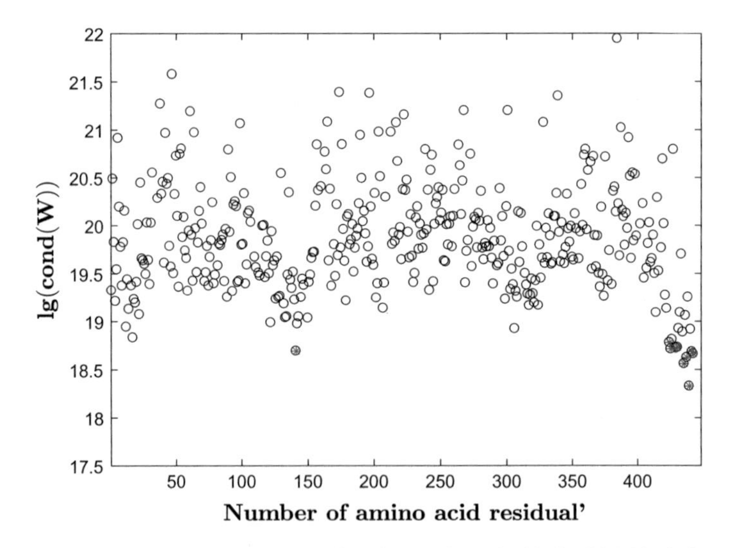

Fig. 5.19 Results of numerical simulation of the interaction of Mdm2 with Mdm2, frame size = 50 for $\varepsilon = 80$. The red dots denote the 10 minimum values of lg(cond(W))

Table 5.5 The ten minimum values of lg(cond(W)) and the amino acid sequences of the detected regions of the Mdm2 protein at a frameshift length equal to 20 a.a., $\varepsilon = 1$

$N^{\underline{0}}$	Amino acid sequence Mdm2	lg(cond(W))
446	PKNGCIVHGKTGHLMACFTC	17.5224
444	GRPKNGCIVHGKTGHLMACF	17.585
448	NGCIVHGKTGHLMACFTCAK	17.628
438	PCVICQGRPKNGCIVHGKTG	17.669
451	IVHGKTGHLMACFTCAKKLK	17.677
452	VHGKTGHLMACFTCAKKLKK	17.688
450	CIVHGKTGHLMACFTCAKKL	17.695
445	RPKNGCIVHGKTGHLMACFT	17.801
442	CQGRPKNGCIVHGKTGHLMA	17.817
168	ETEENSDELSGERQRKRHKS	17.878

lg(cond(W)) is common logarithm of condition number
$N^{\underline{0}}$ is number of amino acid residual

Table 5.6 The ten minimum values of lg(cond(W)) and the amino acid sequences of the detected regions of the Mdm2 protein at a frameshift length equal to 20 a.a., $\varepsilon = 80$

$N^{\underline{0}}$	Amino acid sequence Mdm2	lg(cond(W))
460	MACFTCAKKLKKRNKPCPVC	17.661
452	VHGKTGHLMACFTCAKKLKK	17.738
441	ICQGRPKNGCIVHGKTGHLM	17.758
442	CQGRPKNGCIVHGKTGHLMA	17.760
472	RNKPCPVCRQPIQMIVLTYF	17.781
27	TLVRPKPLLLKLLKSVGAQK	17.814
451	IVHGKTGHLMACFTCAKKLK	17.828
307	TSCNEMNPPLPSHCNRCWAL	17.859
446	PKNGCIVHGKTGHLMACFTC	17.873
435	AIEPCVICQGRPKNGCIVHG	17.885

lg(cond(W)) is common logarithm of condition number
$N^{\underline{0}}$ is number of amino acid residual

proteins is formed at $\varepsilon = 1$ and $\varepsilon = 80$. This result is in good agreement with earlier experimental work, which indicates the oligomerization of the RING domain of the Mdm2 protein and the important role of the C-terminus of Mdm2 in the formation of the Mdm2–Mdm2 homodimer [29–31].

In this section, we gave 10 minimum values of lg(cond(W)) characterizing the interactions of Mdm2 with Mdm2, but we did not exclude the presence of other regions of interaction of Mdm2 with Mdm2 that did not fall in the above list of 10 minimum values of lg(cond(W)) for $\varepsilon = 1$ and $\varepsilon = 80$. The result of the interaction of two Mdm2 proteins of different frameshift lengths from 20 a.a. up to 50 a.a. identifies in all cases the site of the most stable formed complex in the region of the C-terminus of the Mdm2 protein.

Table 5.7 The ten minimum values of lg(cond(W)) and the amino acid sequences of the detected regions of the Mdm2 protein at a frameshift length equal to 30 a.a., $\varepsilon = 1$

$N^{\underline{0}}$	Amino acid sequence Mdm2	lg(cond(W))
435	AIEPCVICQGRPKNGCIVHGKTGHLMACFT	18.020
425	ESVESSLPLNAIEPCVICQGRPKNGCIVHG	18.075
441	ICQGRPKNGCIVHGKTGHLMACFTCAKKLK	18.094
434	NAIEPCVICQGRPKNGCIVHGKTGHLMAC	18.132
426	SVESSLPLNAIEPCVICQGRPKNGCIVHGK	18.154
448	NGCIVHGKTGHLMACFTCAKKLKKRNKPCP	18.206
450	CIVHGKTGHLMACFTCAKKLKKRNKPCPVC	18.220
436	IEPCVICQGRPKNGCIVHGKTGHLMACFTC	18.238
451	IVHGKTGHLMACFTCAKKLKKRNKPCPVCR	18.281
431	LPLNAIEPCVICQGRPKNGCIVHGKTGHLM	18.291

lg(cond(W)) is common logarithm of condition number
$N^{\underline{0}}$ is number of amino acid residual

Table 5.8 The ten minimum values of lg(cond(W)) and the amino acid sequences of the detected regions of the Mdm2 protein at a frameshift length equal to 30 a.a., $\varepsilon = 80$

$N^{\underline{0}}$	Amino acid sequence Mdm2	lg(cond(W))
435	LNAIEPCVICQGRPKNGCIVHGKTGHLMAC	17.994
446	PKNGCIVHGKTGHLMACFTCAKKLKKRNKP	18.069
448	NGCIVHGKTGHLMACFTCAKKLKKRNKPCP	18.094
450	CIVHGKTGHLMACFTCAKKLKKRNKPCPVC	18.098
428	ESSLPLNAIEPCVICQGRPKNGCIVHGKTG	18.117
425	ESVESSLPLNAIEPCVICQGRPKNGCIVHG	18.126
430	SLPLNAIEPCVICQGRPKNGCIVHGKTGHL	18.160
443	QGRPKNGCIVHGKTGHLMACFTCAKKLKKR	18.171
441	ICQGRPKNGCIVHGKTGHLMACFTCAKKLK	18.308
437	EPCVICQGRPKNGCIVHGKTGHLMACFTCA	18.336

lg(cond(W)) is common logarithm of condition number
$N^{\underline{0}}$ is number of amino acid residual

It should also be taken into account that the Algorithm 2 developed by us allows one to analyze the interactions of amino acid residues between two proteins with only approximately symmetrical sequence numbers. We note that if we analyze two identical proteins, we will obtain an analysis of the interaction between identical amino acid sequences. In order to obtain data containing information on the interaction between different regions of the polypeptide chains of proteins, it is better to use Algorithm 1, to extract each protein region of interest from one protein and to determine the stability of each complex formed by it with different regions of the second protein.

To obtain more reliable and qualitative data, it is recommended to take segments of polypeptide sequences of the same size. If we analyze the interactions of sections

Table 5.9 The ten minimum values of lg(cond(W)) and the amino acid sequences of the detected regions of the Mdm2 protein at a frameshift length equal to 40 a.a., $\varepsilon = 1$

$N^{\underline{0}}$	Amino acid sequence Mdm2	lg(cond(W))
442	CQGRPKNGCIVHGKTGHLMACFTCAKKLKKRNKPCPVCRQ	18.321
427	VESSLPLNAIEPCVICQGRPKNGCIVHGKTGHLMACFTCA	18.369
437	EPCVICQGRPKNGCIVHGKTGHLMACFTCAKKLKKRNKPC	18.407
453	HGKTGHLMACFTCAKKLKKRNKPCPVCRQPIQMIVLTYFP	18.425
430	SLPLNAIEPCVICQGRPKNGCIVHGKTGHLMACFTCAKKL	18.531
426	SVESSLPLNAIEPCVICQGRPKNGCIVHGKTGHLMACFTC	18.575
440	VICQGRPKNGCIVHGKTGHLMACFTCAKKLKKRNKPCPVC	18.617
434	NAIEPCVICQGRPKNGCIVHGKTGHLMACFTCAKKLKKRN	18.621
450	CIVHGKTGHLMACFTCAKKLKKRNKPCPVCRQPIQMIVLT	18.626
144	QEEKPSSSHLVSRPSTSSRRRAISETEENSDELSGERQRK	18.664

lg(cond(W)) is common logarithm of condition number
$N^{\underline{0}}$ is number of amino acid residual

Table 5.10 The ten minimum values of lg(cond(W)) and the amino acid sequences of the detected regions of the Mdm2 protein at a frameshift length equal to 40 a.a., $\varepsilon = 80$

$N^{\underline{0}}$	Amino acid sequence Mdm2	lg(cond(W))
433	LNAIEPCVICQGRPKNGCIVHGKTGHLMACFTCAKKLKKR	18.176
427	VESSLPLNAIEPCVICQGRPKNGCIVHGKTGHLMACFTCA	18.226
437	EPCVICQGRPKNGCIVHGKTGHLMACFTCAKKLKKRNKPC	18.270
445	RPKNGCIVHGKTGHLMACFTCAKKLKKRNKPCPVCRQPIQ	18.288
434	NAIEPCVICQGRPKNGCIVHGKTGHLMACFTCAKKLKKRN	18.370
440	VICQGRPKNGCIVHGKTGHLMACFTCAKKLKKRNKPCPVC	18.375
453	HGKTGHLMACFTCAKKLKKRNKPCPVCRQPIQMIVLTYFP	18.390
440	ICQGRPKNGCIVHGKTGHLMACFTCAKKLKKRNKPCPVCR	18.436
450	CIVHGKTGHLMACFTCAKKLKKRNKPCPVCRQPIQMIVLT	18.444
428	ESSLPLNAIEPCVICQGRPKNGCIVHGKTGHLMACFTCAK	18.453

lg(cond(W)) is common logarithm of condition number
$N^{\underline{0}}$ is number of amino acid residual

of different lengths, an error can be introduced in the data, since we compare the values of lg(cond(W)) obtained for matrices of different dimensions.

5.7.2 Numerical Calculation of the Interaction of Polypeptide Sequences of the Protein Nap1

In this section, we consider the results of a numerical calculation of the interaction of two identical polypeptide chains of the protein Nap1 from the N-terminus to the C-terminus according to Algorithm 2 along two polypeptide sequences of the protein

Table 5.11 The ten minimum values of lg(cond(W)) and the amino acid sequences of the detected regions of the Mdm2 protein at a frameshift length equal to 50 a.a., $\varepsilon = 1$

$N^{\underline{0}}$	Amino acid sequence Mdm2	lg(cond(W))
433	LNAIEPCVICQGRPKNGCIVHGKTGHLMACFTCAKKLKKRNKPCPVCRQP	18.453
440	VICQGRPKNGCIVHGKTGHLMACFTCAKKLKKRNKPCPVCRQPIQMIVLT	18.507
423	KEESVESSLPLNAIEPCVICQGRPKNGCIVHGKTGHLMACFTCAKKLKKR	18.587
308	SCNEMNPPLPSHCNRCWALRENWLPEDKGKDKGEISEKAKLENSTQAEEG	18.701
438	PCVICQGRPKNGCIVHGKTGHLMACFTCAKKLKKRNKPCPVCRQPIQMIV	18.742
437	EPCVICQGRPKNGCIVHGKTGHLMACFTCAKKLKKRNKPCPVCRQPIQMI	18.867
428	ESSLPLNAIEPCVICQGRPKNGCIVHGKTGHLMACFTCAKKLKKRNKPCP	18.869
140	VQELQEEKPSSSHLVSRPSTSSRRRAISETEENSDELSGERQRKRHKSDS	18.891
11	TDGAVTTSQIPASEQETLVRPKPLLLKLLKSVGAQKDTYTMKEVLFYLGQ	18.918
127	RCHLEGGSDQKDLVQELQEEKPSSSHLVSRPSTSSRRRAISETEENSDEL	18.924

lg(cond(W)) is common logarithm of condition number
$N^{\underline{0}}$ is number of amino acid residual

Table 5.12 The ten minimum values of lg(cond(W)) and the amino acid sequences of the detected regions of the Mdm2 protein at a frameshift length equal to 50 a.a., $\varepsilon = 80$

$N^{\underline{0}}$	Amino acid sequence Mdm2	lg(cond(W))
440	VICQGRPKNGCIVHGKTGHLMACFTCAKKLKKRNKPCPVCRQPIQMIVLT	18.326
436	IEPCVICQGRPKNGCIVHGKTGHLMACFTCAKKLKKRNKPCPVCRQPIQM	18.560
438	PCVICQGRPKNGCIVHGKTGHLMACFTCAKKLKKRNKPCPVCRQPIQMIV	18.624
443	QGRPKNGCIVHGKTGHLMACFTCAKKLKKRNKPCPVCRQPIQMIVLTYFP	18.662
442	CQGRPKNGCIVHGKTGHLMACFTCAKKLKKRNKPCPVCRQPIQMIVLTYF	18.684
142	ELQEEKPSSSHLVSRPSTSSRRRAISETEENSDELSGERQRKRHKSDSIS	18.693
426	SVESSLPLNAIEPCVICQGRPKNGCIVHGKTGHLMACFTCAKKLKKRNKP	18.712
430	SLPLNAIEPCVICQGRPKNGCIVHGKTGHLMACFTCAKKLKKRNKPCPVC	18.720
431	LPLNAIEPCVICQGRPKNGCIVHGKTGHLMACFTCAKKLKKRNKPCPVCR	18.730
425	ESVESSLPLNAIEPCVICQGRPKNGCIVHGKTGHLMACFTCAKKLKKRNK	18.779

lg(cond(W)) is common logarithm of condition number
$N^{\underline{0}}$ is number of amino acid residual

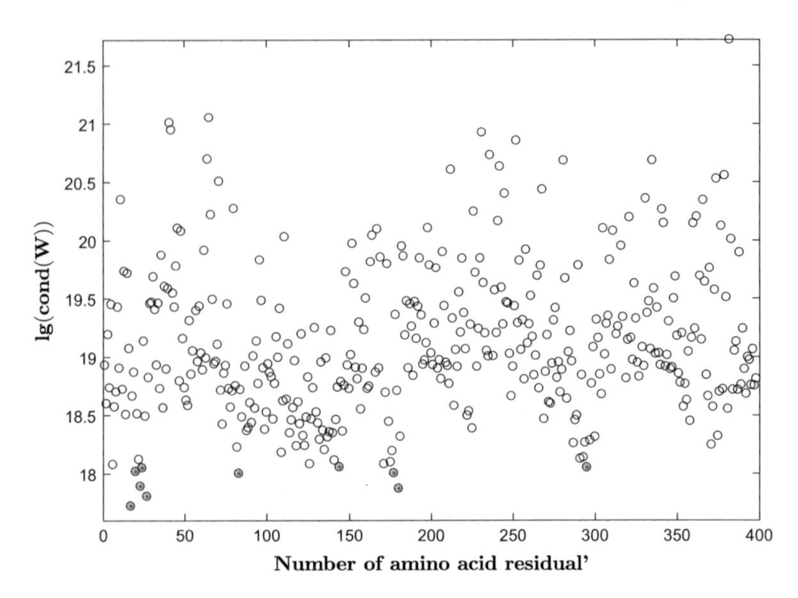

Fig. 5.20 Results of numerical simulation of the interaction of Nap1 with Nap1, frame size equal to 20 a.a., $\varepsilon = 1$. The red dots denote the 10 minimum values of lg(cond(W))

Nap1. We give a step-by-step detailed description of the numerical computations obtained with brief conclusions for each calculation in order to give a clear idea for further modelling of the interaction of other proteins. Note that the formation of the Nap1 homodimer Nap1–Nap1 is due to two identical binding sites in domain 1 of each protein from 74 a.a. to 180 a.a. in opposite directions [25]. The main dimerization site of two proteins belongs to the $\alpha 2$-helix from 90 a.a. to 140 a.a. of each histone chaperone Nap1 protein.

We now turn to the analysis of numerical calculations of the interaction of two Nap1 and Nap1 proteins at a frameshift length of 20 a.a. The results of the numerical interaction are shown in Fig. 5.20 for $\varepsilon = 1$ and in Fig. 5.21 for $\varepsilon = 80$.

As can be seen from the graphs Figs. 5.20 and 5.21, with a frameshift equal to 20 a.a. there are many areas of interaction between proteins Nap1 which are characterized by a small value of lg(cond(W)). In this case, the ten minimum values of lg(cond(W)) obtained by the interaction of Nap1–Nap1 with a frameshift equal to 20 a.a. are given in Table 5.13 for $\varepsilon = 1$ and in Table 5.14 for $\varepsilon = 80$.

As can be seen from the tables above, we were unable to obtain a qualitative interpretation of the numerical calculations of the interaction of the two Nap1 proteins with the formation of the dimer Nap1–Nap1, since according to the available data, the formation of this dimer is carried out with the participation of the amino acid sequence from 74 a.a. to 180 a.a. [25]. Thus, a 20 a.a. frameshift is not suitable for a qualitative calculation of the interaction between the two proteins Nap1.

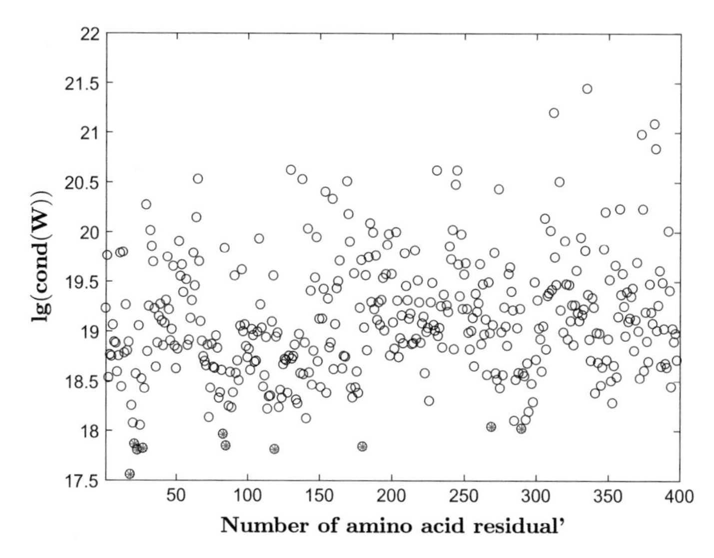

Fig. 5.21 Results of numerical simulation of the interaction of Nap1 with Nap1, frame size equal to 20 a.a., $\varepsilon = 80$. The red dots denote the 10 minimum values of lg(cond(W))

Table 5.13 The ten minimum values of lg(cond(W)) and the amino acid sequences of the detected regions of the Nap1 protein at a frameshift length equal to 20 a.a., $\varepsilon = 1$

$N^{\underline{0}}$	Amino acid sequence Nap1	lg(cond(W))
18	NAPTPHNTPASVLNPSYLKN	17.723
28	SVLNPSYLKNGNPVRAQAQE	17.806
181	EQVKGIPSFWLTALENLPIV	17.872
24	NTPASVLNPSYLKNGNPVRA	17.892
84	GYVGGLPKNVKEKLLSLKTL	18.001
178	SEEEQVKGIPSFWLTALENL	18.003
21	TPHNTPASVLNPSYLKNGNP	18.019
296	KTTKQVRTIEKITPIESFFN	18.049
25	TPASVLNPSYLKNGNPVRAQ	18.049
145	QPKPEQIAKGQEIVESLNE	18.055

lg(cond(W)) is common logarithm of condition number
$N^{\underline{0}}$ is number of amino acid residual

Let us turn to the numerical calculation results of the interaction of two proteins Nap1 at a frameshift length equal to 30 a.a. The results of the numerical interaction are shown in Fig. 5.22 for $\varepsilon = 1$ and in Fig. 5.23 for $\varepsilon = 80$.

As can be seen from the presented graphs, with a frameshift equal to 30 a.a. there are many areas of interaction between proteins Nap1 that are characterized by a small value of lg(cond(w)). In this case, the ten minimum values of lg(cond(W)) obtained by the interaction of two whole Nap1 proteins with a frameshift equal to 30 a.a. are given in Table 5.15 for $\varepsilon = 1$ and in Table 5.16 for $\varepsilon = 80$.

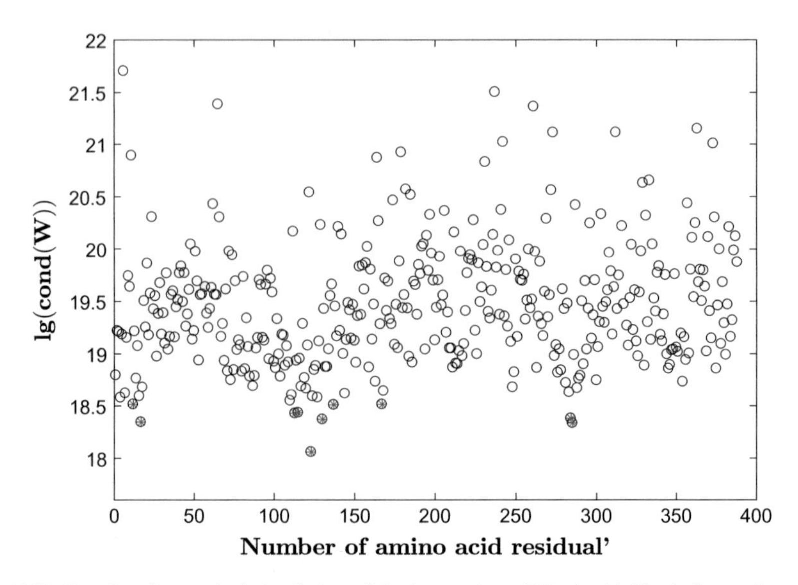

Fig. 5.22 Results of numerical simulation of the interaction of Nap1 with Nap1, frame size equal to 30 a.a., $\varepsilon = 1$. The red dots denote the 10 minimum values of lg(cond(W))

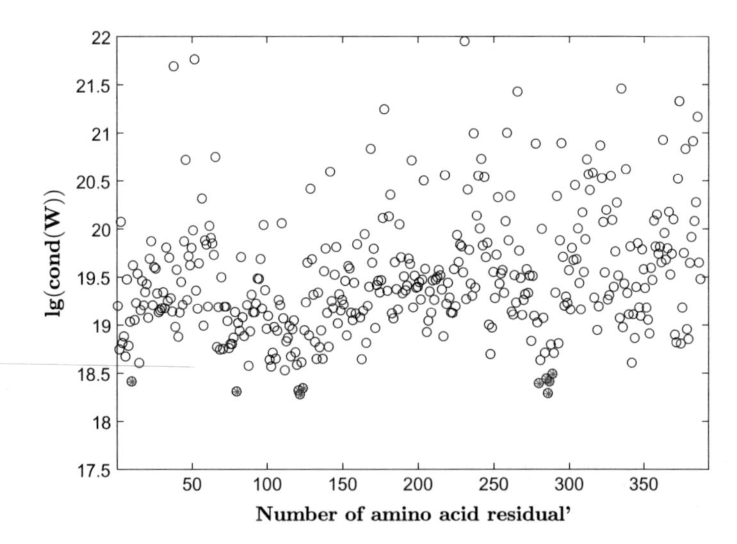

Fig. 5.23 Results of numerical simulation of the interaction of Nap1 with Nap1, frame size equal to 30 a.a., $\varepsilon = 80$. The red dots denote the 10 minimum values of lg(cond(W))

As can be seen from the above results, a larger number of values of lg(cond (W)) fell into the region of domain 1, and an additional cluster was formed from the minimum values in the region 281 a.a. at $\varepsilon = 80$. Despite the fact that half of the values obtained fall within the region of domain 1, the rest of the scattered minimum values of lg(cond(W)) fall on different parts of the polypeptide sequence

Table 5.14 The ten minimum values of lg(cond(W)) and the amino acid sequences of the detected regions of the Nap1 protein at a frameshift length equal to 20 a.a., $\varepsilon = 80$

$N^{\underline{0}}$	Amino acid sequence Nap1	lg(cond(W))
19	APTPHNTPASVLNPSYLKNG	17.557
24	NTPASVLNPSYLKNGNPVRA	17.801
120	ELENKFLQKYKPIWEQRSRI	17.810
28	SVLNPSYLKNGNPVRAQAQE	17.819
181	EQVKGIPSFWLTALENLPIV	17.837
86	VGGLPKNVKEKLLSLKTLQS	17.845
22	PHNTPASVLNPSYLKNGNPV	17.862
84	GYVGGLPKNVKEKLLSLKTL	17.962
291	RKQRNKTTKQVRTIEKITPI	18.020
27	AEGCEISWKDNAHNVTVDLE	18.039

lg(cond(W)) is common logarithm of condition number
$N^{\underline{0}}$ is number of amino acid residual

Table 5.15 The ten minimum values of lg(cond(W)) and the amino acid sequences of the detected regions of the Nap1 protein at a frameshift length equal to 30 a.a., $\varepsilon = 1$

$N^{\underline{0}}$	Amino acid sequence Nap1	lg(cond(W))
124	KFLQKYKPIWEQRSRIISGQEQPKPEQIAK	18.058
286	VDLEMRKQRNKTTKQVRTIEKITPIESFFN	18.334
18	NAPTPHNTPASVLNPSYLKNGNPVRAQAQE	18.345
131	PIWEQRSRIISGQEQPKPEQIAKGQEIVES	18.370
285	TVDLEMRKQRNKTTKQVRTIEKITPIESFF	18.378
114	FQVEMFELENKFLQKYKPIWEQRSRIISGQ	18.428
116	VEMFELENKFLQKYKPIWEQRSRIISGQEQ	18.434
138	RIISGQEQPKPEQIAKGQEIVESLNETELL	18.509
168	VDEEEKAQNDSEEEQVKGIPSFWLTALENL	18.513
13	SMQIDNAPTPHNTPASVLNPSYLKNGNPVR	18.513

lg(cond(W)) is common logarithm of condition number
$N^{\underline{0}}$ is number of amino acid residual

Nap1, which significantly complicates the interpretation of the obtained numerical data, namely, the search for protein sites responsible for dimerization. Numerical calculations obtained at $\varepsilon = 1$ are characterized by the formation of a cluster of values in the region of domain 1, with the smallest value of lg(cond(W)) obtained at 124 a.a. and amounting to 18.058. Analysis of the data obtained with a frameshift of 30 a.a. shows a slightly better result than with a 20 a.a. frameshift.

Let us now turn to the numerical results of the calculation of the interaction of two proteins Nap1 at a frameshift length equal to 40 a.a. The results of the numerical interaction are shown in Fig. 5.24 for $\varepsilon = 1$ and in Fig. 5.25 for $\varepsilon = 80$. As can be seen from the graphs, with a frameshift equal to 40 a.a, two clusters are formed from the minimum values at $\varepsilon = 1$ in the 8 a.a. and 111 a.a. regions, ten minimum are

Table 5.16 The ten minimum values of lg(cond(W)) and the amino acid sequences of the detected regions of the Nap1 protein at a frameshift length equal to 30 a.a., $\varepsilon = 80$

$N^{\underline{0}}$	Amino acid sequence Nap1	lg(cond(W))
123	NKFLQKYKPIWEQRSRIISGQEQPKPEQIA	18.275
287	DLEMRKQRNKTTKQVRTIEKITPIESFFNF	18.287
81	QDSGYVGGLPKNVKEKLLSLKTLQSELFEV	18.306
122	ENKFLQKYKPIWEQRSRIISGQEQPKPEQI	18.315
125	FLQKYKPIWEQRSRIISGQEQPKPEQIAKG	18.340
281	AHNVTVDLEMRKQRNKTTKQVRTIEKITPI	18.390
288	LEMRKQRNKTTKQVRTIEKITPIESFFNFF	18.409
11	KSSMQIDNAPTPHNTPASVLNPSYLKNGNP	18.410
286	VDLEMRKQRNKTTKQVRTIEKITPIESFFN	18.438
290	MRKQRNKTTKQVRTIEKITPIESFFNFFDP	18.487

lg(cond(W)) is common logarithm of condition number
$N^{\underline{0}}$ is number of amino acid residual

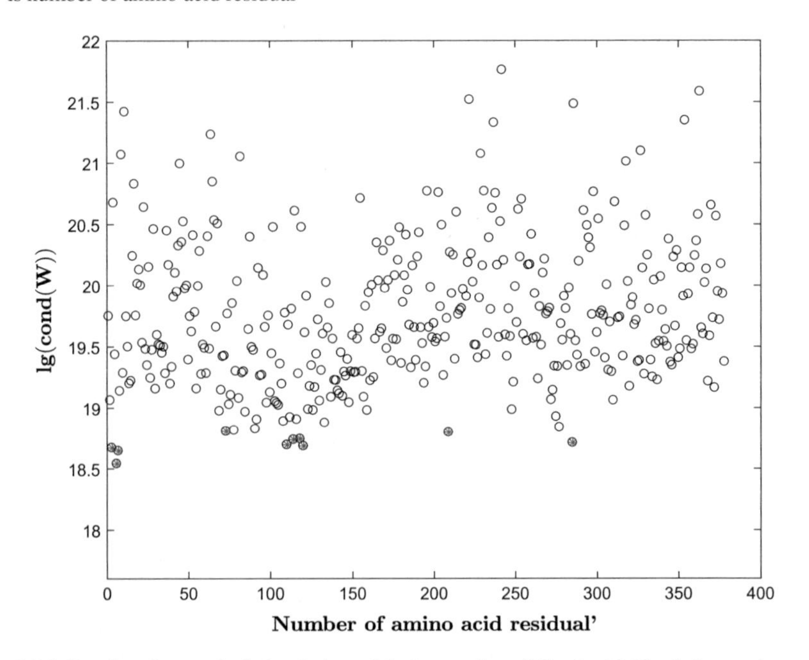

Fig. 5.24 Results of numerical simulation of the interaction of Nap1 with Nap1, frame size equal to 40 a.a., $\varepsilon = 1$. The red dots denote the 10 minimum values of lg(cond(W))

presented in Tables 5.17 and 5.18. The second cluster falls on the the domain 1 region. Analysis of the graph of the numerical results obtained at $\varepsilon = 80$ demonstrates the presence of 6 minimal values in the domain 1 region, whose minimum is obtained 116 a.a. and amounts to 18.520, as well as 3 closely lying values of the magnitude of lg(cond(W)) in the region of the flexible N-terminus of the protein Nap1.

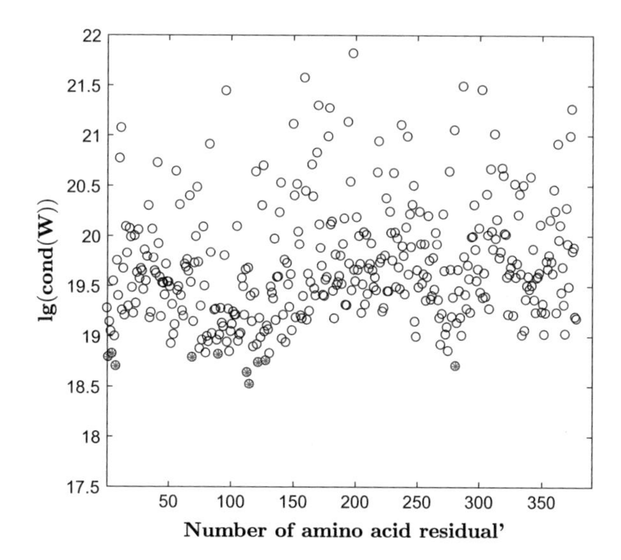

Fig. 5.25 Results of numerical simulation of the interaction of Nap1 with Nap1, frame size equal to 40 a.a., $\varepsilon = 80$. The red dots denote the 10 minimum values of lg(cond(W))

Table 5.17 The ten minimum values of lg(cond(W)) and the amino acid sequences of the detected regions of the Nap1 protein at a frameshift length equal to 40 a.a., $\varepsilon = 1$

$N^{\underline{0}}$	Amino acid sequence Nap1	lg(cond(W))
7	RTKPKSSMQIDNAPTPHNTPASVLNPSYLKNGNPVRAQAQ	18.540
8	TKPKSSMQIDNAPTPHNTPASVLNPSYLKNGNPVRAQAQE	18.645
4	DPIRTKPKSSMQIDNAPTPHNTPASVLNPSYLKNGNPVRA	18.670
121	LENKFLQKYKPIWEQRSRIISGQEQPKPEQIAKGQEIVES	18.684
111	EKEFQVEMFELENKFLQKYKPIWEQRSRIISGQEQPKPEQ	18.693
286	VDLEMRKQRNKTTKQVRTIEKITPIESFFNFFDPPKIQNE	18.710
115	QVEMFELENKFLQKYKPIWEQRSRIISGQEQPKPEQIAKG	18.737
119	FELENKFLQKYKPIWEQRSRIISGQEQPKPEQIAKGQEIV	18.743
210	EVLEYLQDIGLEYLTDGRPGFKLLFRFDSSANPFFTNDI	18.797
74	RLGSLVGQDSGYVGGLPKNVKEKLLSLKTLQSELFEVEKE	18.805

lg(cond(W)) is common logarithm of condition number
$N^{\underline{0}}$ is number of amino acid residual

The results of the numerical calculations presented in Fig. 5.25 and Table 5.18 demonstrate the large number of minimum values of lg(cond(W)) attributable to the domain 1 of the Nap1 protein responsible for the formation of the Nap1–Nap1 dimer at $\varepsilon = 80$.

Let us now turn to the numerical results of the calculation of the interaction of two proteins Nap1 at a frameshift length equal to 50 a.a. The results of the numerical interaction are shown in Fig. 5.26 for $\varepsilon = 1$ and in Fig. 5.27 for $\varepsilon = 80$.

As can be seen from the graphs, with a frameshift equal to 50 a.a. the presence of a cluster of minimum values is observed for $\varepsilon = 1$ in the 75 a.a. and 137 a.a. regions, the minimum of which is obtained at 75 a.a. and amounts to 18.853. Analysis of

Table 5.18 The ten minimum values of lg(cond(W)) and the amino acid sequences of the detected regions of the Nap1 protein at a frameshift length equal to 40 a.a., $\varepsilon = 80$

$N^{\underline{0}}$	Amino acid sequence Nap1	lg(cond(W))
116	VEMFELENKFLQKYKPIWEQRSRIISGQEQPKPEQIAKGQ	18.520
114	FQVEMFELENKFLQKYKPIWEQRSRIISGQEQPKPEQIAK	18.637
9	KPKSSMQIDNAPTPHNTPASVLNPSYLKNGNPVRAQAQEQ	18.700
282	HNVTVDLEMRKQRNKTTKQVRTIEKITPIESFFNFFDPPK	18.700
123	NKFLQKYKPIWEQRSRIISGQEQPKPEQIAKGQEIVESLN	18.8737
129	YKPIWEQRSRIISGQEQPKPEQIAKGQEIVESLNETELLV	18.755
70	SIQDRLGSLVGQDSGYVGGLPKNVKEKLLSLKTLQSELFE	18.787
3	SDPIRTKPKSSMQIDNAPTPHNTPASVLNPSYLKNGNPVR	18.789
91	KNVKEKLLSLKTLQSELFEVEKEFQVEMFELENKFLQKYK	18.818
6	IRTKPKSSMQIDNAPTPHNTPASVLNPSYLKNGNPVRAQA	18.824

lg(cond(W)) is common logarithm of condition number
$N^{\underline{0}}$ is number of amino acid residual

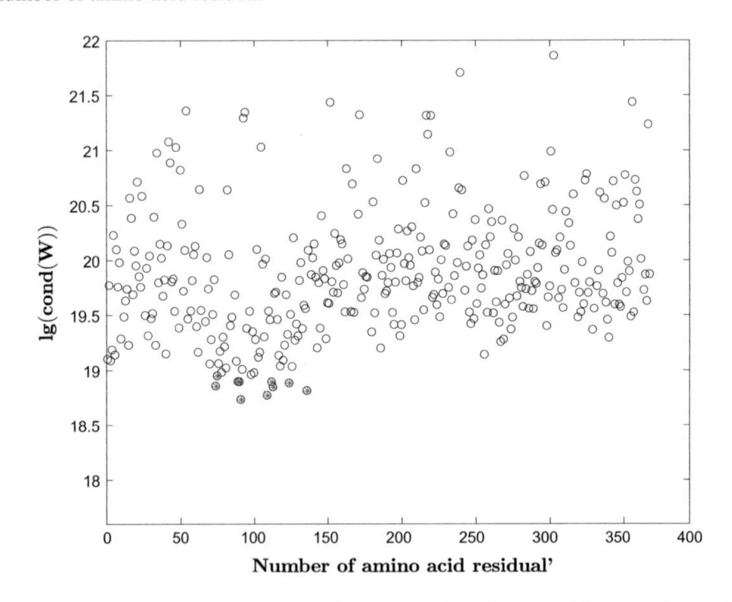

Fig. 5.26 Results of numerical simulation of the interaction of Nap1 with Nap1, frame size equal to 50 a.a., $\varepsilon = 1$. The red dots denote the 10 minimum values of lg(cond(W))

the graph of the numerical results obtained at $\varepsilon = 80$ demonstrates the formation of a cluster from the minimum values in the region from 80 a.a. to 146 a.a., whose minimum is obtained at 116 a.a. and amounts to 18.807, as well as 3 closely lying values of lg(cond(W)) in the region of the flexible N-terminus of the protein Nap1.

The results presented in Figs. 5.26 and 5.27, and in Tables 5.19 and 5.20 show good qualitative agreement with the previous experiment in which domain 1 was

Table 5.19 The ten minimum values of lg(cond(W)) and the amino acid sequences of the detected regions of the Nap1 protein at a frameshift length equal to 50 a.a., $\varepsilon = 1$

$N^{\underline{0}}$	Amino acid sequence Nap1	lg(cond(W))
92	NVKEKLLSLKTLQSELFEVEKEFQVEMFELENKFLQKYKPIWEQRSRIIS	18.732
110	VEKEFQVEMFELENKFLQKYKPIWEQRSRIISGQEQPKPEQIAKGQEIVE	18.769
137	SRIISGQEQPKPEQIAKGQEIVESLNETELLVDEEEKAQNDSEEEQVKGI	18.810
114	FQVEMFELENKFLQKYKPIWEQRSRIISGQEQPKPEQIAKGQEIVESLNE	18.845
75	LGSLVGQDSGYVGGLPKNVKEKLLSLKTLQSELFEVEKEFQVEMFELENK	18.853
125	FLQKYKPIWEQRSRIISGQEQPKPEQIAKGQEIVESLNETELLVDEEEKA	18.881
113	EFQVEMFELENKFLQKYKPIWEQRSRIISGQEQPKPEQIAKGQEIVESLN	18.892
91	KNVKEKLLSLKTLQSELFEVEKEFQVEMFELENKFLQKYKPIWEQRSRII	18.893
90	PKNVKEKLLSLKTLQSELFEVEKEFQVEMFELENKFLQKYKPIWEQRSRI	18.894
76	GSLVGQDSGYVGGLPKNVKEKLLSLKTLQSELFEVEKEFQVEMFELENKF	18.947

lg(cond(W)) is common logarithm of condition number
$N^{\underline{0}}$ is number of amino acid residual

Table 5.20 The ten minimum values of lg(cond(W)) and the amino acid sequences of the detected regions of the Nap1 protein at a frameshift length equal to 50 a.a., $\varepsilon = 80$

$N^{\underline{0}}$	Amino acid sequence Nap1	lg(cond(W))
116	VEMFELENKFLQKYKPIWEQRSRIISGQEQPKPEQIAKGQEIVESLNETE	18.807
105	SELFEVEKEFQVEMFELENKFLQKYKPIWEQRSRIISGQEQPKPEQIAKG	18.863
93	VKEKLLSLKTLQSELFEVEKEFQVEMFELENKFLQKYKPIWEQRSRIISG	18.902
89	LPKNVKEKLLSLKTLQSELFEVEKEFQVEMFELENKFLQKYKPIWEQRSR	18.922
120	ELENKFLQKYKPIWEQRSRIISGQEQPKPEQIAKGQEIVESLNETELLVD	18.977
80	GQDSGYVGGLPKNVKEKLLSLKTLQSELFEVEKEFQVEMFELENKFLQKY	18.989
354	IPRAVDWFTGAALEFEFEEDEEEADEDEDEEEDDDHGLEDDDGESAEEQD	18.999
146	PKPEQIAKGQEIVESLNETELLVDEEEKAQNDSEEEQVKGIPSFWLTALE	19.000
113	EFQVEMFELENKFLQKYKPIWEQRSRIISGQEQPKPEQIAKGQEIVESLN	19.011
270	AEGCEISWKDNAHNVTVDLEMRKQRNKTTKQVRTIEKITPIESFFNFFDP	19.013

lg(cond(W)) is common logarithm of condition number
$N^{\underline{0}}$ is number of amino acid residual

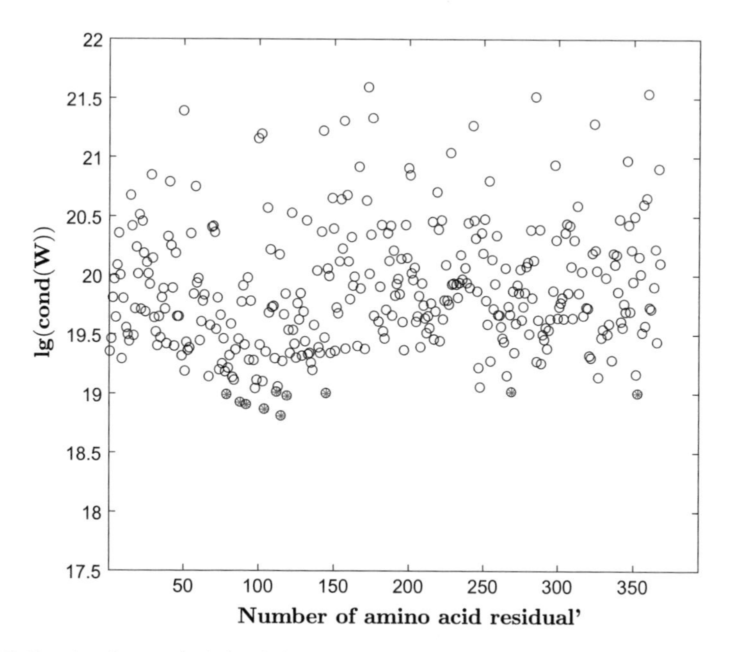

Fig. 5.27 Results of numerical simulation of the interaction of Nap1 with Nap1, frame size equal to 50 a.a., $\varepsilon = 80$. The red dots denote the 10 minimum values of lg(cond(W))

detected, which in turns plays a decisive role in the dimerization of the two proteins Nap1–Nap1 [25].

Let us now turn to the numerical results of the calculation of the interaction of two proteins Nap1 with a frameshift equal to 60 a.a. The results of the numerical interaction are shown in are shown in Fig. 5.28 for $\varepsilon = 1$ and in Fig. 5.29 for $\varepsilon = 80$. As can be seen from the Figs. 5.28 and 5.29, with a frameshift equal to 60 a.a. the presence of a cluster of minimum values is observed for $\varepsilon = 1$ in the region of 82 a.a. and 137 a.a., the minimum of which is obtained at 94 a.a. and amounts to 18.906. Analysis of the graph of numerical results obtained at $\varepsilon = 80$ demonstrates the formation of a cluster from the minimum values in the region from 68 a.a. to 133 a.a., whose minimum is obtained at 85 a.a. and amounts to 18.855.

The results presented in Figs. 5.28 and 5.29 and in Tables 5.21, and 5.22 demonstrate good qualitative agreement with a previous experiment in which domain 1 has been identified, which in turn plays a decisive role in the dimerization of two Nap1–Nap1 proteins [25].

Let us now turn to the numerical results of the calculation of the interaction of two proteins Nap1 at a frameshift length equal to 70 a.a. The results of the numerical interaction are shown in Fig. 5.30 for $\varepsilon = 1$ and in Fig. 5.31 for $\varepsilon = 80$. As can be seen from the Figs. 5.30 and 5.31, with a frameshift equal to 70 a.a., the formation of a cluster from the minimum values is observed for $\varepsilon = 1$ in the region of 70 a.a.$-$83 a.a., whose minimum is obtained at 83 a.a. and amounts to 18.963. Analysis

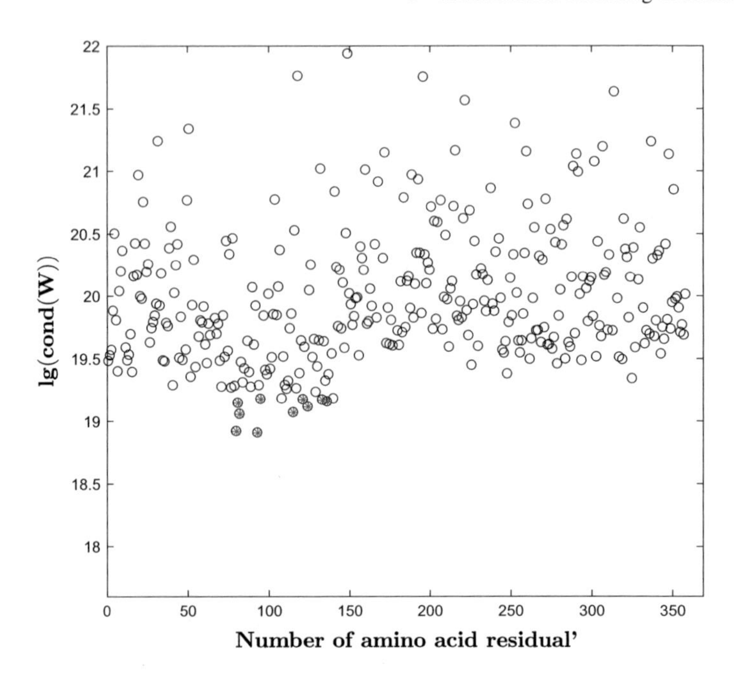

Fig. 5.28 Results of numerical simulation of the interaction of Nap1 with Nap1, frame size equal to 60 a.a., $\varepsilon = 1$. The red dots denote the 10 minimum values of lg(cond(W))

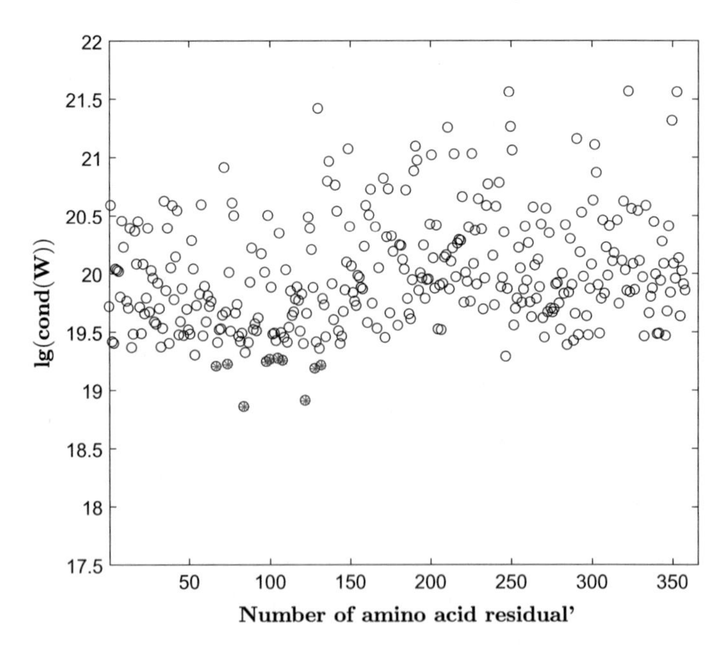

Fig. 5.29 Results of numerical simulation of the interaction of Nap1 with Nap1, frame size equal to 60 a.a., $\varepsilon = 80$. The red dots denote the 10 minimum values of lg(cond(W))

Table 5.21 The ten minimum values of lg(cond(W)) and the amino acid sequences of the detected regions of the Nap1 protein at a frameshift length equal to 60 a.a., $\varepsilon = 1$

$N^{\underline{0}}$	Amino acid sequence Nap1	lg(cond(W))
94	KEKLLSLKTLQSELFEVEKEFQVEMFELENKFLQKYKPIWEQRSRIISGQEQPKPEQIAK	18.906
81	QDSGYVGGLPKNVKEKLLSLKTLQSELFEVEKEFQVEMFELENKFLQKYKPIWEQRSRII	18.917
83	SGYVGGLPKNVKEKLLSLKTLQSELFEVEKEFQVEMFELENKFLQKYKPIWEQRSRIISG	19.055
116	VEMFELENKFLQKYKPIWEQRSRIISGQEQPKPEQIAKGQEIVESLNETELLVDEEEKAQ	19.068
125	FLQKYKPIWEQRSRIISGQEQPKPEQIAKGQEIVESLNETELLVDEEEKAQNDSEEEQVK	19.114
82	DSGYVGGLPKNVKEKLLSLKTLQSELFEVEKEFQVEMFELENKFLQKYKPIWEQRSRIIS	19.144
137	SRIISGQEQPKPEQIAKGQEIVESLNETELLVDEEEKAQNDSEEEQVKGIPSFWLTALEN	19.154
134	EQRSRIISGQEQPKPEQIAKGQEIVESLNETELLVDEEEKAQNDSEEEQVKGIPSFWLTA	19.169
122	ENKFLQKYKPIWEQRSRIISGQEQPKPEQIAKGQEIVESLNETELLVDEEEKAQNDSEEE	19.172
96	KLLSLKTLQSELFEVEKEFQVEMFELENKFLQKYKPIWEQRSRIISGQEQPKPEQIAKGQ	19.175

lg(cond(W)) is common logarithm of condition number
$N^{\underline{0}}$ is number of amino acid residual

Table 5.22 The ten minimum values of lg(cond(W)) and the amino acid sequences of the detected regions of the Nap1 protein at a frameshift length equal to 60 a.a., $\varepsilon = 80$

$N^{\underline{0}}$	Amino acid sequence Nap1	lg(cond(W))
85	YVGGLPKNVKEKLLSLKTLQSELFEVEKEFQVEMFELENKFLQKYKPIWEQRSRIISGQE	18.855
123	NKFLQKYKPIWEQRSRIISGQEQPKPEQIAKGQEIVESLNETELLVDEEEKAQNDSEEEQ	18.908
129	YKPIWEQRSRIISGQEQPKPEQIAKGQEIVESLNETELLVDEEEKAQNDSEEEQVKGIPS	19.182
68	LQSIQDRLGSLVGQDSGYVGGLPKNVKEKLLSLKTLQSELFEVEKEFQVEMFELENKFLQ	19.201
133	WEQRSRIISGQEQPKPEQIAKGQEIVESLNETELLVDEEEKAQNDSEEEQVKGIPSFWLT	19.210
75	LGSLVGQDSGYVGGLPKNVKEKLLSLKTLQSELFEVEKEFQVEMFELENKFLQKYKPIWE	19.219
99	SLKTLQSELFEVEKEFQVEMFELENKFLQKYKPIWEQRSRIISGQEQPKPEQIAKGQEIV	19.241
109	EVEKEFQVEMFELENKFLQKYKPIWEQRSRIISGQEQPKPEQIAKGQEIVESLNETELLV	19.253
101	KTLQSELFEVEKEFQVEMFELENKFLQKYKPIWEQRSRIISGQEQPKPEQIAKGQEIVES	19.261
106	ELFEVEKEFQVEMFELENKFLQKYKPIWEQRSRIISGQEQPKPEQIAKGQEIVESLNETE	19.271

lg(cond(W)) is common logarithm of condition number
$N^{\underline{0}}$ is number of amino acid residual

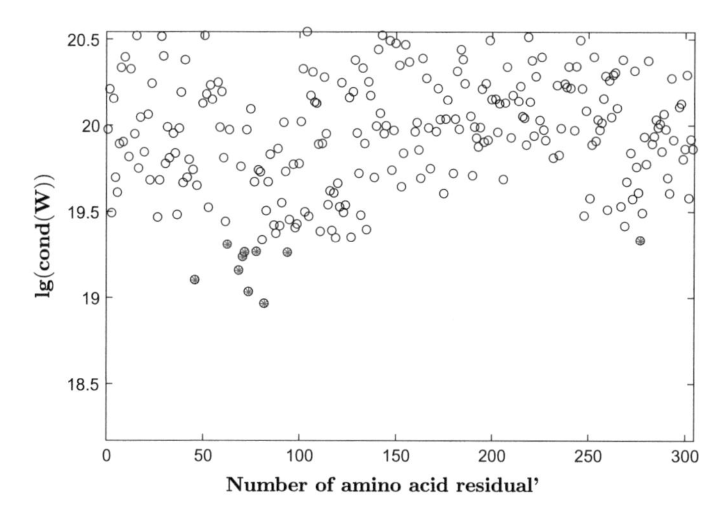

Fig. 5.30 Results of numerical simulation of the interaction of Nap1 with Nap1, frame size equal to 70 a.a., $\varepsilon = 1$. The red dots denote the 10 minimum values of lg(cond(W))

of the graph of the numerical results obtained at $\varepsilon = 80$ demonstrates the formation of a cluster from the minimum values in the region from 74 a.a. to 132 a.a., whose minimum is obtained at 74 a.a. and amounts to 19.012.

Analysis of the numerical results of the numerical experiments carried out with the participation of two identical Nap1 proteins using Algorithm 2, during which frameshifts of different lengths from 20 a.a. up to 70 a.a. determined that when the size of the frame was increased, the results obtained correlated more with the experimental data in which the structure of the Nap1–Nap1 homodimer was analyzed. So with a frameshift length of 20 a.a. there were a large number of possible segments of the polypeptide chain of the protein Nap1 which could form biological complexes with the same amino acid sequences of the second identical protein. However, we cannot say how accurate the data obtained are, because we need to have experimental data on the interaction of different sites between the Nap1 proteins. Also note that the Nap1 protein has a complex three-dimensional native structure, many parts of which are inside the protein and cannot interact with other proteins. With an increase in the length of the frameshift to 30 a.a. we see cluster formation from the minimum values of lg(cond(W)) in several regions of the polypeptide chain of the Nap1 protein, including in the domain 1 region responsible for the homodimerization of the Nap1 protein. With a further increase in the length of the frameshift to 40 a.a., half of the 10 minimum values of lg(cond(W)) fall in domain 1. Analysis of numerical calculations for frameshift dimensions of 50 a.a., 60 a.a., 70 a.a. demonstrated almost identical to the found amino acid sequences of the interacting proteins Nap1 in the domain 1 region. Thus, in finding the interacting regions of proteins Nap1 and Nap1, the best results are those in which a larger-size frameshift was taken: from 50 a.a. to 70 a.a. We explain this by the fact that the interaction between proteins Nap1 is due to an

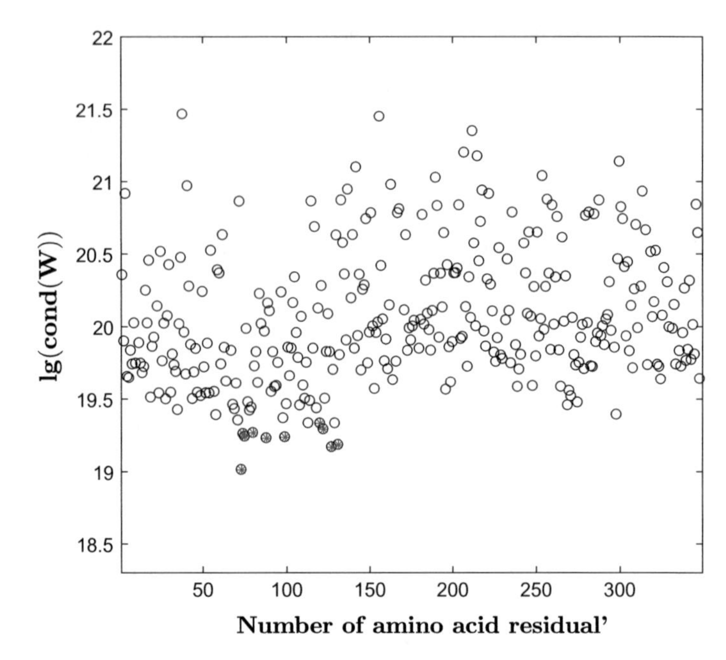

Fig. 5.31 Results of numerical simulation of the interaction of Nap1 with Nap1, frame size equal to 70 a.a., $\varepsilon = 80$. The red dots denote the 10 minimum values of lg(cond(W))

extended site between two proteins, domain 1 is located in the region from 74 a.a. to 180 a.a. [25] .

Thus, the larger size of the frameshift, the closer it allowed us to approach real interactions between proteins. It should be noted that, as an additional criterion determining the binding sites, it is possible to obtain identical binding sites in the form of cluster formation consisting of minimal amino acid residues lg(cond(W)) with a change in the length of the frameshift (Tables 5.21–5.24).

5.7.3 Numerical Calculation of the Interaction of P53 *with Mdm2*

We calculated the interaction between the proteins P53 and Mdm2 according to Algorithm 1 and Algorithm 2.

In this case, we calculated the interaction of the two proteins P53 and Mdm2 separately from the proteins Nap1 and Mdm2, since the nature of the formation of the P53−Mdm2 dimer does not correspond to the characteristics given above, for which the calculation can be applied according to Algorithm 1 or Algorithm 2, since the formation of the biological complex P53−Mdm2 is due to the short amino acid sequence of the P53 protein, which binds to the hydrophobic groove of Mdm2. From

Table 5.23 The ten minimum values of lg(cond(W)) and the amino acid sequences of the detected regions of the Nap1 protein at a frameshift length equal to 70 a.a., $\varepsilon = 1$

$N^{\underline{0}}$	Amino acid sequence Nap1	lg(cond(W))
83	SGYVGGLPKNVKEKLLSLKTLQSELFEVEKEFQVEMFELENKFL QKYKPIWEQRSRIISGQEQPKPEQIA	18.963
75	LGSLVGQDSGYVGGLPKNVKEKLLSLKTLQSELFEVEKEFQVEMF ELENKFLQKYKPIWEQRSRIISGQE	19.031
45	EQDDKIGTINEEDILANQPLLLQSIQDRLGSLVGQDSGYVGGLPKNV KEKLLSLKTLQSELFEVEKEFQV	19.099
70	SIQDRLGSLVGQDSGYVGGLPKNVKEKLLSLKTLQSELFEVEK EFQVEMFELENKFLQKYKPIWEQRSRI	19.155
72	QDRLGSLVGQDSGYVGGLPKNVKEKLLSLKTLQSELFEVEKEFQV EMFELENKFLQKYKPIWEQRSRIIS	19.234
95	EKLLSLKTLQSELFEVEKEFQVEMFELENKFLQKYKPIWEQRS RIISGQEQPKPEQIAKGQEIVESLNET	19.259
73	DRLGSLVGQDSGYVGGLPKNVKEKLLSLKTLQSELFEVEKEFQVE MFELENKFLQKYKPIWEQRSRIISG	19.261
79	VGQDSGYVGGLPKNVKEKLLSLKTLQSELFEVEKEFQVEMFELEN KFLQKYKPIWEQRSRIISGQEQPKP	19.264
64	QPLLLQSIQDRLGSLVGQDSGYVGGLPKNVKEKLLSLKTLQSELFEVE KEFQVEMFELENKFLQKYKPIW	19.305
278	KDNAHNVTVDLEMRKQRNKTTKQVRTIEKITPIESFFNFFDPPKI QNEDQDEELEEDLEERLALDYSIGE	19.331

lg(cond(W)) is common logarithm of condition number
$N^{\underline{0}}$ is number of amino acid residual

Table 5.24 The ten minimum values of lg(cond(W)) and the amino acid sequences of the detected regions of the Nap1 protein at a frameshift length equal to 70 a.a., $\varepsilon = 80$

$N^{\underline{0}}$	Amino acid sequence Nap1	lg(cond(W))
74	RLGSLVGQDSGYVGGLPKNVKEKLLSLKTLQSELFEV EKEFQVEMFELENKFLQKYKPIWEQRSRIISGQ	19.012
128	KYKPIWEQRSRIISGQEQPKPEQIAKGQEIVESLNETELLV DEEEKAQNDSEEEQVKGIPSFWLTALENL	19.168
132	IWEQRSRIISGQEQPKPEQIAKGQEIVESLNETELLVDE EEKAQNDSEEEQVKGIPSFWLTALENLPIVC	19.184
89	LPKNVKEKLLSLKTLQSELFEVEKEFQVEMFELENKFL QKYKPIWEQRSRIISGQEQPKPEQIAKGQEIV	19.229
100	LKTLQSELFEVEKEFQVEMFELENKFLQKYKPIWEQRSRIISG QEQPKPEQIAKGQEIVESLNETELLVD	19.236
76	GSLVGQDSGYVGGLPKNVKEKLLSLKTLQSELFEVEKEFQVE MFELENKFLQKYKPIWEQRSRIISGQEQ	19.241
75	LGSLVGQDSGYVGGLPKNVKEKLLSLKTLQSELFEVEKEFQVE MFELENKFLQKYKPIWEQRSRIISGQE	19.259
81	QDSGYVGGLPKNVKEKLLSLKTLQSELFEVEKEFQVEMFEL ENKFLQKYKPIWEQRSRIISGQEQPKPEQ	19.266
123	NKFLQKYKPIWEQRSRIISGQEQPKPEQIAKGQEIVESLNETELLV DEEEKAQNDSEEEQVKGIPSFWLT	19.290
121	LENKFLQKYKPIWEQRSRIISGQEQPKPEQIAKGQEIVESLNETELL VDEEEKAQNDSEEEQVKGIPSFW	19.331

lg(cond(W)) is common logarithm of condition number
$N^{\underline{0}}$ is number of amino acid residual

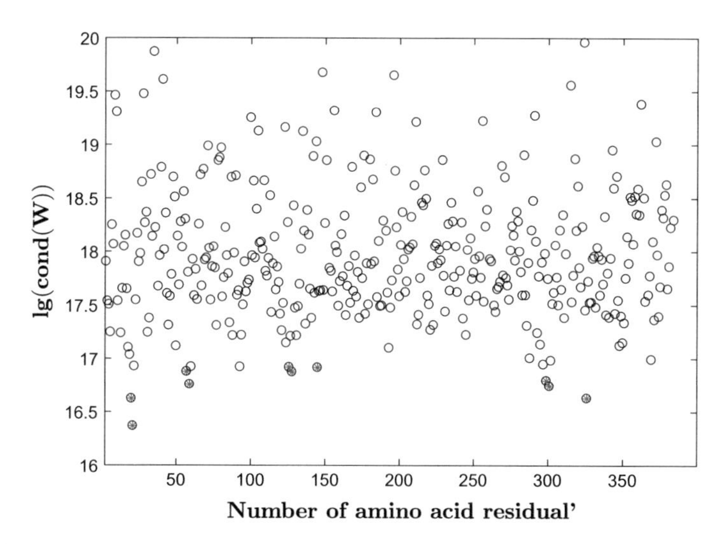

Fig. 5.32 Results of numerical simulation of the interaction of Mdm2 with P53, frame size equal to 10 a.a., $\varepsilon = 1$. The red dots denote the 10 minimum values of lg(cond(W))

the side of the P53 protein, amino acid residues from about 15 a.a. to 30 a.a., arranged consecutively one after the other, and from the side of the protein Mdm2 L54, L57, G58, I61, M62, Y67, Q72, V75, F91, V93, H96, I99, Y100 [23].

In the formation of the dimer P53–Mdm2, the N-termini of the two proteins are involved, but the interacting amino acid residues are not symmetric with respect to the N-terminus of the proteins. However, for example, we calculated the interaction of two given P53 and Mdm2 proteins according to Algorithm 2 and Algorithm 1. Let us now turn to the numerical results of the interaction of the proteins p53 and Mdm2 according to Algorithm 2 with a 10 a.a. frameshift and $\varepsilon = 1$. The results of the numerical calculation are shown in Fig. 5.32. As we can see from the above Fig. 5.32, the four minimum values of lg(cond(W)) fall on the Mdm2 protein domain of the P53-binding domain. The ten minimum values of lg(cond(W)) are given in the Table 5.25. for $\varepsilon = 1$.

As can be seen from the Table 5.25, the first two minimum values of lg(cond(W)) are 21 a.a. and 20 a.a., which are located in the P53-binding domain of the protein Mdm2 [32, 33].

However, it is difficult to accurately identify the binding site of the P53 and Mdm2 protein, and a better experimental or theoretical approach will be required.

Results of numerical calculations of the interaction between proteins P53 and Mdm2 according to Algorithm 2 with a 15 a.a. frameshift at $\varepsilon = 1$ was completed. The numerical results are shown in Fig. 5.33. In Fig. 5.33, we see three clusters, the largest accumulation of minimum values fall in the P53-binding domain of the Mdm2 protein, the second cluster of the 2nd minimum values lies in the regions of 125 a.a. and 126 a.a., the third cluster of third minimum values lies in the region of 300 a.a.

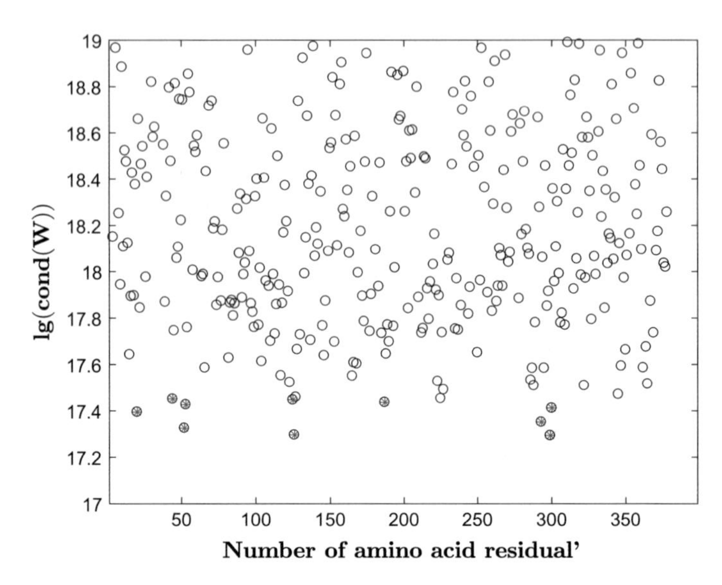

Fig. 5.33 Results of numerical simulation of the interaction of Mdm2 with P53, frame size equal to 15 a.a., $\varepsilon = 1$. The red dots denote the 10 minimum values of lg(cond(W))

Table 5.25 The ten minimum values of lg(cond(W)) and the amino acid sequences of the detected regions regions of the protein P53 in interaction with Mdm2 at a frameshift length equal to 10 a.a., $\varepsilon = 1$

$N^{\underline{0}}$	Amino acid sequence Mdm2	Amino acid sequence P53	lg(cond(W))
21	ASEQETLVRP	DLWKLLPENN	16.367
20	PASEQETLVR	SDLWKLLPEN	16.622
326	RENWLPEDKG	EYFTLQIRGR	16.627
301	DYWKCTSCNE	PGSTKRALPN	16.741
59	QYIMTKRLYD	GPDEAPRMPE	16.755
299	LADYWKCTSC	LPPGSTKRAL	16.793
128	HLEGGSDQKD	PALNKMFCQL	16.874
57	LGQYIMTKRL	DPGPDEAPRM	16.877
145	EKPSSSHLVS	LWVDSTPPPG	16.914
126	RCHLEGGSDQ	YSPALNKMFC	16.917

lg(cond(W)) is common logarithm of condition number
$N^{\underline{0}}$ is number of amino acid residual

As can be seen from Table 5.26, five values of lg(cond(W)) fall on the P53-binding domain of the Mdm2 protein and the N-terminus of the P53 protein, which is in satisfactory agreement with the previously identified binding sites [32, 33].

Analysis of data demonstrates the minimum values of lg(cond(W)) at the site of the P53-binding domain and the N-terminus of the P53 protein, which is in satisfactory agreement with the previously identified binding sites of Mdm2 and P53 proteins.

Table 5.26 The ten minimum values of lg(cond(W)) and the amino acid sequences of the detected regions regions of the protein P53 in interaction with Mdm2 at a frameshift length equal to 15 a.a., $\varepsilon = 1$

$N^{\underline{o}}$	Amino acid sequence Mdm2	Amino acid sequence P53	lg(cond(W))
299	LADYWKCTSCNEMNP	LPPGSTKRALPNNTS	17.292
126	RCHLEGGSDQKDLVQ	YSPALNKMFCQLAKT	17.295
52	EVLFYLGQYIMTKRL	QWFTEDPGPDEAPRM	17.324
293	EDPEISLADYWKCTS	GEPHHELPPGSTKRA	17.350
20	PASEQETLVRPKPLL	SDLWKLLPENNVLSP	17.394
300	ADYWKCTSCNEMNPP	PPGSTKRALPNNTSS	17.411
53	VLFYLGQYIMTKRLY	WFTEDPGPDEAPRMP	17.426
187	DSISLSFDESLALCV	GLAPPQHLIRVEGNL	17.435
125	NRCHLEGGSDQKDLV	TYSPALNKMFCQLAK	17.446
44	QKDTYTMKEVLFYLG	MLSPDDIEQWFTEDP	17.450

lg(cond(W)) is common logarithm of condition number
$N^{\underline{o}}$ is number of amino acid residual

However, the presence of the remaining minimum values, which are not located on the N-terminus of the two proteins, makes it difficult to accurately identify the binding site, nor can we say anything about the structure of the complex formed upon the formation of the dimer Mdm2−P53.

Let us turn to the analysis of the data obtained characterizing the interactions of P53 and Mdm2 proteins according to Algorithm 1. Numerical calculation of the interaction of the $P53_{(11-30)}$ and Mdm2 proteins according to Algorithm 1 for $\varepsilon = 80$.

We selected a short sequence of protein $P53_{(11-30)}$, which is directly involved in the formation of the biological complex with the protein Mdm2. The results of the numerical calculation are shown in Fig. 5.34.

As can be seen from Fig. 5.34, the greatest number of lowest values of lg(cond(W)) fall on the P53-binding domain of the Mdm2 protein, which is in satisfactory agreement with the previously identified binding sites [32, 33].

As can be seen from the Table 5.27 and graph Fig. 5.34 above, the cluster of the smallest values falls on the P53-binding domain of the Mdm2 protein in the interaction of $P53_{(11-30)}$, and the smallest value is obtained at 25 a.a. This result is in satisfactory agreement with earlier experiments, in which a complex part of the interaction of P53 and Mdm2 proteins [32, 33].

In this section, a numerical simulation of the interaction of proteins P53 and Mdm2 was performed without taking into account the phosphorylation of the N-terminus of the P53 protein and the subsequent influence of phosphorylation processes on the stability of the formed complex P53−Mdm2. In the next chapter, we will present a model for the phosphorylation of the amino acid residues of a single protein and the effect of phosphorylation on the stability of the formed biological complex on the example of the dimer Mdm2−P53.

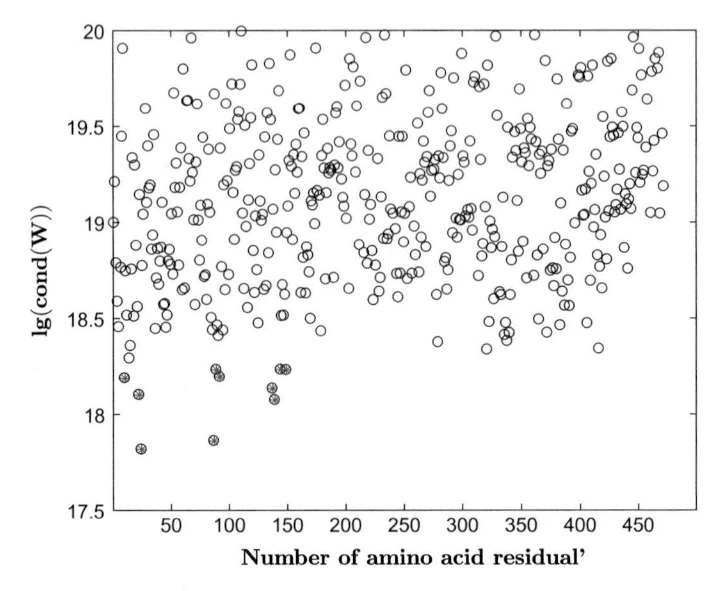

Fig. 5.34 Results of numerical simulation of the interaction of Mdm2 with $P53_{(11-30)}$, $\varepsilon = 80$. The red dots denote the 10 minimum values of lg(cond(W))

Table 5.27 The ten minimum values of lg(cond(W)) and the amino acid sequences of the detected regions regions of the protein $P53_{(11-30)}$ in interaction with Mdm2, $\varepsilon = 80$

$N^{\underline{0}}$	Amino acid sequence MdM2	lg(cond(W))
25	TLVRPKPLLLKLLKSVGAQ	17.817
87	VPSFSVKEHRKIYTMIYRN	17.861
139	QELQEEKPSSSHLVSRPST	18.073
23	QETLVRPKPLLLKLLKSVG	18.100
137	LVQELQEEKPSSSHLVSRP	18.133
11	GAVTTSQIPASEQETLVRP	18.188
92	VKEHRKIYTMIYRNLVVVN	18.194
149	SHLVSRPSTSSRRRAISET	18.231
89	SFSVKEHRKIYTMIYRNLV	18.232
144	EKPSSSHLVSRPSTSSRRR	18.233

lg(cond(W)) is common logarithm of condition number
$N^{\underline{0}}$ is number

5.8 Matlab Script Algorithm 1 for Mathematical Modeling Identification of Active Sites Interaction of Protein Molecules

Input parameters:

1. S_{100}, S_{20} are amino acid sequences of biological complexes ($S_{100} \geq S_{20}$)
2. sh is step shift
3. epsilon is the dielectric constant of the medium.

Output parameters:

lg(cond(W) is the common logarithm of the condition number of the matrix W, where its elements are composed of the electrostatic potential energy which is created based on the interaction between pair of amino acid residues of biological complexes.

Compute:

lg(cond(W) is the common logarithm of the condition number of the matrix W, which will allow a prediction the reactivity of the studied biological complexes.

```
1  clc,clear all
2  format long e
3  %MDM2
4  S_100=['M' 'C' 'N' 'T' 'N' 'M' 'S' 'V' 'P' 'T' 'D'...
5  'G' 'A' 'V' 'T' 'T' 'S' 'Q' 'I' 'P'    'A' 'S' 'E'...
6  'Q' 'E' 'T' 'L' 'V' 'R' 'P'  'K' 'P' 'L' 'L' 'L' 'K'...
7  'L' 'L' 'K' 'S'  'V' 'G' 'A'  'Q' 'K' 'D' 'T' 'Y' 'T' ...
8  'M'  'K' 'E' 'V' 'L' 'F' 'Y' 'L' 'G' 'Q' 'Y'  'I' 'M'...
9  'T' 'K' 'R' 'L' 'Y' 'D' 'E' 'K'  'Q' 'Q'  'H' 'I' 'V'...
10 'Y' 'C' 'S' 'N' 'D'  'L' 'L' 'G' 'D' 'L' 'F' 'G'...
11 'V' 'P' 'S'  'F' 'S' 'V' 'K' 'E' 'H' 'R' 'K' 'I' 'Y'...
12 'T' 'M' 'I' 'Y' 'R' 'N' 'L' 'V' 'V' 'V'  'N' 'Q' 'Q' ...
13 'E' 'S' 'S' 'D' 'S' 'G' 'T'  'S' 'V' 'S' 'E' 'N' 'R' ...
14 'C' 'H' 'L' 'E'  'G' 'G' 'S' 'D' 'Q' 'K' 'D' 'L' 'V' ...
15 'Q'  'E' 'L' 'Q' 'E' 'E' 'K' 'P' 'S' 'S' 'S'  'H' 'L' ...
16 'V' 'S' 'R' 'P' 'S' 'T' 'S' 'S'  'R' 'R' 'R' 'A' 'I'...
17 'S' 'E' 'T' 'E' 'E'  'N' 'S' 'D' 'E' 'L' 'S' 'G' 'E' ...
18 'R' 'Q'  'R' 'K' 'R' 'H' 'K' 'S' 'D' 'S' 'I' 'S'  'L' ...
19 'S' 'F' 'D' 'E' 'S' 'L' 'A' 'L' 'C'  'V' 'I' 'R' 'E'...
20 'I' 'C' 'C' 'E' 'R' 'S'  'S' 'S' 'S' 'E' 'S' 'T' 'G' ...
21 'T' 'P'  'S'  'N' 'P' 'D' 'L' 'D' 'A' 'G' 'V' 'S' 'E' ...
22 'H' 'S' 'G' 'D'  'W' 'L' 'D' 'Q' 'D' 'S'  'V' 'S' 'D'...
23 'Q' 'F' 'S' 'V' 'E' 'F'  'E'  'V' 'E' 'S' 'L' 'D' 'S'...
24 'E' 'D' 'Y' 'S'  'L' 'S' 'E' 'E' 'G' 'Q' 'E' 'L' 'S'...
25 'D'  'E' 'D' 'D' 'E' 'V' 'Y' 'Q' 'V' 'T'  'V'  'Y'...
26 'Q' 'A' 'G' 'E' 'S' 'D' 'T' 'D' 'S'  'F' 'E' 'E' 'D' ...
27 'P' 'E' 'I' 'S' 'L' 'A'  'D' 'Y' 'W' 'K' 'C' 'T' 'S'...
28 'C' 'N' 'E' 'M' 'N' 'P' 'P' 'L' 'P' 'S' 'H' 'C' 'N' ...
29 'R' 'C' 'W' 'A' 'L' 'R' 'E' 'N' 'W' 'L'  'P' 'E' 'D'...
30 'K' 'G' 'K' 'D' 'K' 'G' 'E'  'I' 'S' 'E' 'K' 'A' 'K'...
```

```
31 'L' 'E' 'N' 'S' 'T' 'Q' 'A' 'E' 'E' 'G' 'F' 'D' 'V' ...
32 'P' 'D' 'C' 'K' 'K' 'T' 'I' 'V' 'N' 'D' 'S' 'R' 'E' ...
33 'S' 'C' 'V' 'E' 'E' 'N' 'D' 'D' 'K' 'I' 'T' 'Q' 'A'...
34 'S' 'Q' 'S' 'Q' 'E' 'S' 'E' 'D' 'Y' 'S' 'Q' 'P' 'S'...
35 'T' 'S' 'S' 'S' 'I' 'I' 'Y' 'S' 'S' 'Q' 'E' 'D' ...
36 'V' 'K' 'E' 'F' 'E' 'R' 'E' 'E' 'T' 'Q' 'D' 'K'...
37 'E' 'E' 'S' 'V' 'E' 'S' 'S' 'L' 'P' 'L' 'N' 'A' ...
38 'I' 'E' 'P' 'C' 'V' 'I' 'C' 'Q' 'G' 'R' 'P' 'K'...
39 'N' 'G' 'C' 'I' 'V' 'H' 'G' 'K' 'T' 'G' 'H' 'L'...
40 'M' 'A' 'C' 'F' 'T' 'C' 'A' 'K' 'K' 'L' 'K' 'K' ...
41 'R' 'N' 'K' 'P' 'C' 'P' 'V' 'C' 'R' 'Q' 'P' 'I'...
42 'Q' 'M' 'I' 'V' 'L' 'T' 'Y' 'F' 'P']
43
44 S_20=['M' 'C' 'N' 'T' 'N' 'M' 'S' 'V' 'P' 'T' 'D'...
45 'G' 'A' 'V' 'T' 'T' 'S' 'Q' 'I' 'P' 'A' 'S' 'E'...
46 'Q' 'E' 'T' 'L' 'V' 'R' 'P' 'K' 'P' 'L' 'L' 'L' 'K'...
47 'L' 'L' 'K' 'S' 'V' 'G' 'A' 'Q' 'K' 'D' 'T' 'Y' 'T' ...
48 'M' 'K' 'E' 'V' 'L' 'F' 'Y' 'L' 'G' 'Q' 'Y' 'I' 'M'...
49 'T' 'K' 'R' 'L' 'Y' 'D' 'E' 'K' 'Q' 'Q' 'H' 'I' 'V'...
50 'Y' 'C' 'S' 'N' 'D' 'L' 'L' 'G' 'D' 'L' 'F' 'G'...
51 'V' 'P' 'S' 'F' 'S' 'V' 'K' 'E' 'H' 'R' 'K' 'I' 'Y'...
52 'T' 'M' 'I' 'Y' 'R' 'N' 'L' 'V' 'V' 'V' 'N' 'Q' 'Q' ...
53 'E' 'S' 'S' 'D' 'S' 'G' 'T' 'S' 'V' 'S' 'E' 'N' 'R' ...
54 'C' 'H' 'L' 'E' 'G' 'G' 'S' 'D' 'Q' 'K' 'D' 'L' 'V' ...
55 'Q' 'E' 'L' 'Q' 'E' 'E' 'K' 'P' 'S' 'S' 'S' 'H' 'L' ...
56 'V' 'S' 'R' 'P' 'S' 'T' 'S' 'S' 'R' 'R' 'R' 'A' 'I'...
57 'S' 'E' 'T' 'E' 'E' 'N' 'S' 'D' 'E' 'L' 'S' 'G' 'E' ...
58 'R' 'Q' 'R' 'K' 'R' 'H' 'K' 'S' 'D' 'S' 'I' 'S' 'L' ...
59 'S' 'F' 'D' 'E' 'S' 'L' 'A' 'L' 'C' 'V' 'I' 'R' 'E'...
60 'I' 'C' 'C' 'E' 'R' 'S' 'S' 'S' 'S' 'E' 'S' 'T' 'G' ...
61 'T' 'P' 'S' 'N' 'P' 'D' 'L' 'D' 'A' 'G' 'V' 'S' 'E' ...
62 'H' 'S' 'G' 'D' 'W' 'L' 'D' 'Q' 'D' 'S' 'V' 'S' 'D'...
63 'Q' 'F' 'S' 'V' 'E' 'F' 'E' 'V' 'E' 'S' 'L' 'D' 'S'...
64 'E' 'D' 'Y' 'S' 'L' 'S' 'E' 'E' 'G' 'Q' 'E' 'L' 'S'...
65 'D' 'E' 'D' 'D' 'E' 'V' 'Y' 'Q' 'V' 'T' 'V' 'Y'...
66 'Q' 'A' 'G' 'E' 'S' 'D' 'T' 'D' 'S' 'F' 'E' 'E' 'D' ...
67 'P' 'E' 'I' 'S' 'L' 'A' 'D' 'Y' 'W' 'K' 'C' 'T' 'S'...
68 'C' 'N' 'E' 'M' 'N' 'P' 'P' 'L' 'P' 'S' 'H' 'C' 'N' ...
69 'R' 'C' 'W' 'A' 'L' 'R' 'E' 'N' 'W' 'L' 'P' 'E' 'D'...
70 'K' 'G' 'K' 'D' 'K' 'G' 'E' 'I' 'S' 'E' 'K' 'A' 'K'...
71 'L' 'E' 'N' 'S' 'T' 'Q' 'A' 'E' 'E' 'G' 'F' 'D' 'V' ...
72 'P' 'D' 'C' 'K' 'K' 'T' 'I' 'V' 'N' 'D' 'S' 'R' 'E' ...
73 'S' 'C' 'V' 'E' 'E' 'N' 'D' 'D' 'K' 'I' 'T' 'Q' 'A'...
74 'S' 'Q' 'S' 'Q' 'E' 'S' 'E' 'D' 'Y' 'S' 'Q' 'P' 'S'...
75 'T' 'S' 'S' 'S' 'I' 'I' 'Y' 'S' 'S' 'Q' 'E' 'D' ...
76 'V' 'K' 'E' 'F' 'E' 'R' 'E' 'E' 'T' 'Q' 'D' 'K'...
77 'E' 'E' 'S' 'V' 'E' 'S' 'S' 'L' 'P' 'L' 'N' 'A' ...
78 'I' 'E' 'P' 'C' 'V' 'I' 'C' 'Q' 'G' 'R' 'P' 'K'...
79 'N' 'G' 'C' 'I' 'V' 'H' 'G' 'K' 'T' 'G' 'H' 'L'...
80 'M' 'A' 'C' 'F' 'T' 'C' 'A' 'K' 'K' 'L' 'K' 'K' ...
81 'R' 'N' 'K' 'P' 'C' 'P' 'V' 'C' 'R' 'Q' 'P' 'I'...
```

```
82   'Q' 'M' 'I' 'V' 'L' 'T' 'Y' 'F'  'P']
83  sh=1;
84  epsilon=1;
85  %-----------------------------------------------
86  len_S20=length(S_20);
87  len_S100=length(S_100);
88  N1=1*len_S100;
89  n_el=10;
90  del_len=len_S100-len_S20;
91  X=[];
92  Out=[];
93  V=[];
94  Z=[];
95  F=[];
96  br=ceil(del_len/sh)-1;
97  for ii=0:br+1
98      if ii~=br+1
99          X=[S_100(ii*sh+1:ii*sh+1+len_S20-1)];
100     else
101         X=[S_100(del_len+1:len_S100)];
102     end
103         S_1=X;
104         num=ii;
105 N=length(S_1);
106 M=length(S_20);
107 S_2=S_20;
108 Q1=[];
109 Q2=[];
110 R1=[];
111 R2=[];
112 [S_1,S_2,Q1,Q2,R1,R2,h]=potential(S_1,S_2,N1,N,M);
113 [A]=electrostatic(Q1,Q2, R1,R2,h,M,N,N1,epsilon);
114 [cond2]=condmy(A)
115 Out=[Out; X];
116 F=[F {num, S_1,S_2,(real(cond2))}'];
117 end
118 len_X=length(X);
119 len_Out=length(Out);
120 F;
121 barX=cell2mat(F(1,:));
122 barY=cell2mat(F(4,:));
123 SortF = sortrows(F',4);
124 barX_sort=cell2mat(SortF(:,1));
125 barY_sort=cell2mat(SortF(:,4));
126 minelem=[SortF(1:n_el,1) SortF(1:n_el,2) SortF(1:n_el,3)...
127     SortF(1:n_el,4)]
128 figure();
129 bar(barX,barY)
130 hold on
131 for i=1:n_el
132     bar(cell2mat(SortF(i,1)),cell2mat(SortF(i,4)),'red')
```

```
133 end
134 set(0,'DefaultTextInterpreter', 'latex');
135 set(0,'DefaultTextFontSize',14,...
136 'DefaultTextFontName','Arial Cyr');
137 xlabel('\bf Numer aminoacid residual');
138 set(0,'DefaultTextFontSize',14,...
139 'DefaultTextFontName','Arial Cyr');
140 ylabel('lg(cond(W))');
141 figure();
142 plot(barX,barY,'ok')
143 hold on
144 for i=1:n_el
145     plot(cell2mat(SortF(i,1)),cell2mat(SortF(i,4)),'*r')
146 end
147 set(0,'DefaultTextInterpreter', 'latex');
148 set(0,'DefaultTextFontSize',14,'DefaultTextFontName',...
149     'Arial Cyr');
150 xlabel('\bf Numer aminoacid residual');
151 set(0,'DefaultTextFontSize',14,...
152 'DefaultTextFontName','Arial Cyr');
153 ylabel('lg(cond(W))');
154 
155 function [S_1,S_2,Q1,Q2,R1,R2,h]=potential(S_1,S_2,N1,N,M);
156 N=length(S_1);
157 M=length(S_2);
158 Q1=[];
159 Q2=[];
160 R1=[];
161 R2=[];
162 for i=1:length(S_1);
163 for j=1:length(S_2);
164 if (S_1(i)=='D' & S_2(j)=='E')| (S_1(i)=='E' & S_2(j)=='D');
165 Q1(i,j)= 0.16e-19;
166 Q2(i,j)= 0.16e-19;
167 else
168 if (S_1(i)=='D' & S_2(j)=='D');
169 Q1(i,j)= 0.07e-19;
170 Q2(i,j)= 0.07e-19;
171 else
172 if (S_1(i)=='D' & S_2(j) =='C')|(S_1(i)=='C' & S_2(j) =='D');
173 Q1(i,j)= 0.05e-19;
174 Q2(i,j)= 0.05e-19;
175 else
176 if (S_1(i)=='D' &S_2(j)=='N')|(S_1(i)=='N' &S_2(j)=='D')|...
177 (S_1(i)=='D' &S_2(j)=='F')|(S_1(i)=='D' &S_2(j)=='Y')|...
178 (S_1(i)=='D' &S_2(j)=='Q')|(S_1(i)=='D' &S_2(j)=='S')|...
179 (S_1(i)=='F' &S_2(j)=='D')|(S_1(i)=='Y' &S_2(j)=='D')|...
180 (S_1(i)=='Q' &S_2(j)=='D')|(S_1(i)=='S' &S_2(j)=='D');
181 Q1(i,j)= 0.57e-19;
182 Q2(i,j)= 0.57e-19;
183 else
184 if ((S_1(i)=='D' & S_2(j)=='M')|(S_1(i)=='D' & S_2(j)=='T')|...
```

```
185 (S_1(i)=='D' & S_2(j)=='I')|(S_1(i)=='D' & S_2(j)=='G')|...
186 (S_1(i)=='D' & S_2(j)=='V')|(S_1(i)=='D' & S_2(j)=='W')|...
187 (S_1(i)=='D' & S_2(j)=='L')|(S_1(i)=='D' & S_2(j)=='A')|...
188 (S_1(i)=='M' & S_2(j)=='D')|(S_1(i)=='T' & S_2(j)=='D')|...
189 (S_1(i)=='I' & S_2(j)=='D')|(S_1(i)=='G' & S_2(j)=='D')|...
190 (S_1(i)=='V' & S_2(j)=='D')|(S_1(i)=='W' & S_2(j)=='D')|...
191 (S_1(i)=='L' & S_2(j)=='D')|(S_1(i)=='A' & S_2(j)=='D'));
192 Q1(i,j)= 0.64e-19;
193 Q2(i,j)= 0.64e-19;
194 else
195 if ((S_1(i)=='D'& S_2(j)=='P')|(S_1(i)=='P'& S_2(j)=='D'));
196 Q1(i,j)= 0.78e-19;
197 Q2(i,j)= 0.78e-19;
198 else
199 if ((S_1(i)=='D' & S_2(j)=='H')|(S_1(i)=='H'& S_2(j)=='D'));
200 Q1(i,j)= 0.99e-19;
201 Q2(i,j)= 0.99e-19;
202 else
203 if ((S_1(i)=='D'& S_2(j)=='K')|(S_1(i)=='K'& S_2(j)=='D'));
204 Q1(i,j)= 1.4e-19;
205 Q2(i,j)= 1.4e-19;
206 else
207 if ((S_1(i)=='D' & S_2(j)=='R')|(S_1(i)=='R'& S_2(j)=='D'));
208 Q1(i,j)= 1.59e-19;
209 Q2(i,j)= 1.59e-19;
210 else
211 if ((S_1(i)=='E'&S_2(j)=='E'));
212 Q1(i,j)= 0.16e-19;
213 Q2(i,j)= 0.16e-19;
214 else
215 if ((S_1(i)=='E' & S_2(j)=='C')|(S_1(i)=='E' & S_2(j)=='F')|...
216 (S_1(i)=='E' & S_2(j)=='N')|(S_1(i)=='C' & S_2(j)=='E')|...
217 (S_1(i)=='F' & S_2(j)=='E')|(S_1(i)=='N' & S_2(j)=='E'));
218 Q1(i,j)= 0.55e-19;
219 Q2(i,j)= 0.55e-19;
220 else
221  if ((S_1(i)=='E' & S_2(j)=='Q')|(S_1(i)=='E' & S_2(j)=='Y')|...
222  (S_1(i)=='E' & S_2(j)=='S')|(S_1(i)=='E' & S_2(j)=='M')|...
223  (S_1(i)=='E' & S_2(j)=='T')|(S_1(i)=='E' & S_2(j)=='I')|...
224   (S_1(i)=='E' & S_2(j)=='G')|(S_1(i)=='E' & S_2(j)=='V')|...
225   (S_1(i)=='E' & S_2(j)=='W')|(S_1(i)=='E' & S_2(j)=='L')|...
226   (S_1(i)=='E' & S_2(j)=='A')|(S_1(i)=='Q' & S_2(j)=='E')|...
227   (S_1(i)=='Y' & S_2(j)=='E')| (S_1(i)=='S' & S_2(j)=='E')|...
228   (S_1(i)=='M' & S_2(j)=='E')|(S_1(i)=='T' & S_2(j)=='E')|...
229   (S_1(i)=='I' & S_2(j)=='E')|(S_1(i)=='G' & S_2(j)=='E')|...
230   (S_1(i)=='V' & S_2(j)=='E')|(S_1(i)=='W' & S_2(j)=='E')|...
231   (S_1(i)=='L' & S_2(j)=='E')|(S_1(i)=='A' & S_2(j)=='E'));
232 Q1(i,j)= 0.64e-19;
233 Q2(i,j)= 0.64e-19;
234 else
235 if ((S_1(i)=='E' & S_2(j)=='P' )|(S_1(i)=='P' & S_2(j)=='E'));
236 Q1(i,j)= 0.78e-19;
```

```
237 Q2(i,j)= 0.78e-19;
238 else
239 if ((S_1(i)=='E' & S_2(j)=='H')|(S_1(i)=='H' &S_2(j)=='E'));
240 Q1(i,j)= 0.99e-19;
241 Q2(i,j)= 0.99e-19;
242 else
243 if (S_1(i)=='E'& S_2(j)=='K')| (S_1(i)=='K'& S_2(j)=='E');
244 Q1(i,j)= 1.34e-19;
245 Q2(i,j)= 1.34e-19;
246 else
247 if (S_1(i)=='E' & S_2(j)=='R')|(S_1(i)=='R' & S_2(j)=='E');
248 Q1(i,j)= 1.58e-19;
249 Q2(i,j)= 1.58e-19;
250 else
251 if (S_1(i)=='C' & S_2(j)=='C')|(S_1(i)=='C' & S_2(j)=='F')|...
252 (S_1(i)=='C' & S_2(j)=='Q')|(S_1(i)=='C'& S_2(j)=='Y')|...
253 (S_1(i)=='C' & S_2(j)=='S')|(S_1(i)=='C' & S_2(j)=='M')|...
254 (S_1(i)=='C' & S_2(j)=='T')|(S_1(i)=='C' & S_2(j)=='I')|...
255 (S_1(i)=='C' & S_2(j)=='G')|(S_1(i)=='C' & S_2(j)=='V')|...
256 (S_1(i)=='C' & S_2(j)=='W')|(S_1(i)=='C' & S_2(j)=='L')|...
257 (S_1(i)=='C' & S_2(j)=='L')|(S_1(i)=='C' & S_2(j)=='A')|...
258 (S_1(i)=='F' & S_2(j)=='C')|(S_1(i)=='Q' & S_2(j)=='C')|...
259 (S_1(i)=='Y'& S_2(j)=='C')|(S_1(i)=='S' & S_2(j)=='C')|...
260 (S_1(i)=='M' & S_2(j)=='C')|(S_1(i)=='T' & S_2(j)=='C')|...
261 (S_1(i)=='I' & S_2(j)=='C')|(S_1(i)=='G' & S_2(j)=='C')|...
262 (S_1(i)=='V' & S_2(j)=='C')|(S_1(i)=='W' & S_2(j)=='C')|...
263 (S_1(i)=='L' & S_2(j)=='C')|( S_1(i)=='A' & S_2(j)=='C');
264 Q1(i,j)=0.74e-19;
265 Q2(i,j)=0.74e-19;
266 else
267 if (S_1(i)=='C' & S_2(j)=='H')| (S_1(i)=='H' & S_2(j)=='C');
268 Q1(i,j)= 0.99e-19;
269 Q2(i,j)= 0.99e-19;
270 else
271 if (S_1(i)=='C' & S_2(j)=='K')|(S_1(i)=='K' & S_2(j)=='C');
272 Q1(i,j)= 1.34e-19;
273 Q2(i,j)= 1.34e-19;
274 else
275 if (S_1(i)=='C' & S_2(j)=='R')|(S_1(i)=='R' & S_2(j)=='C');
276 Q1(i,j)= 1.59e-19;
277 Q2(i,j)= 1.59e-19;
278 else
279 if (S_1(i)=='N' & S_2(j)=='N')|(S_1(i)=='N' & S_2(j)=='F')|...
280 (S_1(i)=='N' & S_2(j)=='Q')|(S_1(i)=='N' & S_2(j)=='Y')|...
281 (S_1(i)=='N' & S_2(j)=='S')|(S_1(i)=='N'& S_2(j)=='M')|...
282 (S_1(i)=='F' & S_2(j)=='N')|(S_1(i)=='Q' & S_2(j)=='N')|...
283 (S_1(i)=='Y' & S_2(j)=='N')|(S_1(i)=='S' & S_2(j)=='N')|...
284 (S_1(i)=='M'& S_2(j)=='N');
285 Q1(i,j)=0.74e-19;
286 Q2(i,j)=0.74e-19;
287 else
```

```
288 if (S_1(i)=='N' & S_2(j)=='H')|(S_1(i)=='H' & S_2(j)=='N')
289 Q1(i,j)= 0.99e-19;
290 Q2(i,j)= 0.99e-19;
291 else
292 if(S_1(i)=='N' & S_2(j)=='K')|(S_1(i)=='K' & S_2(j)=='N');
293 Q1(i,j)= 1.05e-19;
294 Q2(i,j)= 1.05e-19;
295 else
296 if (S_1(i)=='N' & S_2(j)=='R')|(S_1(i)=='R' & S_2(j)=='N');
297 Q1(i,j)= 1.1e-19;
298 Q2(i,j)= 1.1e-19;
299 else
300 if ((S_1(i)=='F' & S_2(j)=='F')|(S_1(i)=='F' & S_2(j)=='Q'));
301 Q1(i,j)=0.74e-19;
302 Q2(i,j)=0.74e-19;
303 else
304 if ((S_1(i)=='F' & S_2(j)=='Y')|(S_1(i)=='F' & S_2(j)=='S')|...
305 (S_1(i)=='F' & S_2(j)=='M')|(S_1(i)=='Q' & S_2(j)=='F')|...
306 (S_1(i)=='Y' & S_2(j)=='F'));
307 Q1(i,j)=0.74e-19;
308 Q2(i,j)=0.74e-19;
309 else
310 if (S_1(i)=='S' & S_2(j)=='F')|(S_1(i)=='M' & S_2(j)=='F');
311 Q1(i,j)=0.74e-19;
312 Q2(i,j)=0.74e-19;
313 else
314 if (S_1(i)=='F' & S_2(j)=='H')|(S_1(i)=='H' & S_2(j)=='F');
315 Q1(i,j)= 0.99e-19;
316 Q2(i,j)= 0.99e-19;
317 else
318 if (S_1(i)=='F' & S_2(j)=='K')|(S_1(i)=='K' & S_2(j)=='F');
319 Q1(i,j)= 1.05e-19;
320 Q2(i,j)= 1.05e-19;
321 else
322 if (S_1(i)=='F' & S_2(j)=='R')|(S_1(i)=='R' & S_2(j)=='F');
323 Q1(i,j)= 1.1e-19;
324 Q2(i,j)= 1.1e-19;
325 else
326 % Q
327 if (S_1(i)=='Q' & S_2(j)=='H')|(S_1(i)=='H' & S_2(j)=='Q');
328 Q1(i,j)= 0.99e-19;
329 Q2(i,j)= 0.99e-19;
330 else
331 if (S_1(i)=='Q' & S_2(j)=='K')|(S_1(i)=='K' & S_2(j)=='Q');
332 Q1(i,j)= 1.05e-19;
333 Q2(i,j)= 1.05e-19;
334 else
335 if (S_1(i)=='Q' & S_2(j)=='R')|(S_1(i)=='R' & S_2(j)=='Q');
336 Q1(i,j)= 1.1e-19;
337 Q2(i,j)= 1.1e-19;
338 else
339 % Y
340 if (S_1(i)=='Q' & S_2(j)=='H')|(S_1(i)=='H' & S_2(j)=='Q');
```

```
341 Q1(i,j)= 0.99e-19;
342 Q2(i,j)= 0.99e-19;
343 else
344 if (S_1(i)=='Y' & S_2(j)=='K')|(S_1(i)=='K' & S_2(j)=='Y')
345 Q1(i,j)= 1.05e-19;
346 Q2(i,j)= 1.05e-19;
347 else
348 if (S_1(i)=='Y' & S_2(j)=='R')|(S_1(i)=='R' & S_2(j)=='Y');
349 Q1(i,j)= 1.1e-19;
350 Q2(i,j)= 1.1e-19;
351 else
352 if (S_1(i)=='S' & S_2(j)=='H')|(S_1(i)=='H' & S_2(j)=='S');
353 Q1(i,j)= 0.99e-19;
354 Q2(i,j)= 0.99e-19;
355 else
356 if (S_1(i)=='S' & S_2(j)=='K')|(S_1(i)=='K' & S_2(j)=='S');
357 Q1(i,j)= 1e-19;
358 Q2(i,j)= 1e-19;
359 else
360 if (S_1(i)=='S' & S_2(j)=='R')|(S_1(i)=='R' & S_2(j)=='S');
361 Q1(i,j)= 1.1e-19;
362 Q2(i,j)= 1.1e-19;
363 else
364 if (S_1(i)=='M' & S_2(j)=='H')|(S_1(i)=='H' & S_2(j)=='M');
365 Q1(i,j)= 0.99e-19;
366 Q2(i,j)= 0.99e-19;
367 else
368 if (S_1(i)=='M' & S_2(j)=='K')|(S_1(i)=='K' & S_2(j)=='M');
369 Q1(i,j)= 1e-19;
370 Q2(i,j)= 1c-19;
371 else
372 if (S_1(i)=='M' & S_2(j)=='R')|(S_1(i)=='R' & S_2(j)=='M');
373 Q1(i,j)= 1.1e-19;
374 Q2(i,j)= 1.1e-19;
375 else
376 if (S_1(i)=='T' & S_2(j)=='H')|(S_1(i)=='H' & S_2(j)=='T');
377 Q1(i,j)= 0.99e-19;
378 Q2(i,j)= 0.99e-19;
379 else
380 if (S_1(i)=='T' & S_2(j)=='K')|(S_1(i)=='K' & S_2(j)=='T');
381 Q1(i,j)= 1e-19;
382 Q2(i,j)= 1e-19;
383 else
384 if (S_1(i)=='T' & S_2(j)=='R')|(S_1(i)=='R' & S_2(j)=='T');
385 Q1(i,j)= 1.05e-19;
386 Q2(i,j)= 1.05e-19;
387 else
388 if (S_1(i)=='I' & S_2(j)=='H')|(S_1(i)=='H' & S_2(j)=='I');
389 Q1(i,j)= 0.99e-19;
390 Q2(i,j)= 0.99e-19;
391 else
392 if (S_1(i)=='I' & S_2(j)=='K')|(S_1(i)=='K' & S_2(j)=='I');
393 Q1(i,j)= 1e-19;
```

```
394 Q2(i,j)= 1e-19;
395 else
396 if (S_1(i)=='I' & S_2(j)=='R')|(S_1(i)=='R' & S_2(j)=='I');
397 Q1(i,j)= 1.05e-19;
398 Q2(i,j)= 1.05e-19;
399 else
400 if (S_1(i)=='G' & S_2(j)=='H')|(S_1(i)=='H' & S_2(j)=='G');
401 Q1(i,j)= 0.99e-19;
402 Q2(i,j)= 0.99e-19;
403 else
404 if (S_1(i)=='G' & S_2(j)=='K')|(S_1(i)=='K' & S_2(j)=='G');
405 Q1(i,j)= 1e-19;
406 Q2(i,j)= 1e-19;
407 else
408 if (S_1(i)=='G' & S_2(j)=='R')|(S_1(i)=='R' & S_2(j)=='G');
409 Q1(i,j)= 1.05e-19;
410 Q2(i,j)= 1.05e-19;
411 else
412 if (S_1(i)=='V' & S_2(j)=='H')|(S_1(i)=='H' & S_2(j)=='V');
413 Q1(i,j)= 0.99e-19;
414 Q2(i,j)= 0.99e-19;
415 else
416 if (S_1(i)=='V' & S_2(j)=='K')|(S_1(i)=='K' & S_2(j)=='V');
417 Q1(i,j)= 1e-19;
418 Q2(i,j)= 1e-19;
419 else
420 if (S_1(i)=='V' & S_2(j)=='R')|(S_1(i)=='R' & S_2(j)=='V');
421 Q1(i,j)= 1.05e-19;
422 Q2(i,j)= 1.05e-19;
423 else
424 if (S_1(i)=='W' & S_2(j)=='H')|(S_1(i)=='H' & S_2(j)=='W');
425 Q1(i,j)= 0.99e-19;
426 Q2(i,j)= 0.99e-19;
427 else
428 if (S_1(i)=='W' & S_2(j)=='K')|(S_1(i)=='K' & S_2(j)=='W');
429 Q1(i,j)= 1e-19;
430 Q2(i,j)= 1e-19;
431 else
432 if (S_1(i)=='W' & S_2(j)=='R')|(S_1(i)=='R' & S_2(j)=='W');
433 Q1(i,j)= 1.05e-19;
434 Q2(i,j)= 1.05e-19;
435 else
436 if (S_1(i)=='L' & S_2(j)=='H')|(S_1(i)=='H' & S_2(j)=='L');
437 Q1(i,j)= 0.99e-19;
438 Q2(i,j)= 0.99e-19;
439 else
440 if (S_1(i)=='L' & S_2(j)=='K')|(S_1(i)=='K' & S_2(j)=='L');
441 Q1(i,j)= 1e-19;
442 Q2(i,j)= 1e-19;
443 else
444 if (S_1(i)=='L' & S_2(j)=='R')|(S_1(i)=='R' & S_2(j)=='L');
445 Q1(i,j)= 1.05e-19;
446 Q2(i,j)= 1.05e-19;
```

```
447 else
448 if (S_1(i)=='A' & S_2(j)=='H')|(S_1(i)=='H' & S_2(j)=='A');
449 Q1(i,j)= 0.99e-19;
450 Q2(i,j)= 0.99e-19;
451 else
452 if (S_1(i)=='A' & S_2(j)=='K')|(S_1(i)=='K' & S_2(j)=='A');
453 Q1(i,j)= 1e-19;
454 Q2(i,j)= 1e-19;
455 else
456 if (S_1(i)=='A' & S_2(j)=='R')|(S_1(i)=='R' & S_2(j)=='A');
457 Q1(i,j)= 1.05e-19;
458 Q2(i,j)= 1.05e-19;
459 else
460 if (S_1(i)=='P' & S_2(j)=='H')|(S_1(i)=='H' & S_2(j)=='P');
461 Q1(i,j)= 0.99e-19;
462 Q2(i,j)= 0.99e-19;
463 else
464 if (S_1(i)=='P' & S_2(j)=='K')|(S_1(i)=='K' & S_2(j)=='P');
465 Q1(i,j)= 0.82e-19;
466 Q2(i,j)= 0.82e-19;
467 else
468 if (S_1(i)=='P' & S_2(j)=='R')|(S_1(i)=='R' & S_2(j)=='P');
469 Q1(i,j)= 0.96e-19;
470 Q2(i,j)= 0.96e-19;
471 else
472 if (S_1(i)=='H' & S_2(j)=='H');
473 Q1(i,j)= 0.82e-19;
474 Q2(i,j)= 0.82e-19;
475 else
476 if (S_1(i)=='H' & S_2(j)=='K')|(S_1(i)=='K' & S_2(j)=='H');
477 Q1(i,j)= 0.82e-19;
478 Q2(i,j)= 0.82e-19;
479 else
480 if (S_1(i)=='H' & S_2(j)=='R')|(S_1(i)=='R' & S_2(j)=='H');
481 Q1(i,j)= 0.74e-19;
482 Q2(i,j)= 0.74e-19;
483 else
484 if (S_1(i)=='K' & S_2(j)=='K');
485 Q1(i,j)= 0.54e-19;
486 Q2(i,j)= 0.54e-19;
487 else
488 if (S_1(i)=='K' & S_2(j)=='R')|(S_1(i)=='R' & S_2(j)=='K');
489 Q1(i,j)= 0.41e-19;
490 Q2(i,j)= 0.41e-19;
491 else
492 if (S_1(i)=='R' & S_2(j)=='R');
493 Q1(i,j)= 0.16e-19;
494 Q2(i,j)= 0.16e-19;
495 else
496 Q1(i,j)= 0.824e-19;
497 Q2(i,j)= 0.824e-19;
```

```
498 end
499 end
500 end
501 end
502 end
503 end
504 end
505 end
506 end
507 end
508 end
509 end
510 end
511 end
512 end
513 end
514 end
515 end
516 end
517 end
518 end
519 end
520 end
521 end
522 end
523 end
524 end
525 end
526 end
527 end
528 end
529 end
530 end
531 end
532 end
533 end
534 end
535 end
536 end
537 end
538 end
539 end
540 end
541 end
542 end
543 end
544 end
545 end
546 end
547 end
548 end
549 end
```

```
550 end
551 end
552 end
553 end
554 end
555 end
556 end
557 end
558 end
559 end
560 end
561 end
562 end
563 end
564 end
565 end
566 end
567 end
568 end
569 end
570 end
571 end
572  Q3=[];
573  Q4=[];
574  R1=[];
575  R2=[];
576 for i=1:length(S_1);
577 if (S_1(i)=='A');
578 R1(i)=0.6e-9;
579 else
580 if (S_1(i)=='R');
581 R1(i)=0.809e-9;
582 else
583 if (S_1(i)=='N');
584 R1(i)=0.682e-9;
585 else
586 if (S_1(i)=='D');
587 R1(i)=0.665e-9;
588 else
589 if (S_1(i)=='C');
590 R1(i)=0.629e-9;
591 else
592 if (S_1(i)=='Q');
593 R1(i)=0.725e-9;
594 else
595 if (S_1(i)=='E');
596 R1(i)=0.714e-9;
597 else
598 if (S_1(i)=='G');
599 R1(i)=0.537e-9;
600 else
601 if (S_1(i)=='H');
```

```
602 R1(i)=0.732e-9;
603 else
604 if (S_1(i)=='I');
605 R1(i)=0.735e-9;
606 else
607 if (S_1(i)=='L');
608 R1(i)=0.734e-9;
609 else
610 if (S_1(i)=='K');
611 R1(i)=0.737e-9;
612 else
613 if (S_1(i)=='M');
614 R1(i)=0.741e-9;
615 else
616 if (S_1(i)=='F');
617 R1(i)=0.781e-9;
618 else
619 if (S_1(i)=='P');
620 R1(i)=0.672e-9;
621 else
622 if (S_1(i)=='S');
623 R1(i)=0.615e-9;
624 else
625 if (S_1(i)=='T');
626 R1(i)=0.659e-9;
627 else
628 if (S_1(i)=='W');
629 R1(i)=0.826e-9;
630 else
631 if (S_1(i)=='Y');
632 R1(i)=0.781e-9;
633 else
634 if (S_1(i)=='V');
635 R1(i)=0.694e-9;
636 end
637 end
638 end
639 end
640 end
641 end
642 end
643 end
644 end
645 end
646 end
647 end
648 end
649 end
650 end
651 end
652 end
653 end
```

```
654 end
655 end
656 for j=1:length(S_2);
657 if (S_2(j)=='A');
658 R2(j)=0.6e-9;
659 else
660 if (S_2(j)=='R');
661 R2(j)= 0.809e-9;
662 else
663 if (S_2(j)=='N');
664 R2(j)=0.682e-9;
665 else
666 if (S_2(j)=='D');
667 R2(j)=0.665e-9;
668 else
669 if (S_2(j)=='C');
670 R2(j)=0.629e-9;
671 else
672 if (S_2(j)=='Q');
673 R2(j)=0.725e-9;
674 else
675 if (S_2(j)=='E');
676 R2(j)=0.714e-9;
677 else
678 if (S_2(j)=='G');
679 R2(j)=0.537e-9;
680 else
681 if (S_2(j)=='H');
682 R2(j)=0.732e-9;
683 else
684 if (S_2(j)=='I');
685 R2(j)=0.735e-9;
686 else
687 if(S_2(j)=='L');
688 R2(j)=0.734e-9;
689 else
690 if (S_2(j)=='K')
691 R2(j)=0.737e-9;
692 else
693 if (S_2(j)=='M')
694 R2(j)=0.741e-9;
695 else
696 if (S_2(j)=='F')
697 R2(j)=0.781e-9;
698 else
699 if (S_2(j)=='P');
700 R2(j)=0.672e-9;
701 else
702 if (S_2(j)=='S');
703 R2(j)=0.615e-9;
704 else
705 if (S_2(j)=='T');
706 R2(j)=0.659e-9;
```

```
707 else
708 if (S_2(j)=='W');
709 R2(j)=0.826e-9;
710 else
711 if (S_2(j)=='Y');
712 R2(j)=0.781e-9;
713 else
714 if (S_2(j)=='V');
715 R2(j)=0.694e-9;
716 end
717 end
718 end
719 end
720 end
721 end
722 end
723 end
724 end
725 end
726 end
727 end
728 end
729 end
730 end
731 end
732 end
733 end
734 end
735 end
736 end
737 end
738  Ra=0.6e-9;
739  Rr=0.809e-9;
740  Rn=0.682e-9;
741  Rd=0.665e-9;
742  Rc=0.629e-9;
743  Rq=0.725e-9;
744  Re=0.714e-9;
745  Rg=0.725e-9;
746  Rh=0.732e-9;
747  Ri=0.735e-9;
748  Rl=0.734e-9;
749  Rk=0.737e-9;
750  Rm=0.741e-9;
751  Rf=0.781e-9;
752  Rp=0.672e-9;
753  Rs=0.615e-9;
754  Rt=0.659e-9;
755  Rw=0.826e-9;
756  Ry=0.781e-9;
757  Rv=0.694e-9;
```

```
758 for i=1:length(S_1);
759 for j=1:length(S_2);
760 if (S_1(i)=='R'& S_2(j)=='D');
761    h(i,j)=.15*10^(-9)+Rr+Rd;
762 else
763 if (S_1(i)=='R'& S_2(j)=='E');
764     h(i,j)=.15*10^(-9)+Rr+Re;
765 else
766 if (S_1(i)=='D'& S_2(j)=='R');
767     h(i,j)=.15*10^(-9)+Rd+Rr;
768 else
769 if (S_1(i)=='D'& S_2(j)=='H');
770    h(i,j)=.15*10^(-9)+Rd+Rh;
771 else
772 if (S_1(i)=='D'& S_2(j)=='R');
773     h(i,j)=.15*10^(-9)+Rd+Rr;
774 else
775 if (S_1(i)=='D'& S_2(j)=='H');
776     h(i,j)=.15*10^(-9)+Rd+Rh;
777 else
778 if (S_1(i)=='D'& S_2(j)=='K');
779     h(i,j)=.15*10^(-9)+Rd+Rk;
780 else
781 if (S_1(i)=='E')& (S_2(j)=='R');
782     h(i,j)=.15*10^(-9)+Re+Rr;
783 else
784 if (S_1(i)=='E'& S_2(j)=='H');
785    h(i,j)=.15*10^(-9)+Re+Rh;
786 else
787 if (S_1(i)=='E'& S_2(j)=='K');
788    h(i,j)=.15*10^(-9)+Re+Rk;
789 else
790 if (S_1(i)=='H'& S_2(j)=='D')
791     h(i,j)=.15*10^(-9)+Rh+Rd;
792 else
793 if (S_1(i)=='H'& S_2(j)=='E')
794     h(i,j)=.15*10^(-9)+Rh+Re;
795 else
796 if (S_1(i)=='R'& S_2(j)=='R')
797     h(i,j)=.4*10^(-9)+Rr+Rr;
798 else
799 if (S_1(i)=='R'& S_2(j)=='H')
800    h(i,j)=.4*10^(-9)+Rr+Rh;
801 else
802 if (S_1(i)=='R'& S_2(j)=='H')
803     h(i,j)=.4*10^(-9)+Rr+Rh;
804 else
805 if (S_1(i)=='R'& S_2(j)=='K')
806     h(i,j)=.4*10^(-9)+Rr+Rk;
807 else
808 if (S_1(i)=='D'& S_2(j)=='E');
809     h(i,j)=.4*10^(-9)+Rd+Re;
```

```
810 else
811 if (S_1(i)=='D'& S_2(j)=='D');
812     h(i,j)=.4*10^(-9)+Rd+Rd;
813 else
814 if (S_1(i)=='H'& S_2(j)=='R')
815     h(i,j)=.4*10^(-9)+Rh+Rr;
816 else
817 if (S_1(i)=='H'& S_2(j)=='H')
818     h(i,j)=.4*10^(-9)+Rh+Rh;
819 else
820 if (S_1(i)=='H'& S_2(j)=='K')
821      h(i,j)=.4*10^(-9)+Rh+Rk;
822 else
823 if (S_1(i)=='K'& S_2(j)=='R')
824    h(i,j)=.4*10^(-9)+Rk+Rr;
825 else
826 if (S_1(i)=='K'& S_2(j)=='H')
827     h(i,j)=.4*10^(-9)+Rk+Rh;
828 else
829 if (S_1(i)=='K'& S_2(j)=='K')
830     h(i,j)=.4*10^(-9)+Rk+Rk;
831 else
832 if (S_1(i)=='N'& S_2(j)=='Q')
833    h(i,j)=.25*10^(-9)+Rn+Rq;
834 else
835 if (S_1(i)=='N'& S_2(j)=='S')
836    h(i,j)=.25*10^(-9)+Rn+Rs;
837 else
838 if (S_1(i)=='N'& S_2(j)=='Y')
839     h(i,j)=.25*10^(-9)+Rn+Ry;
840 else
841 if (S_1(i)=='Q'& S_2(j)=='S')|(S_1(i)=='Q')& (S_2(j)=='Y');
842     h(i,j)=.25*10^(-9)+Rq+Rs;
843 else
844 if (S_1(i)=='Q')& (S_2(j)=='Y');
845      h(i,j)=.25*10^(-9)+Rq+Ry;
846 else
847 if (S_1(i)=='S'& S_2(j)=='Y');
848     h(i,j)=.25*10^(-9)+Rs+Ry;
849 else
850     h(i,j)=1.76*10^(-9);
851 end
852 end
853 end
854 end
855 end
856 end
857 end
858 end
859 end
860 end
861 end
```

```
862 end
863 end
864 end
865 end
866 end
867 end
868 end
869 end
870 end
871 end
872 end
873 end
874 end
875 end
876 end
877 end
878 end
879 end
880 end
881 end
882 end
883 %---------------------------------------------------
884 function[A]=electrostatic(Q1,Q2, R1,R2,h,M,N,N1,epsilon)
885 for i=1:N
886     for j=1:M
887         if R1(i)>R2(j)
888             gamma(i,j)=R1(i)/R2(j);
889         else
890             if  R1(i)<R2(j)
891                 gamma(i,j)=R2(j)/R1(i);
892               else if R1(i)==R2(j);
893      gamma(i,j)=R2(j)/R1(i);
894          end
895             end
896         end
897         if h(i,j)>(R1(i)+R2(j))
898             r(i,j)=h(i,j)/(R1(i)+R2(j));
899         else if  h(i,j)<=(R1(i)+R2(j))
900             r(i,j)=(R1(i)+R2(j))/h(i,j);
901         end
902         end
903     y(i,j)=(((r(i,j)^2*(1+gamma(i,j))^2)-...
904     (1+(gamma(i,j))^2))/(2*gamma(i,j)));
905     beta(i,j)=acosh(y(i,j));
906     z(i,j)=exp(-beta(i,j));
907     S12=0;
908     S22=0;
909     S11=0;
910     for k=1:N1
911 
912         S_1(k)=(z(i,j)^k)/(((1-z(i,j)^(2*k)))*...
913         ((gamma(i,j)+y(i,j))-(y(i,j)^2-1)^(1/2)*...
```

```
914             (1+z(i,j)^(2*k))/(1-z(i,j)^(2*k))));
915             S11=S11+S_1(k);
916             S_2(k)=(z(i,j)^(2*k))/(1-(z(i,j)^(2*k)));
917             S12=S12+S_2(k);
918             S_3(k)=(z(i,j)^k)/(((1-z(i,j)^(2*k)))*((1-gamma(i,j)*...
919             y(i,j))-gamma(i,j)*(y(i,j)^2-1)^(1/2)*...
920             (1+z(i,j)^(2*k))/(1-z(i,j)^(2*k))));
921             S22=S22+S_3(k);
922         end
923         epsilon0=8.85418781762*10^(-12);
924         c11(i,j)=(2*gamma(i,j)*((y(i,j)^2-1)^(1/2))).*S11;
925         c22(i,j)=(2*gamma(i,j)*((y(i,j)^2-1)^(1/2))).*S22;
926         c12(i,j)=-((2*gamma(i,j)*((y(i,j)^2-1))^(1/2))/(r(i,j)*...
927         (1+gamma(i,j)))).*S12;
928         delta(i,j)=((c11(i,j)*c22(i,j)-c12(i,j)^2));
929          k=1/(4*pi*epsilon0);
930         k1=1/(4*pi*epsilon* epsilon0);
931             alpha(i,j)=Q2(i,j)/Q1(i,j);
932         if R1(i)>R2(j)
933             gamma(i,j)=R1(i)/R2(j);
934       W1(i,j)=((1/k1)*R2(j)*gamma(i,j))*...
935       ((1+gamma(i,j))/(2*alpha(i,j)))*...
936       ((alpha(i,j)^2*c11(i,j)-2*alpha(i,j)*...
937       c12(i,j)+c22(i,j))/delta(i,j));
938             else if (R1(i)<R2(j))
939                 gamma(i,j)=R2(j)/R1(i);
940      W1(i,j)=((1/k1)*R1(i)*gamma(i,j))*...
941       ((1+gamma(i,j))/(2*alpha(i,j)))*...
942     ((alpha(i,j)^2*c11(i,j)-2*alpha(i,j)*...
943     c12(i,j)+c22(i,j))/delta(i,j));
944      else if R1(i)==R2(j);
945     W1(i,j)=((1/k1)*R1(i)*gamma(i,j))*...
946     ((1+gamma(i,j))/(2*alpha(i,j)))*...
947     ((alpha(i,j)^2*c11(i,j)-2*alpha(i,j)*...
948     c12(i,j)+c22(i,j))/delta(i,j));
949     end
950     end
951     end
952     W2(i,j)=k*((Q1(i,j)*Q2(i,j)))/((R1(i)+R2(j)));
953     A1(i,j)=W1(i,j);
954     A2(i,j)=W2(i,j);
955     A(i,j)=A1(i,j)/A2(i,j);
956     end
957     end
958     return
959
960     function[cond2]=condmy(A)
961     [U,S,V]=SVD_2(A);
962     lambda_max=max(diag(S));
963     lambda_min=min(diag(S));
964     cond_1=((lambda_max)/(lambda_min));
965     cond2=(log(cond_1))/(log(10));
```

```
966  return
967
968  function [Uout,Sout,Vout] = SVD_2(A)
969        m = size(A,1);
970        n = size(A,2);
971        U = eye(m);
972        V = eye(n);
973        e = eps*fro(A);
974        while (sum(abs(A(~eye(m,n)))) > e)
975        for i = 1:n
976        for j = i+1:n
977            [J1,J2] = jacobi(A,m,n,i,j);
978            A = mtimes(J1,mtimes(A,J2));
979            U = mtimes(U,J1');
980            V = mtimes(J2',V);
981        end
982        for j = n+1:m
983            J1 = jacobi2(A,m,n,i,j);
984            A = mtimes(J1,A);
985            U = mtimes(U,J1');
986        end
987        end
988        end
989        S = A;
990        if (nargout < 3)
991           Uout = diag(S);
992        else
993            Uout = U; Sout = times(S,eye(m,n)); Vout = V;
994        end
995        end
996      function [J1,J2] = jacobi(A,m,n,i,j)
997         B = [A(i,i), A(i,j); A(j,i), A(j,j)];
998         [U,S,V] = tinySVD(B); %
999         J1 = eye(m);
1000        J1(i,i) = U(1,1);
1001        J1(j,j) = U(2,2);
1002        J1(i,j) = U(2,1);
1003        J1(j,i) = U(1,2);
1004        J2 = eye(n);
1005        J2(i,i) = V(1,1);
1006        J2(j,j) = V(2,2);
1007        J2(i,j) = V(2,1);
1008        J2(j,i) = V(1,2);
1009     end
1010     function J1 = jacobi2(A,m,n,i,j)
1011        B = [A(i,i), 0; A(j,i), 0];
1012        [U,S,V] = tinySVD(B);
1013        J1 = eye(m);
1014        J1(i,i) = U(1,1);
1015        J1(j,j) = U(2,2);
1016        J1(i,j) = U(2,1);
1017        J1(j,i) = U(1,2);
```

```
1018        end
1019        function [Uout,Sout,Vout] = tinySVD(A)
1020    t = rdivide((minus(A(1,2),A(2,1))),(plus(A(1,1),A(2,2))));
1021          c = rdivide(1,sqrt(1+t^2));
1022          s = times(t,c);
1023          R = [c,-s;s,c];
1024          M = mtimes(R,A);
1025          [U,S,V] = tinySymmetricSVD(M);
1026          U = mtimes(R',U);
1027          if (nargout < 3)
1028              Uout = diag(S);
1029          else
1030              Uout = U; Sout = S; Vout = V;
1031          end
1032          end
1033        function [Uout,Sout,Vout] = tinySymmetricSVD(A)
1034          if (A(2,1) == 0)
1035             S = A;
1036             U = eye(2);
1037             V = U;
1038          else
1039             w = A(1,1);
1040             y = A(2,1);
1041             z = A(2,2);
1042             ro = rdivide(minus(z,w),times(2,y));
1043    t2 = rdivide(sign(ro),plus(abs(ro),sqrt(plus(times(ro,ro),1))));
1044             t = t2;
1045             c = rdivide(1,sqrt(plus(1,times(t,t))));
1046             s = times(t,c);
1047             U = [c, -s; s, c];
1048             V = [c,  s;-s, c];
1049             S = mtimes(U,mtimes(A,V));
1050             U = U';
1051             V = V';
1052          end
1053          [U,S,V] = fixSVD(U,S,V);
1054          if (nargout < 3)
1055              Uout = diag(S);
1056          else
1057              Uout = U; Sout = S; Vout = V;
1058          end
1059          end
1060        function [U,S,V] = fixSVD(U,S,V)
1061          Z = [sign(S(1,1)),0; 0,sign(S(2,2))]; %
1062          U = mtimes(U,Z);
1063          S = mtimes(Z,S);
1064          if (S(1,1) < S(2,2))
1065              P = [0,1;1,0];
1066              U = mtimes(U,P);
1067              S = mtimes(P,mtimes(S,P));
1068              V = mtimes(P,V);
1069          end
```

```
1070        end
1071     function f = fro(M)
1072        f = sqrt(sum(sum(times(M,M))));
1073     end
1074     function s = sign(x)
1075        if (x > 0)
1076            s = 1;
1077        else
1078            s = -1;
1079        end
1080        end
```

5.9 Matlab Script Algorithm 2 for Mathematical Modeling Identification of Active Sites Interaction of Protein Molecules

Input parameters:

1. S_{100}, S_{20} are amino acid sequences of biological complexes ($S_{100} \geq S_{20}$).
2. sh0 is the initial shift
3. sh1 is the length of the frame
4. sh2 is the frame step
5. epsilon is the dielectric constant of the medium

Output parameters:

lg(cond(W) is the common logarithm of the condition number of the matrix W, where its elements are composed of the electrostatic potential energy which is created based on the interaction between pair of amino acid residues of biological complexes.

Compute:

lg(cond(W)) is the common logarithm of the condition number of the matrix W, which will allow a prediction the reactivity of the studied biological complexes.

```
1  clear all
2  clc
3  format long e
4  %Nap1
5  S_100=[ 'M' 'S' 'D' 'P' 'I' 'R' 'T' 'K' 'P'...
6  'K' 'S' 'S' 'M' 'Q' 'I' 'D' 'N' 'A' 'P' ...
7  'T' 'P' 'H' 'N' 'T' 'P' 'A' 'S' 'V' 'L' ...
8  'N' 'P' 'S' 'Y' 'L' 'K' 'N' 'G' 'N' 'P' ...
9  'V' 'R' 'A' 'Q' 'A' 'Q' 'E' 'Q' 'D' 'D' ...
10 'K' 'I' 'G' 'T' 'I' 'N' 'E' 'E' 'D' 'I' ...
11 'L' 'A' 'N' 'Q' 'P' 'L' 'L' 'L' 'Q' 'S' ...
12 'I' 'Q' 'D' 'R' 'L' 'G' 'S' 'L' 'V' 'G'...
13 'Q' 'D' 'S' 'G' 'Y' 'V' 'G' 'G' 'L' 'P' ...
14 'K' 'N' 'V' 'K' 'E' 'K' 'L' 'L' 'S' 'L'...
15 'K' 'T' 'L' 'Q' 'S' 'E' 'L' 'F' 'E' 'V' ...
16 'E' 'K' 'E' 'F' 'Q' 'V' 'E' 'M' 'F' 'E' ...
17 'L' 'E' 'N' 'K' 'F' 'L' 'Q' 'K' 'Y' 'K' ...
18 'P' 'I' 'W' 'E' 'Q' 'R' 'S' 'R' 'I' 'I'...
19 'S' 'G' 'Q' 'E' 'Q' 'P' 'K' 'P' 'E' 'Q'...
20 'I' 'A' 'K' 'G' 'Q' 'E' 'I' 'V' 'E' 'S' ...
21 'L' 'N' 'E' 'T' 'E' 'L' 'L' 'V' 'D' 'E' ...
22 'E' 'E' 'K' 'A' 'Q' 'N' 'D' 'S' 'E' 'E' ...
23 'E' 'Q' 'V' 'K' 'G' 'I' 'P' 'S' 'F' 'W' ...
24 'L' 'T' 'A' 'L' 'E' 'N' 'L' 'P' 'I' 'V' ...
25 'C' 'D' 'T' 'I' 'T' 'D' 'R' 'D' 'A' 'E' ...
26 'V' 'L' 'E' 'Y' 'L' 'Q' 'D' 'I' 'G' 'L'...
27 'E' 'Y' 'L' 'T' 'D' 'G' 'R' 'P' 'G' 'F' ...
28 'K' 'L' 'L' 'F' 'R' 'F' 'D' 'S' 'S' 'A' ...
29 'N' 'P' 'F' 'F' 'T' 'N' 'D' 'I' 'L' 'C' ...
30 'K' 'T' 'Y' 'F' 'Y' 'Q' 'K' 'E' 'L' 'G' ...
31 'Y' 'S' 'G' 'D' 'F' 'I' 'Y' 'D' 'H' 'A' ...
32 'E' 'G' 'C' 'E' 'I' 'S' 'W' 'K' 'D' 'N'...
33 'A' 'H' 'N' 'V' 'T' 'V' 'D' 'L' 'E' 'M' ...
34 'R' 'K' 'Q' 'R' 'N' 'K' 'T' 'T' 'K' 'Q' ...
35 'V' 'R' 'T' 'I' 'E' 'K' 'I' 'T' 'P' 'I'...
36 'E' 'S' 'F' 'F' 'N' 'F' 'F' 'D' 'P' 'P'...
37 'K' 'I' 'Q' 'N' 'E' 'D' 'Q' 'D' 'E' 'E' ...
38 'L' 'E' 'E' 'D' 'L' 'E' 'E' 'R' 'L' 'A' ...
39 'L' 'D' 'Y' 'S' 'I' 'G' 'E' 'Q' 'L' 'K' ...
40 'D' 'K' 'L' 'I' 'P' 'R' 'A' 'V' 'D' 'W' ...
41 'F' 'T' 'G' 'A' 'A' 'L' 'E' 'F' 'E' 'F'...
42 'E' 'E' 'D' 'E' 'E' 'E' 'A' 'D' 'E' 'D' ...
43 'E' 'D' 'E' 'E' 'E' 'D' 'D' 'D' 'H' 'G' ...
44 'L' 'E' 'D' 'D' 'D' 'G' 'E' 'S' 'A' 'E' ...
45 'E' 'Q' 'D' 'D' 'F' 'A' 'G' 'R' 'P' 'E' ...
46 'Q' 'A' 'P' 'E' 'C' 'K' 'Q' 'S' ]
47
48 S_20=[ 'M' 'S' 'D' 'P' 'I' 'R' 'T' 'K' 'P'...
49 'K' 'S' 'S' 'M' 'Q' 'I' 'D' 'N' 'A' 'P' ...
50 'T' 'P' 'H' 'N' 'T' 'P' 'A' 'S' 'V' 'L' ...
51 'N' 'P' 'S' 'Y' 'L' 'K' 'N' 'G' 'N' 'P' ...
52 'V' 'R' 'A' 'Q' 'A' 'Q' 'E' 'Q' 'D' 'D' ...
```

```
53 'K' 'I' 'G' 'T' 'I' 'N' 'E' 'E' 'D' 'I' ...
54 'L' 'A' 'N' 'Q' 'P' 'L' 'L' 'L' 'Q' 'S' ...
55 'I' 'Q' 'D' 'R' 'L' 'G' 'S' 'L' 'V' 'G'...
56 'Q' 'D' 'S' 'G' 'Y' 'V' 'G' 'G' 'L' 'P' ...
57 'K' 'N' 'V' 'K' 'E' 'K' 'L' 'L' 'S' 'L'...
58 'K' 'T' 'L' 'Q' 'S' 'E' 'L' 'F' 'E' 'V' ...
59 'E' 'K' 'E' 'F' 'Q' 'V' 'E' 'M' 'F' 'E' ...
60 'L' 'E' 'N' 'K' 'F' 'L' 'Q' 'K' 'Y' 'K' ...
61 'P' 'I' 'W' 'E' 'Q' 'R' 'S' 'R' 'I' 'I'...
62 'S' 'G' 'Q' 'E' 'Q' 'P' 'K' 'P' 'E' 'Q'...
63 'I' 'A' 'K' 'G' 'Q' 'E' 'I' 'V' 'E' 'S' ...
64 'L' 'N' 'E' 'T' 'E' 'L' 'L' 'V' 'D' 'E' ...
65 'E' 'E' 'K' 'A' 'Q' 'N' 'D' 'S' 'E' 'E' ...
66 'E' 'Q' 'V' 'K' 'G' 'I' 'P' 'S' 'F' 'W' ...
67 'L' 'T' 'A' 'L' 'E' 'N' 'L' 'P' 'I' 'V' ...
68 'C' 'D' 'T' 'I' 'T' 'D' 'R' 'D' 'A' 'E' ...
69 'V' 'L' 'E' 'Y' 'L' 'Q' 'D' 'I' 'G' 'L'...
70 'E' 'Y' 'L' 'T' 'D' 'G' 'R' 'P' 'G' 'F' ...
71 'K' 'L' 'L' 'F' 'R' 'F' 'D' 'S' 'S' 'A' ...
72 'N' 'P' 'F' 'F' 'T' 'N' 'D' 'I' 'L' 'C' ...
73 'K' 'T' 'Y' 'F' 'Y' 'Q' 'K' 'E' 'L' 'G' ...
74 'Y' 'S' 'G' 'D' 'F' 'I' 'Y' 'D' 'H' 'A' ...
75 'E' 'G' 'C' 'E' 'I' 'S' 'W' 'K' 'D' 'N'...
76 'A' 'H' 'N' 'V' 'T' 'V' 'D' 'L' 'E' 'M' ...
77 'R' 'K' 'Q' 'R' 'N' 'K' 'T' 'T' 'K' 'Q' ...
78 'V' 'R' 'T' 'I' 'E' 'K' 'I' 'T' 'P' 'I'...
79 'E' 'S' 'F' 'F' 'N' 'F' 'F' 'D' 'P' 'P'...
80 'K' 'I' 'Q' 'N' 'E' 'D' 'Q' 'D' 'E' 'E' ...
81 'L' 'E' 'E' 'D' 'L' 'E' 'E' 'R' 'L' 'A' ...
82 'L' 'D' 'Y' 'S' 'I' 'G' 'E' 'Q' 'L' 'K' ...
83 'D' 'K' 'L' 'I' 'P' 'R' 'A' 'V' 'D' 'W' ...
84 'F' 'T' 'G' 'A' 'A' 'L' 'E' 'F' 'E' 'F'...
85 'E' 'E' 'D' 'E' 'E' 'E' 'A' 'D' 'E' 'D' ...
86 'E' 'D' 'E' 'E' 'E' 'D' 'D' 'D' 'H' 'G' ...
87 'L' 'E' 'D' 'D' 'D' 'G' 'E' 'S' 'A' 'E' ...
88 'E' 'Q' 'D' 'D' 'F' 'A' 'G' 'R' 'P' 'E' ...
89 'Q' 'A' 'P' 'E' 'C' 'K' 'Q' 'S' ]
90 sh0=0;
91 sh1=70;
92 sh2=1;
93 n_el=10;
94 epsilon=1;
95 %-----------------------------------------------------------
96 len_S20=length(S_20);
97 len_S100=length(S_100);
98 N1=.1*len_S100;
99 del_len=len_S100-len_S20;
100 X=[];
101 Out=[];
102 F=[];
103 br=ceil(((len_S20-sh0)-(sh1-1))/sh2);
```

```
104 ost=len_S20-sh0-br*sh2-(sh1-sh2);
105 if ost~=0
106     OSTATOK_1=[S_20(len_S20-ost+1:len_S20)];
107     OSTATOK_2=[S_100(len_S20-ost+1:len_S20)];
108 end
109 for i=1:br
110      U_S20=[S_20(sh2*i+sh0-(sh2-1):sh2*i+sh0-(sh2-1)+(sh1-1))];
111     X=[S_100(sh2*i+sh0-(sh2-1):sh2*i+sh0-(sh2-1)+(sh1-1))];
112       S_1=X;
113     num=i;
114 N=length(S_1);
115 M=sh1;
116 S_2=U_S20;
117 [S_1,S_2,Q1,Q2,R1,R2,h]=potential(S_1,S_2,N1,N,M);
118 [A]=electrostatic(Q1,Q2, R1,R2,h,M,N,N1,epsilon);
119 [cond2]=condmy(A);
120 Out=[Out; X];
121 F=[F {num, S_1,S_2,(real(cond2))}'];
122 end
123 len_X=length(X);
124 len_Out=length(Out);
125 F;
126 barX=cell2mat(F(1,:));
127 barY=cell2mat(F(4,:));
128 SortF = sortrows(F',4);
129 barX_sort=cell2mat(SortF(:,1));
130 barY_sort=cell2mat(SortF(:,4));
131 minelem=[SortF(1:n_el,1) SortF(1:n_el,2) SortF(1:n_el,3)...
132  SortF(1:n_el,4)]
133 figure();
134 bar(barX,barY)
135 hold on
136 for i=1:n_el
137     bar(cell2mat(SortF(i,1)),cell2mat(SortF(i,4)),'red');
138 end
139 set(0,'DefaultTextInterpreter', 'latex');
140 set(0,'DefaultTextFontSize',14,...
141 'DefaultTextFontName','Arial Cyr');
142 xlabel('\bf Numer aminoacid residual');
143 set(0,'DefaultTextFontSize',14,...
144 'DefaultTextFontName','Arial Cyr');
145 ylabel('lg(cond(W))');
146 figure();
147 plot(barX,barY,'ok')
148 hold on
149 for i=1:n_el
150     plot(cell2mat(SortF(i,1)),cell2mat(SortF(i,4)),'*r')
151 end
152 set(0,'DefaultTextInterpreter', 'latex');
153 set(0,'DefaultTextFontSize',14,...
154 'DefaultTextFontName','Arial Cyr');
155 xlabel('\bf Numer aminoacid residual');
```

```
156 set(0,'DefaultTextFontSize',14,...
157 'DefaultTextFontName','Arial Cyr');
158 ylabel('lg(cond(W))');
159 [S_1,S_2,Q1,Q2,R1,R2,h]=potential(S_1,S_2,N1,N,M);
160 [A]=electrostatic(Q1,Q2, R1,R2,h,M,N,N1,epsilon);
161 [cond2]=condmy(A)
162 Out=[Out; X];
163 F=[F {num, S_1,S_2,(real(cond2))}'];
164 end
165 len_X=length(X);
166 len_Out=length(Out);
167 F;
168 barX=cell2mat(F(1,:));
169 barY=cell2mat(F(4,:));
170 SortF = sortrows(F',4);
171 barX_sort=cell2mat(SortF(:,1));
172 barY_sort=cell2mat(SortF(:,4));
173 minelem=[SortF(1:n_el,1) SortF(1:n_el,2) SortF(1:n_el,3)...
174     SortF(1:n_el,4)]
175 figure();
176 bar(barX,barY)
177 hold on
178 for i=1:n_el
179     bar(cell2mat(SortF(i,1)),cell2mat(SortF(i,4)),'red')
180 end
181 set(0,'DefaultTextInterpreter', 'latex');
182 set(0,'DefaultTextFontSize',14,'DefaultTextFontName',...
183     'Arial Cyr');
184 xlabel('\bf Numer aminoacid residual');
185 set(0,'DefaultTextFontSize',14,...
186 'DefaultTextFontName','Arial Cyr');
187 ylabel('lg(cond(W))');
188 figure();
189 plot(barX,barY,'ok')
190 hold on
191 for i=1:n_el
192     plot(cell2mat(SortF(i,1)),cell2mat(SortF(i,4)),'*r')
193 end
194 set(0,'DefaultTextInterpreter', 'latex');
195 set(0,'DefaultTextFontSize',14,'DefaultTextFontName',...
196     'Arial Cyr');
197 xlabel('\bf Numer aminoacid residual');
198 set(0,'DefaultTextFontSize',14,...
199 'DefaultTextFontName','Arial Cyr');
200 ylabel('lg(cond(W))');
201 %-----------------------------------------------------
202 function [S_1,S_2,Q1,Q2,R1,R2,h]=potential(S_1,S_2,N1,N,M);
203 N=length(S_1);
204 M=length(S_2);
205 Q1=[];
206 Q2=[];
207 R1=[];
```

```
208 R2=[];
209 for i=1:length(S_1);
210 for j=1:length(S_2);
211 if (S_1(i)=='D' & S_2(j)=='E')| (S_1(i)=='E' & S_2(j)=='D');
212 Q1(i,j)= 0.16e-19;
213 Q2(i,j)= 0.16e-19;
214 else
215 if (S_1(i)=='D' & S_2(j)=='D');
216 Q1(i,j)= 0.07e-19;
217 Q2(i,j)= 0.07e-19;
218 else
219 if (S_1(i)=='D' & S_2(j) =='C')|(S_1(i)=='C' & S_2(j) =='D');
220 Q1(i,j)= 0.05e-19;
221 Q2(i,j)= 0.05e-19;
222 else
223 if (S_1(i)=='D' &S_2(j)=='N')|(S_1(i)=='N' &S_2(j)=='D')|...
224 (S_1(i)=='D' &S_2(j)=='F')|(S_1(i)=='D' &S_2(j)=='Y')|...
225 (S_1(i)=='D' &S_2(j)=='Q')|(S_1(i)=='D' &S_2(j)=='S')|...
226 (S_1(i)=='F' &S_2(j)=='D')|(S_1(i)=='Y' &S_2(j)=='D')|...
227 (S_1(i)=='Q' &S_2(j)=='D')|(S_1(i)=='S' &S_2(j)=='D');
228 Q1(i,j)= 0.57e-19;
229 Q2(i,j)= 0.57e-19;
230 else
231 if ((S_1(i)=='D' & S_2(j)=='M')|(S_1(i)=='D' & S_2(j)=='T')|...
232 (S_1(i)=='D' & S_2(j)=='I')|(S_1(i)=='D' & S_2(j)=='G')|...
233 (S_1(i)=='D' & S_2(j)=='V')|(S_1(i)=='D' & S_2(j)=='W')|...
234 (S_1(i)=='D' & S_2(j)=='L')|(S_1(i)=='D' & S_2(j)=='A')|...
235 (S_1(i)=='M' & S_2(j)=='D')|(S_1(i)=='T' & S_2(j)=='D')|...
236 (S_1(i)=='I' & S_2(j)=='D')|(S_1(i)=='G' & S_2(j)=='D')|...
237 (S_1(i)=='V' & S_2(j)=='D')|(S_1(i)=='W' & S_2(j)=='D')|...
238 (S_1(i)=='L' & S_2(j)=='D')|(S_1(i)=='A' & S_2(j)=='D'));
239 Q1(i,j)= 0.64e-19;
240 Q2(i,j)= 0.64e-19;
241 else
242 if ((S_1(i)=='D'& S_2(j)=='P')|(S_1(i)=='P'& S_2(j)=='D'));
243 Q1(i,j)= 0.78e-19;
244 Q2(i,j)= 0.78e-19;
245 else
246 if ((S_1(i)=='D' & S_2(j)=='H')|(S_1(i)=='H'& S_2(j)=='D'));
247 Q1(i,j)= 0.99e-19;
248 Q2(i,j)= 0.99e-19;
249 else
250 if ((S_1(i)=='D'& S_2(j)=='K')|(S_1(i)=='K'& S_2(j)=='D'));
251 Q1(i,j)= 1.4e-19;
252 Q2(i,j)= 1.4e-19;
253 else
254 if ((S_1(i)=='D' & S_2(j)=='R')|(S_1(i)=='R'& S_2(j)=='D'));
255 Q1(i,j)= 1.59e-19;
256 Q2(i,j)= 1.59e-19;
257 else
258 if ((S_1(i)=='E'&S_2(j)=='E'));
259 Q1(i,j)= 0.16e-19;
```

```
260 Q2(i,j)= 0.16e-19;
261 else
262  if ((S_1(i)=='E'  & S_2(j)=='C')|(S_1(i)=='E' & S_2(j)=='F')|...
263 (S_1(i)=='E' & S_2(j)=='N')|(S_1(i)=='C'  & S_2(j)=='E')|...
264 (S_1(i)=='F' & S_2(j)=='E')|(S_1(i)=='N' & S_2(j)=='E'));
265 Q1(i,j)= 0.55e-19;
266 Q2(i,j)= 0.55e-19;
267 else
268 if  ((S_1(i)=='E' & S_2(j)=='Q')|(S_1(i)=='E' & S_2(j)=='Y')|...
269  (S_1(i)=='E' & S_2(j)=='S')|(S_1(i)=='E' & S_2(j)=='M')|...
270  (S_1(i)=='E' & S_2(j)=='T')|(S_1(i)=='E' & S_2(j)=='I')|...
271   (S_1(i)=='E' & S_2(j)=='G')|(S_1(i)=='E' & S_2(j)=='V')|...
272   (S_1(i)=='E' & S_2(j)=='W')|(S_1(i)=='E' & S_2(j)=='L')|...
273   (S_1(i)=='E' & S_2(j)=='A')|(S_1(i)=='Q' & S_2(j)=='E')|...
274   (S_1(i)=='Y' & S_2(j)=='E')| (S_1(i)=='S' & S_2(j)=='E')|...
275   (S_1(i)=='M' & S_2(j)=='E')|(S_1(i)=='T' & S_2(j)=='E')|...
276   (S_1(i)=='I' & S_2(j)=='E')|(S_1(i)=='G' & S_2(j)=='E')|...
277   (S_1(i)=='V' & S_2(j)=='E')|(S_1(i)=='W' & S_2(j)=='E')|...
278   (S_1(i)=='L' & S_2(j)=='E')|(S_1(i)=='A' & S_2(j)=='E'));
279 Q1(i,j)= 0.64e-19;
280 Q2(i,j)= 0.64e-19;
281 else
282 if ((S_1(i)=='E' & S_2(j)=='P' )|(S_1(i)=='P' & S_2(j)=='E'));
283 Q1(i,j)= 0.78e-19;
284 Q2(i,j)= 0.78e-19;
285 else
286 if ((S_1(i)=='E' & S_2(j)=='H')|(S_1(i)=='H' &S_2(j)=='E'));
287 Q1(i,j)= 0.99e-19;
288 Q2(i,j)= 0.99e-19;
289 else
290 if (S_1(i)=='E'& S_2(j)=='K')| (S_1(i)=='K'& S_2(j)=='E');
291 Q1(i,j)= 1.34e-19;
292 Q2(i,j)= 1.34e-19;
293 else
294 if (S_1(i)=='E' & S_2(j)=='R')|(S_1(i)=='R' & S_2(j)=='E');
295 Q1(i,j)= 1.58e-19;
296 Q2(i,j)= 1.58e-19;
297 else
298 if (S_1(i)=='C' & S_2(j)=='C')|(S_1(i)=='C' & S_2(j)=='F')|...
299 (S_1(i)=='C' & S_2(j)=='Q')|(S_1(i)=='C'& S_2(j)=='Y')|...
300 (S_1(i)=='C' & S_2(j)=='S')|(S_1(i)=='C' & S_2(j)=='M')|...
301 (S_1(i)=='C' & S_2(j)=='T')|(S_1(i)=='C' & S_2(j)=='I')|...
302 (S_1(i)=='C' & S_2(j)=='G')|(S_1(i)=='C' & S_2(j)=='V')|...
303 (S_1(i)=='C' & S_2(j)=='W')|(S_1(i)=='C' & S_2(j)=='L')|...
304 (S_1(i)=='C' & S_2(j)=='L')|(S_1(i)=='C' & S_2(j)=='A')|...
305 (S_1(i)=='F' & S_2(j)=='C')|(S_1(i)=='Q' & S_2(j)=='C')|...
306 (S_1(i)=='Y'& S_2(j)=='C')|(S_1(i)=='S' & S_2(j)=='C')|...
307 (S_1(i)=='M' & S_2(j)=='C')|(S_1(i)=='T' & S_2(j)=='C')|...
308 (S_1(i)=='I' & S_2(j)=='C')|(S_1(i)=='G' & S_2(j)=='C')|...
309 (S_1(i)=='V' & S_2(j)=='C')|(S_1(i)=='W' & S_2(j)=='C')|...
310 (S_1(i)=='L' & S_2(j)=='C')|( S_1(i)=='A' & S_2(j)=='C');
311 Q1(i,j)=0.74e-19;
```

```
312 Q2(i,j)=0.74e-19;
313 else
314 if (S_1(i)=='C' & S_2(j)=='H')| (S_1(i)=='H' & S_2(j)=='C');
315 Q1(i,j)= 0.99e-19;
316 Q2(i,j)= 0.99e-19;
317 else
318 if (S_1(i)=='C' & S_2(j)=='K')|(S_1(i)=='K' & S_2(j)=='C');
319 Q1(i,j)= 1.34e-19;
320 Q2(i,j)= 1.34e-19;
321 else
322 if (S_1(i)=='C' & S_2(j)=='R')|(S_1(i)=='R' & S_2(j)=='C');
323 Q1(i,j)= 1.59e-19;
324 Q2(i,j)= 1.59e-19;
325 else
326 if (S_1(i)=='N' & S_2(j)=='N')|(S_1(i)=='N' & S_2(j)=='F')|...
327 (S_1(i)=='N' & S_2(j)=='Q')|(S_1(i)=='N' & S_2(j)=='Y')|...
328 (S_1(i)=='N' & S_2(j)=='S')|(S_1(i)=='N'& S_2(j)=='M')|...
329 (S_1(i)=='F' & S_2(j)=='N')|(S_1(i)=='Q' & S_2(j)=='N')|...
330 (S_1(i)=='Y' & S_2(j)=='N')|(S_1(i)=='S' & S_2(j)=='N')|...
331 (S_1(i)=='M'& S_2(j)=='N');
332 Q1(i,j)=0.74e-19;
333 Q2(i,j)=0.74e-19;
334 else
335 if (S_1(i)=='N' & S_2(j)=='H')|(S_1(i)=='H' & S_2(j)=='N')
336 Q1(i,j)= 0.99e-19;
337 Q2(i,j)= 0.99e-19;
338 else
339 if(S_1(i)=='N' & S_2(j)=='K')|(S_1(i)=='K' & S_2(j)=='N');
340 Q1(i,j)= 1.05e-19;
341 Q2(i,j)= 1.05e-19;
342 else
343 if (S_1(i)=='N' & S_2(j)=='R')|(S_1(i)=='R' & S_2(j)=='N');
344 Q1(i,j)= 1.1e-19;
345 Q2(i,j)= 1.1e-19;
346 else
347 if ((S_1(i)=='F' & S_2(j)=='F')|(S_1(i)=='F' & S_2(j)=='Q'));
348 Q1(i,j)=0.74e-19;
349 Q2(i,j)=0.74e-19;
350 else
351 if ((S_1(i)=='F' & S_2(j)=='Y')|(S_1(i)=='F' & S_2(j)=='S')|...
352 (S_1(i)=='F' & S_2(j)=='M')|(S_1(i)=='Q' & S_2(j)=='F')|...
353 (S_1(i)=='Y' & S_2(j)=='F'));
354 Q1(i,j)=0.74e-19;
355 Q2(i,j)=0.74e-19;
356 else
357 if (S_1(i)=='S' & S_2(j)=='F')|(S_1(i)=='M' & S_2(j)=='F');
358 Q1(i,j)=0.74e-19;
359 Q2(i,j)=0.74e-19;
360 else
361 if (S_1(i)=='F' & S_2(j)=='H')|(S_1(i)=='H' & S_2(j)=='F');
362 Q1(i,j)= 0.99e-19;
363 Q2(i,j)= 0.99e-19;
```

```
364 else
365 if (S_1(i)=='F' & S_2(j)=='K')|(S_1(i)=='K' & S_2(j)=='F');
366 Q1(i,j)= 1.05e-19;
367 Q2(i,j)= 1.05e-19;
368 else
369 if (S_1(i)=='F' & S_2(j)=='R')|(S_1(i)=='R' & S_2(j)=='F');
370 Q1(i,j)= 1.1e-19;
371 Q2(i,j)= 1.1e-19;
372 else
373 if (S_1(i)=='Q' & S_2(j)=='H')|(S_1(i)=='H' & S_2(j)=='Q');
374 Q1(i,j)= 0.99e-19;
375 Q2(i,j)= 0.99e-19;
376 else
377 if (S_1(i)=='Q' & S_2(j)=='K')|(S_1(i)=='K' & S_2(j)=='Q');
378 Q1(i,j)= 1.05e-19;
379 Q2(i,j)= 1.05e-19;
380 else
381 if (S_1(i)=='Q' & S_2(j)=='R')|(S_1(i)=='R' & S_2(j)=='Q');
382 Q1(i,j)= 1.1e-19;
383 Q2(i,j)= 1.1e-19;
384 else
385 if (S_1(i)=='Q' & S_2(j)=='H')|(S_1(i)=='H' & S_2(j)=='Q');
386 Q1(i,j)= 0.99e-19;
387 Q2(i,j)= 0.99e-19;
388 else
389 if (S_1(i)=='Y' & S_2(j)=='K')|(S_1(i)=='K' & S_2(j)=='Y')
390 Q1(i,j)= 1.05e-19;
391 Q2(i,j)= 1.05e-19;
392 else
393 if (S_1(i)=='Y' & S_2(j)=='R')|(S_1(i)=='R' & S_2(j)=='Y');
394 Q1(i,j)= 1.1e-19;
395 Q2(i,j)= 1.1e-19;
396 else
397 if (S_1(i)=='S' & S_2(j)=='H')|(S_1(i)=='H' & S_2(j)=='S');
398 Q1(i,j)= 0.99e-19;
399 Q2(i,j)= 0.99e-19;
400 else
401 if (S_1(i)=='S' & S_2(j)=='K')|(S_1(i)=='K' & S_2(j)=='S');
402 Q1(i,j)= 1e-19;
403 Q2(i,j)= 1e-19;
404 else
405 if (S_1(i)=='S' & S_2(j)=='R')|(S_1(i)=='R' & S_2(j)=='S');
406 Q1(i,j)= 1.1e-19;
407 Q2(i,j)= 1.1e-19;
408 else
409 if (S_1(i)=='M' & S_2(j)=='H')|(S_1(i)=='H' & S_2(j)=='M');
410 Q1(i,j)= 0.99e-19;
411 Q2(i,j)= 0.99e-19;
412 else
413 if (S_1(i)=='M' & S_2(j)=='K')|(S_1(i)=='K' & S_2(j)=='M');
414 Q1(i,j)= 1e-19;
415 Q2(i,j)= 1e-19;
```

```
416 else
417 if (S_1(i)=='M' & S_2(j)=='R')|(S_1(i)=='R' & S_2(j)=='M');
418 Q1(i,j)= 1.1e-19;
419 Q2(i,j)= 1.1e-19;
420 else
421 if (S_1(i)=='T' & S_2(j)=='H')|(S_1(i)=='H' & S_2(j)=='T');
422 Q1(i,j)= 0.99e-19;
423 Q2(i,j)= 0.99e-19;
424 else
425 if (S_1(i)=='T' & S_2(j)=='K')|(S_1(i)=='K' & S_2(j)=='T');
426 Q1(i,j)= 1e-19;
427 Q2(i,j)= 1e-19;
428 else
429 if (S_1(i)=='T' & S_2(j)=='R')|(S_1(i)=='R' & S_2(j)=='T');
430 Q1(i,j)= 1.05e-19;
431 Q2(i,j)= 1.05e-19;
432 else
433 if (S_1(i)=='I' & S_2(j)=='H')|(S_1(i)=='H' & S_2(j)=='I');
434 Q1(i,j)= 0.99e-19;
435 Q2(i,j)= 0.99e-19;
436 else
437 if (S_1(i)=='I' & S_2(j)=='K')|(S_1(i)=='K' & S_2(j)=='I');
438 Q1(i,j)= 1e-19;
439 Q2(i,j)= 1e-19;
440 else
441 if (S_1(i)=='I' & S_2(j)=='R')|(S_1(i)=='R' & S_2(j)=='I');
442 Q1(i,j)= 1.05e-19;
443 Q2(i,j)= 1.05e-19;
444 else
445 if (S_1(i)=='G' & S_2(j)=='H')|(S_1(i)=='H' & S_2(j)=='G');
446 Q1(i,j)= 0.99e-19;
447 Q2(i,j)= 0.99e-19;
448 else
449 if (S_1(i)=='G' & S_2(j)=='K')|(S_1(i)=='K' & S_2(j)=='G');
450 Q1(i,j)= 1e-19;
451 Q2(i,j)= 1e-19;
452 else
453 if (S_1(i)=='G' & S_2(j)=='R')|(S_1(i)=='R' & S_2(j)=='G');
454 Q1(i,j)= 1.05e-19;
455 Q2(i,j)= 1.05e-19;
456 else
457 if (S_1(i)=='V' & S_2(j)=='H')|(S_1(i)=='H' & S_2(j)=='V');
458 Q1(i,j)= 0.99e-19;
459 Q2(i,j)= 0.99e-19;
460 else
461 if (S_1(i)=='V' & S_2(j)=='K')|(S_1(i)=='K' & S_2(j)=='V');
462 Q1(i,j)= 1e-19;
463 Q2(i,j)= 1e-19;
464 else
465 if (S_1(i)=='V' & S_2(j)=='R')|(S_1(i)=='R' & S_2(j)=='V');
466 Q1(i,j)= 1.05e-19;
467 Q2(i,j)= 1.05e-19;
```

```
468 else
469 if (S_1(i)=='W' & S_2(j)=='H')|(S_1(i)=='H' & S_2(j)=='W');
470 Q1(i,j)= 0.99e-19;
471 Q2(i,j)= 0.99e-19;
472 else
473 if (S_1(i)=='W' & S_2(j)=='K')|(S_1(i)=='K' & S_2(j)=='W');
474 Q1(i,j)= 1e-19;
475 Q2(i,j)= 1e-19;
476 else
477 if (S_1(i)=='W' & S_2(j)=='R')|(S_1(i)=='R' & S_2(j)=='W');
478 Q1(i,j)= 1.05e-19;
479 Q2(i,j)= 1.05e-19;
480 else
481 if (S_1(i)=='L' & S_2(j)=='H')|(S_1(i)=='H' & S_2(j)=='L');
482 Q1(i,j)= 0.99e-19;
483 Q2(i,j)= 0.99e-19;
484 else
485 if (S_1(i)=='L' & S_2(j)=='K')|(S_1(i)=='K' & S_2(j)=='L');
486 Q1(i,j)= 1e-19;
487 Q2(i,j)= 1e-19;
488 else
489 if (S_1(i)=='L' & S_2(j)=='R')|(S_1(i)=='R' & S_2(j)=='L');
490 Q1(i,j)= 1.05e-19;
491 Q2(i,j)= 1.05e-19;
492 else
493 if (S_1(i)=='A' & S_2(j)=='H')|(S_1(i)=='H' & S_2(j)=='A');
494 Q1(i,j)= 0.99e-19;
495 Q2(i,j)= 0.99e-19;
496 else
497 if (S_1(i)=='A' & S_2(j)=='K')|(S_1(i)=='K' & S_2(j)=='A');
498 Q1(i,j)= 1e-19;
499 Q2(i,j)= 1e-19;
500 else
501 if (S_1(i)=='A' & S_2(j)=='R')|(S_1(i)=='R' & S_2(j)=='A');
502 Q1(i,j)= 1.05e-19;
503 Q2(i,j)= 1.05e-19;
504 else
505 if (S_1(i)=='P' & S_2(j)=='H')|(S_1(i)=='H' & S_2(j)=='P');
506 Q1(i,j)= 0.99e-19;
507 Q2(i,j)= 0.99e-19;
508 else
509 if (S_1(i)=='P' & S_2(j)=='K')|(S_1(i)=='K' & S_2(j)=='P');
510 Q1(i,j)= 0.82e-19;
511 Q2(i,j)= 0.82e-19;
512 else
513 if (S_1(i)=='P' & S_2(j)=='R')|(S_1(i)=='R' & S_2(j)=='P');
514 Q1(i,j)= 0.96e-19;
515 Q2(i,j)= 0.96e-19;
516 else
517 if (S_1(i)=='H' & S_2(j)=='H');
518 Q1(i,j)= 0.82e-19;
519 Q2(i,j)= 0.82e-19;
```

```
520 else
521 if (S_1(i)=='H' & S_2(j)=='K')|(S_1(i)=='K' & S_2(j)=='H');
522 Q1(i,j)= 0.82e-19;
523 Q2(i,j)= 0.82e-19;
524 else
525 if (S_1(i)=='H' & S_2(j)=='R')|(S_1(i)=='R' & S_2(j)=='H');
526 Q1(i,j)= 0.74e-19;
527 Q2(i,j)= 0.74e-19;
528 else
529 if (S_1(i)=='K' & S_2(j)=='K');
530 Q1(i,j)= 0.54e-19;
531 Q2(i,j)= 0.54e-19;
532 else
533 if (S_1(i)=='K' & S_2(j)=='R')|(S_1(i)=='R' & S_2(j)=='K');
534 Q1(i,j)= 0.41e-19;
535 Q2(i,j)= 0.41e-19;
536 else
537 if (S_1(i)=='R' & S_2(j)=='R');
538 Q1(i,j)= 0.16e-19;
539 Q2(i,j)= 0.16e-19;
540 else
541 Q1(i,j)= 0.824e-19;
542 Q2(i,j)= 0.824e-19;
543 end
544 end
545 end
546 end
547 end
548 end
549 end
550 end
551 end
552 end
553 end
554 end
555 end
556 end
557 end
558 end
559 end
560 end
561 end
562 end
563 end
564 end
565 end
566 end
567 end
568 end
569 end
570 end
571 end
```

```
572 end
573 end
574 end
575 end
576 end
577 end
578 end
579 end
580 end
581 end
582 end
583 end
584 end
585 end
586 end
587 end
588 end
589 end
590 end
591 end
592 end
593 end
594 end
595 end
596 end
597 end
598 end
599 end
600 end
601 end
602 end
603 end
604 end
605 end
606 end
607 end
608 end
609 end
610 end
611 end
612 end
613 end
614 end
615 end
616 end
617  Q3=[];
618  Q4=[];
619  R1=[];
620  R2=[];
621 for i=1:length(S_1);
622 if (S_1(i)=='A');
623 R1(i)=0.6e-9;
```

```
624 else
625 if (S_1(i)=='R');
626 R1(i)=0.809e-9;
627 else
628 if (S_1(i)=='N');
629 R1(i)=0.682e-9;
630 else
631 if (S_1(i)=='D');
632 R1(i)=0.665e-9;
633 else
634 if (S_1(i)=='C');
635 R1(i)=0.629e-9;
636 else
637 if (S_1(i)=='Q');
638 R1(i)=0.725e-9;
639 else
640 if (S_1(i)=='E');
641 R1(i)=0.714e-9;
642 else
643 if (S_1(i)=='G');
644 R1(i)=0.537e-9;
645 else
646 if (S_1(i)=='H');
647 R1(i)=0.732e-9;
648 else
649 if (S_1(i)=='I');
650 R1(i)=0.735e-9;
651 else
652 if (S_1(i)=='L');
653 R1(i)=0.734e-9;
654 else
655 if (S_1(i)=='K');
656 R1(i)=0.737e-9;
657 else
658 if (S_1(i)=='M');
659 R1(i)=0.741e-9;
660 else
661 if (S_1(i)=='F');
662 R1(i)=0.781e-9;
663 else
664 if (S_1(i)=='P');
665 R1(i)=0.672e-9;
666 else
667 if (S_1(i)=='S');
668 R1(i)=0.615e-9;
669 else
670 if (S_1(i)=='T');
671 R1(i)=0.659e-9;
672 else
673 if (S_1(i)=='W');
674 R1(i)=0.826e-9;
675 else
```

```
676 if (S_1(i)=='Y');
677 R1(i)=0.781e-9;
678 else
679 if (S_1(i)=='V');
680 R1(i)=0.694e-9;
681 end
682 end
683 end
684 end
685 end
686 end
687 end
688 end
689 end
690 end
691 end
692 end
693 end
694 end
695 end
696 end
697 end
698 end
699 end
700 end
701 for j=1:length(S_2);
702 if (S_2(j)=='A');
703 R2(j)=0.6e-9;
704 else
705 if (S_2(j)=='R');
706 R2(j)= 0.809e-9;
707 else
708 if (S_2(j)=='N');
709 R2(j)=0.682e-9;
710 else
711 if (S_2(j)=='D');
712 R2(j)=0.665e-9;
713 else
714 if (S_2(j)=='C');
715 R2(j)=0.629e-9;
716 else
717 if (S_2(j)=='Q');
718 R2(j)=0.725e-9;
719 else
720 if (S_2(j)=='E');
721 R2(j)=0.714e-9;
722 else
723 if (S_2(j)=='G');
724 R2(j)=0.537e-9;
725 else
726 if (S_2(j)=='H');
727 R2(j)=0.732e-9;
```

```
728 else
729 if (S_2(j)=='I');
730 R2(j)=0.735e-9;
731 else
732 if(S_2(j)=='L');
733 R2(j)=0.734e-9;
734 else
735 if (S_2(j)=='K')
736 R2(j)=0.737e-9;
737 else
738 if (S_2(j)=='M')
739 R2(j)=0.741e-9;
740 else
741 if (S_2(j)=='F')
742 R2(j)=0.781e-9;
743 else
744 if (S_2(j)=='P');
745 R2(j)=0.672e-9;
746 else
747 if (S_2(j)=='S');
748 R2(j)=0.615e-9;
749 else
750 if (S_2(j)=='T');
751 R2(j)=0.659e-9;
752 else
753 if (S_2(j)=='W');
754 R2(j)=0.826e-9;
755 else
756 if (S_2(j)=='Y');
757 R2(j)=0.781e-9;
758 else
759 if (S_2(j)=='V');
760 R2(j)=0.694e-9;
761 end
762 end
763 end
764 end
765 end
766 end
767 end
768 end
769 end
770 end
771 end
772 end
773 end
774 end
775 end
776 end
777 end
778 end
779 end
```

```
780 end
781 end
782 end
783  Ra=0.6e-9;
784  Rr=0.809e-9;
785  Rn=0.682e-9;
786  Rd=0.665e-9;
787  Rc=0.629e-9;
788  Rq=0.725e-9;
789  Re=0.714e-9;
790  Rg=0.725e-9;
791  Rh=0.732e-9;
792  Ri=0.735e-9;
793  Rl=0.734e-9;
794  Rk=0.737e-9;
795  Rm=0.741e-9;
796  Rf=0.781e-9;
797  Rp=0.672e-9;
798  Rs=0.615e-9;
799  Rt=0.659e-9;
800  Rw=0.826e-9;
801  Ry=0.781e-9;
802  Rv=0.694e-9;
803 for i=1:length(S_1);
804 for j=1:length(S_2);
805 if (S_1(i)=='R'& S_2(j)=='D');
806    h(i,j)=.15*10^(-9)+Rr+Rd;
807 else
808 if (S_1(i)=='R'& S_2(j)=='E');
809     h(i,j)=.15*10^(-9)+Rr+Re;
810 else
811 if (S_1(i)=='D'& S_2(j)=='R');
812     h(i,j)=.15*10^(-9)+Rd+Rr;
813 else
814 if (S_1(i)=='D'& S_2(j)=='H');
815    h(i,j)=.15*10^(-9)+Rd+Rh;
816 else
817 if (S_1(i)=='D'& S_2(j)=='R');
818     h(i,j)=.15*10^(-9)+Rd+Rr;
819 else
820 if (S_1(i)=='D'& S_2(j)=='H');
821     h(i,j)=.15*10^(-9)+Rd+Rh;
822 else
823 if (S_1(i)=='D'& S_2(j)=='K');
824     h(i,j)=.15*10^(-9)+Rd+Rk;
825 else
826 if (S_1(i)=='E')& (S_2(j)=='R');
827     h(i,j)=.15*10^(-9)+Re+Rr;
828 else
829 if (S_1(i)=='E'& S_2(j)=='H');
830    h(i,j)=.15*10^(-9)+Re+Rh;
831 else
```

```
832 if (S_1(i)=='E'& S_2(j)=='K');
833    h(i,j)=.15*10^(-9)+Re+Rk;
834 else
835 if (S_1(i)=='H'& S_2(j)=='D')
836     h(i,j)=.15*10^(-9)+Rh+Rd;
837 else
838 if (S_1(i)=='H'& S_2(j)=='E')
839     h(i,j)=.15*10^(-9)+Rh+Re;
840 else
841 if (S_1(i)=='R'& S_2(j)=='R')
842     h(i,j)=.4*10^(-9)+Rr+Rr;
843 else
844 if (S_1(i)=='R'& S_2(j)=='H')
845    h(i,j)=.4*10^(-9)+Rr+Rh;
846 else
847 if (S_1(i)=='R'& S_2(j)=='H')
848     h(i,j)=.4*10^(-9)+Rr+Rh;
849 else
850 if (S_1(i)=='R'& S_2(j)=='K')
851     h(i,j)=.4*10^(-9)+Rr+Rk;
852 else
853 if (S_1(i)=='D'& S_2(j)=='E');
854     h(i,j)=.4*10^(-9)+Rd+Re;
855 else
856 if (S_1(i)=='D'& S_2(j)=='D');
857     h(i,j)=.4*10^(-9)+Rd+Rd;
858 else
859 if (S_1(i)=='H'& S_2(j)=='R')
860     h(i,j)=.4*10^(-9)+Rh+Rr;
861 else
862 if (S_1(i)=='H'& S_2(j)=='H')
863     h(i,j)=.4*10^(-9)+Rh+Rh;
864 else
865 if (S_1(i)=='H'& S_2(j)=='K')
866      h(i,j)=.4*10^(-9)+Rh+Rk;
867 else
868 if (S_1(i)=='K'& S_2(j)=='R')
869    h(i,j)=.4*10^(-9)+Rk+Rr;
870 else
871 if (S_1(i)=='K'& S_2(j)=='H')
872     h(i,j)=.4*10^(-9)+Rk+Rh;
873 else
874 if (S_1(i)=='K'& S_2(j)=='K')
875     h(i,j)=.4*10^(-9)+Rk+Rk;
876 else
877 if (S_1(i)=='N'& S_2(j)=='Q')
878    h(i,j)=.25*10^(-9)+Rn+Rq;
879 else
880 if (S_1(i)=='N'& S_2(j)=='S')
881    h(i,j)=.25*10^(-9)+Rn+Rs;
882 else
883 if (S_1(i)=='N'& S_2(j)=='Y')
```

```
884         h(i,j)=.25*10^(-9)+Rn+Ry;
885     else
886     if (S_1(i)=='Q'& S_2(j)=='S')|(S_1(i)=='Q')& (S_2(j)=='Y');
887         h(i,j)=.25*10^(-9)+Rq+Rs;
888     else
889     if (S_1(i)=='Q')& (S_2(j)=='Y');
890          h(i,j)=.25*10^(-9)+Rq+Ry;
891     else
892     if (S_1(i)=='S'& S_2(j)=='Y');
893         h(i,j)=.25*10^(-9)+Rs+Ry;
894     else
895         h(i,j)=1.76*10^(-9);
896     end
897     end
898     end
899     end
900     end
901     end
902     end
903     end
904     end
905     end
906     end
907     end
908     end
909     end
910     end
911     end
912     end
913     end
914     end
915     end
916     end
917     end
918     end
919     end
920     end
921     end
922     end
923     end
924     end
925     end
926     end
927     end
928
929     function[A]=electrostatic(Q1,Q2, R1,R2,h,M,N,N1,epsilon)
930     for i=1:N
931         for j=1:M
932             if R1(i)>R2(j)
933                 gamma(i,j)=R1(i)/R2(j);
934             else
935                 if  R1(i)<R2(j)
```

```
936                     gamma(i,j)=R2(j)/R1(i);
937                 else if R1(i)==R2(j);
938       gamma(i,j)=R2(j)/R1(i);
939              end
940                   end
941            end
942            if h(i,j)>(R1(i)+R2(j))
943                r(i,j)=h(i,j)/(R1(i)+R2(j));
944            else if  h(i,j)<=(R1(i)+R2(j))
945                r(i,j)=(R1(i)+R2(j))/h(i,j);
946            end
947            end
948        y(i,j)=(((r(i,j)^2*(1+gamma(i,j))^2)-...
949        (1+(gamma(i,j))^2))/(2*gamma(i,j)));
950        beta(i,j)=acosh(y(i,j));
951        z(i,j)=exp(-beta(i,j));
952        S12=0;
953        S22=0;
954        S11=0;
955        for k=1:N1
956 
957            S_1(k)=(z(i,j)^k)/(((1-z(i,j)^(2*k)))*...
958            ((gamma(i,j)+y(i,j))-(y(i,j)^2-1)^(1/2)*...
959            (1+z(i,j)^(2*k))/(1-z(i,j)^(2*k))));
960            S11=S11+S_1(k);
961            S_2(k)=(z(i,j)^(2*k))/(1-(z(i,j)^(2*k)));
962            S12=S12+S_2(k);
963     S_3(k)=(z(i,j)^k)/(((1-z(i,j)^(2*k)))*((1-gamma(i,j)*...
964            y(i,j))-gamma(i,j)*(y(i,j)^2-1)^(1/2)*...
965            (1+z(i,j)^(2*k))/(1-z(i,j)^(2*k))));
966            S22=S22+S_3(k);
967        end
968        epsilon0=8.85418781762*10^(-12);
969        c11(i,j)=(2*gamma(i,j)*((y(i,j)^2-1)^(1/2))).*S11;
970        c22(i,j)=(2*gamma(i,j)*((y(i,j)^2-1)^(1/2))).*S22;
971        c12(i,j)=-((2*gamma(i,j)*((y(i,j)^2-1))^(1/2))/(r(i,j)*...
972        (1+gamma(i,j)))).*S12;
973        delta(i,j)=((c11(i,j)*c22(i,j)-c12(i,j)^2));
974         k=1/(4*pi*epsilon0);
975        k1=1/(4*pi*epsilon* epsilon0);
976            alpha(i,j)=Q2(i,j)/Q1(i,j);
977        if R1(i)>R2(j)
978            gamma(i,j)=R1(i)/R2(j);
979     W1(i,j)=((1/k1)*R2(j)*gamma(i,j))*...
980     ((1+gamma(i,j))/(2*alpha(i,j)))*...
981     ((alpha(i,j)^2*c11(i,j)-2*alpha(i,j)*...
982     c12(i,j)+c22(i,j))/delta(i,j));
983            else if (R1(i)<R2(j))
984                gamma(i,j)=R2(j)/R1(i);
985    W1(i,j)=((1/k1)*R1(i)*gamma(i,j))*...
986     ((1+gamma(i,j))/(2*alpha(i,j)))*...
987   ((alpha(i,j)^2*c11(i,j)-2*alpha(i,j)*...
```

```
988  c12(i,j)+c22(i,j))/delta(i,j));
989   else if R1(i)==R2(j);
990  W1(i,j)=((1/k1)*R1(i)*gamma(i,j))*...
991  ((1+gamma(i,j))/(2*alpha(i,j)))*...
992  ((alpha(i,j)^2*c11(i,j)-2*alpha(i,j)*...
993  c12(i,j)+c22(i,j))/delta(i,j));
994  end
995  end
996  end
997  W2(i,j)=k*((Q1(i,j)*Q2(i,j)))/((R1(i)+R2(j)));
998  A1(i,j)=W1(i,j);
999  A2(i,j)=W2(i,j);
1000 A(i,j)=A1(i,j)/A2(i,j);
1001 end
1002 end
1003 return
1004
1005 function[cond2]=condmy(A)
1006 [U,S,V]=SVD_2(A);
1007 lambda_max=max(diag(S));
1008 lambda_min=min(diag(S));
1009 cond_1=((lambda_max)/(lambda_min));
1010 cond2=(log(cond_1))/(log(10));
1011 return
1012
1013 function [Uout,Sout,Vout] = SVD_2(A)
1014        m = size(A,1);
1015        n = size(A,2);
1016        U = eye(m);
1017        V = eye(n);
1018        e = eps*fro(A);
1019        while (sum(abs(A(~eye(m,n)))) > e)
1020        for i = 1:n
1021        for j = i+1:n
1022            [J1,J2] = jacobi(A,m,n,i,j);
1023            A = mtimes(J1,mtimes(A,J2));
1024            U = mtimes(U,J1');
1025            V = mtimes(J2',V);
1026        end
1027        for j = n+1:m
1028            J1 = jacobi2(A,m,n,i,j);
1029            A = mtimes(J1,A);
1030            U = mtimes(U,J1');
1031        end
1032        end
1033        end
1034        S = A;
1035        if (nargout < 3)
1036           Uout = diag(S);
1037        else
1038            Uout = U; Sout = times(S,eye(m,n)); Vout = V;
1039        end
```

```
1040        end
1041      function [J1,J2] = jacobi(A,m,n,i,j)
1042         B = [A(i,i), A(i,j); A(j,i), A(j,j)];
1043         [U,S,V] = tinySVD(B); %
1044         J1 = eye(m);
1045         J1(i,i) = U(1,1);
1046         J1(j,j) = U(2,2);
1047         J1(i,j) = U(2,1);
1048         J1(j,i) = U(1,2);
1049         J2 = eye(n);
1050         J2(i,i) = V(1,1);
1051         J2(j,j) = V(2,2);
1052         J2(i,j) = V(2,1);
1053         J2(j,i) = V(1,2);
1054      end
1055      function J1 = jacobi2(A,m,n,i,j)
1056         B = [A(i,i), 0; A(j,i), 0];
1057         [U,S,V] = tinySVD(B);
1058         J1 = eye(m);
1059         J1(i,i) = U(1,1);
1060         J1(j,j) = U(2,2);
1061         J1(i,j) = U(2,1);
1062         J1(j,i) = U(1,2);
1063      end
1064      function [Uout,Sout,Vout] = tinySVD(A)
1065   t = rdivide((minus(A(1,2),A(2,1))),(plus(A(1,1),A(2,2))));
1066        c = rdivide(1,sqrt(1+t^2));
1067        s = times(t,c);
1068        R = [c,-s;s,c];
1069        M = mtimes(R,A);
1070        [U,S,V] = tinySymmetricSVD(M);
1071        U = mtimes(R',U);
1072        if (nargout < 3)
1073            Uout = diag(S);
1074        else
1075            Uout = U; Sout = S; Vout = V;
1076        end
1077        end
1078      function [Uout,Sout,Vout] = tinySymmetricSVD(A)
1079        if (A(2,1) == 0)
1080           S = A;
1081           U = eye(2);
1082           V = U;
1083        else
1084           w = A(1,1);
1085           y = A(2,1);
1086           z = A(2,2);
1087           ro = rdivide(minus(z,w),times(2,y));
1088    t2 = rdivide(sign(ro),plus(abs(ro),sqrt(plus(times(ro,ro),1))));
1089           t = t2;
1090           c = rdivide(1,sqrt(plus(1,times(t,t))));
1091           s = times(t,c);
```

```
1092          U = [c, -s; s, c];
1093          V = [c,  s;-s, c];
1094          S = mtimes(U,mtimes(A,V));
1095          U = U';
1096          V = V';
1097       end
1098       [U,S,V] = fixSVD(U,S,V);
1099       if (nargout < 3)
1100           Uout = diag(S);
1101       else
1102           Uout = U; Sout = S; Vout = V;
1103       end
1104       end
1105    function [U,S,V] = fixSVD(U,S,V)
1106       Z = [sign(S(1,1)),0; 0,sign(S(2,2))]; %
1107       U = mtimes(U,Z);
1108       S = mtimes(Z,S);
1109       if (S(1,1) < S(2,2))
1110           P = [0,1;1,0];
1111           U = mtimes(U,P);
1112           S = mtimes(P,mtimes(S,P));
1113           V = mtimes(P,V);
1114       end
1115       end
1116    function f = fro(M)
1117       f = sqrt(sum(sum(times(M,M))));
1118    end
1119    function s = sign(x)
1120        if (x > 0)
1121            s = 1;
1122        else
1123            s = -1;
1124        end
1125        end
```

References

1. System Computer Biology. Monograph. Novosibirsk: Publishing House of the SB RAS (2008), 769 p
2. M.J. Betts, M.J. Sternberg, An analysis of conformational changes on proteinprotein association: implications for predictive docking. Protein Eng. **12**, 271–283 (1999)
3. T.V. Pyrkov, I.V. Ozerov, E.D. Balitskaya, R.G. Efremov, Molecular docking: the role of non-valence interactions in the formation of protein complexes with nucleotides and peptides. Bioorganic Chem. **36**(4), 482–492 (2010)
4. The Universal Protein Resource http://www.uniprot.org/
5. D. Lane, A. Levine, P53 Research: the past thirty years and the next thirty years Cold. Spring. Harb. Perspect. Biol. **2**(12) (2010)
6. S. Nag, J. Qin, K.S. Srivenugopal, M. Wang, R. Zhanga, The MDM2-p53 pathway revisited. J. Biomed. Res. **27**(4), 254–271 (2013)
7. D.P. Lane, L.V. Crawford, T antigen is bound to a host protein in SV40-transformed cells. Nature **278**, 261–263 (1979)

8. C.J. Sherr, F. McCormick, The RB and p53 pathways in cancer. Cancer Cell **2**(2), 103–112 (2002)
9. T. Ozaki, A. Nakagawara, Role of p53 in cell death and human cancers. Cancers(Basel) **3**(1), 994–1013 (2011)
10. J.T. Zilfou, S.W. Lowe, Tumor Suppressive Functions of p53 Cold. Spring. Harb. Perspect. Biol. **1**(5) (2009)
11. Y. Qian, X. Chen, Senescence regulation by the p53 protein family. Methods Mol. Biol. **965**, 37–61 (2013)
12. Y. Liu, M. Kulesz-Martin, p53 protein at the hub of cellular DNA damage response pathways through sequence-specific and non-sequence-specific DNA binding. Carcinogenesis **22**(6), 851–860 (2001)
13. M. Hassan, H. Watari, A. AbuAlmaaty, Y. Ohba, N. Sakuragi, Apoptosis and molecular targeting therapy in cancer. Biomed. Res. Int. **2014** (2014)
14. J. Loughery, M. Cox, L.M. Smith, D.W. Meek, Critical role for p53-serine 15 phosphorylation in stimulating transactivation at p53-responsive promoters. Nucleic. Acids. Res. **42**(12), 7666–7680 (2014)
15. P.A. Lazo, Reverting p53 activation after recovery of cellular stress to resume with cell cycle progression. Cell Signal **33**, 49–58 (2017)
16. G.P. Zambetti (ed.), *The p53 Tumor Suppressor Pathway and Cancer* (Springer, Berlin, 2005)
17. M.A. McCoy, J.J. Gesell, M.M. Senior, D.F. Wyss, Flexible lid to the p53-binding domain of human Mdm2: implications for p53 regulation. Proc. Natl. Acad. Sci. U. S. A. **100**(4), 1645–1648 (2003)
18. H. Liang, H. Atkins, et al., Genomic organisation of the human MDM2 oncogene and relationship to its alternatively spliced mRNAs. Gene **338**(2), 217–223
19. T. Hamzehloie, M. Mojarrad, M. Hasanzadeh-Nazarabadi, S. Shekouhi, The role of tumor protein 53 mutations in common human cancers and targeting the murine double minute 2P53 interaction for cancer therapy. Iran. J. Med. Sci. **37**(1), 3–8 (2012)
20. Y. Zhao, H. Yu, W. Hu, The regulation of MDM2 oncogene and its impact on human cancers. Acta Biochim. Biophys. Sin. (Shanghai) **46**(3), 180–189 (2014)
21. P. Chene, Inhibition of the p53-MDM2 Interaction: targeting a protein-protein interface. Mol. Cancer. Res. **2**(1), 20–28 (2004)
22. U.M. Mol, O. Petrenko, Molecular dynamic simulation insights into the normal state and restoration of p53 function. Mol. Cancer. Res. **1**(14), 1001–1008 (2003)
23. P.H. Kussie, S. Gorina, V. Marechal, B. Elenbaas, J. Moreau, A.J. Levine, N.P. Pavletich, Structure of the MDM2 oncoprotein bound to the p53 tumor suppressor transactivation domain. Science **274**(5289), 948–953 (1996)
24. P.L. Leslie, H. Ke, Y. Zhang, The MDM2 RING domain and central acidic domain play distinct roles in MDM2 protein homodimerization and MDM2-MDMX protein heterodimerization. J. Biol. Chem. **290**(20), 12941–12950 (2015)
25. Y.J. Park, K. Luger, The structure of nucleosome assembly protein 1. Proc. Natl. Acad. Sci. USA **103**(5), 1248–1253 (2006)
26. S.J. McBryant, Y.J. Park, S.M. Abernathy, P.J. Laybourn, J.K. Nyborg, K. Luger, Preferential binding of the histone $(H3\text{-}H4)_2$ tetramer by NAP1 is mediated by the amino-terminal histone tails. J. Biol. Chem. **278**(45), 44574–44583 (2003)
27. J. Zlatanova, C. Seebart, M. Tomschik, Nap1: taking a closer look at a juggler protein of extraordinary skills. FASEB J. **21**(7), 1294–1310 (2007)
28. https://www.rcsb.org/structure/5G2E
29. L.L. Patrick, H. Ke, Z. Yanping, The MDM2 RING domain and central acidic domain play distinct roles in MDM2 protein homodimerization and MDM2-MDMX protein heterodimerization. J. Biol. Chem. **290**(20), 12941–12950 (2015)
30. S. Uldrijan, W.J. Pannekoek, K.H. Vousden, An essential function of the extreme C-terminus of MDM2 can be provided by MDMX. EMBO J. **26**, 102–112 (2007)
31. M.V. Poyurovsky, C. Priest, A. Kentsis, K.L. Borden, Z.Q. Pan, N. Pavletich, C. Prives, The Mdm2 RING domain C-terminus is required for supramolecular assembly and ubiquitin ligase activity. EMBO J. **26**, 90–101 (2007)

32. Y. Zhao, A. Aguilar, D. Bernard, S. Wang, Small molecule inhibitors of MDM2-p53 and MDMX-p53 interactions as new cancer therapeutics. J. Med. Chem. **58**(3), 1038–1052 (2015)
33. J. Chen, V. Marechal, A.J. Levine, Mapping of the p53 and mdm-2 interaction domains. Mol. Cell. Biol. **13**, 4107–4114 (1993)

Chapter 6
Mathematical Modelling of the Effect of Phosphorylation on the Stability of the Formation of Biological Complexes P53–Mdm2 and P53–P300

Abstract This chapter presents a physical model of phosphorylation the amino acid residues of the polypeptide chain of a protein on the formation of biological complexes with other proteins.

6.1 Introduction

In this chapter, we develop a physical model of the effect of phosphorylation of the amino acid residues of the polypeptide chain of a protein on the formation of biological complexes with other proteins, for example, the phosphorylation of the flexible N-terminus of the p53 protein by two amino acid residues 18a.a. and 20a.a., as well as an analysis of the stability of the biological complexes P53–Mdm2 and P53–P300 formed before and after phosphorylation. We took short sites of P53, Mdm2, and P300 proteins in the calculations and analyzed the stability of the heterodimers formed by them.

The authors suggest that a developed model of phosphorylation of the main(key) amino acid residues of the protein will allow us to predict an increase or decrease in the stability of the formed protein complexes before and after phosphorylation with the participation of other non-phosphorylated proteins. The test is supposed to be performed using small sequences of proteins that are directly involved in the formation of the protein complex.

So, in our work, we took sites of the proteins $P53_{(1-22)}$, $P53_{(10-51)}$, $P300_{(1726-1806)}$, $Mdm2_{(25-104)}$, $Mdm2_{(51-104)}$, since according to previous studies [1–4], the N-terminus of the P53 protein is directly involved in complexation with the P300 and Mdm proteins, the Mdm2 protein forms a protein complex with the P53 protein also in the N-terminal region, and the protein region P300 is actively involved in the formation of the dimer with the protein P53 stored in the domain region Taz2 from 1726a.a. to 1806a.a.

T. Koshlan and K. Kulikov, *Mathematical Modeling of Protein Complexes*, Biological and Medical Physics, Biomedical Engineering,
https://doi.org/10.1007/978-3-319-98304-2_6

Thus, we selected P53, P300 and Mdm2 protein sites in the region responsible for the formation of the heterodimers P53–P300 and P53–Mdm2.

In this chapter, we investigate the interaction of the flexible weakly ordered N-terminus of the P53 protein with the N-terminus of the Mdm2 protein and the Taz2 domain of the P300 protein, taking into account the phosphorylation effect. The presence of the flexible N-terminus of the P53 protein leads to the fact that it can take various conformations when forming a biological complex with proteins such as Mdm2 or P300.

The phosphorylation of some amino acid residues of the N-terminus of the P53 protein leads to a marked change in the affinity of the interaction of the P53 protein with the P300 and Mdm2 proteins.

This affinity change in the phosphorylation of selected amino acid residues at the flexible N-terminus of the P53 protein is of great importance in the fate of the cell, leading to the stabilization of the P53 protein, the arrest of the cell cycle or apoptosis [5].

6.2 Protein Phosphorylation

Protein phosphorylation is the most common form of their regulatory post-translational modification.

Phosphorylation can stimulate or inhibit the catalytic activity of enzymes, the affinity with which the protein binds to other molecules, its intracellular localization and its ability for further covalent modifications. It can also alter its stability.

Most eukaryotic cells are phosphorylated with the participation of protein kinases; their dephosphorylation is catalyzed by phosphoprotein phosphatases.

Protein kinases transfer a phosphate group with ATP to the residues of Ser, Thr, Tyr in protein substrates [6].

Let us consider in more detail some earlier work describing the mechanisms of phosphorylation and the effects of the phosphorylation of the N-terminus of the P53 protein on the interaction with the proteins Mdm2 and P300.

In [5], it was reported that the p53 TAD is phosphorylated by a number of activated kinases and is critical for the many protein-protein interactions that either modulate the stability and subcellular localization of p53 or affect its function as a transcription factor. When unbound, the TAD is unstructured [7], but it adopts a helical conformation upon complex formation. The minimal Mdm2-binding region resides fully within TAD1, which forms a helix encompassing residues 19–25. In cells, complex formation results in ubiquitination of the P53 REG domain by the C-terminal E3 ligase domain of Mdm2, leading to nuclear export and degradation of p53. Since Mdm2 is a transcriptional target of P53, these two proteins form anegative feedback loop that controls P53 levels in the absence of stress and during the return to homeostasis following stress. The importance of TAD phosphorylation in the regulation of P53 function has led to numerous in vitro studies examining the effects of P53 phosphorylations on interactions with its binding partners. For the interaction with Mdm2, which is primarily stabilized by the hydrophobic effect, phosphorylation prevents complex formation. In contrast, TAD phosphorylation enhances binding to CBP/P300. Thus, phosphorylation couples relief of negative regulation with enhancement of transcriptional activation.

In the absence of cellular stress, most of the serines and threonines of the p53 TAD are unphosphorylated. In particular, the absence of phosphorylation of Thr18 allows tight binding of Mdm2 to suppress P53 activity by enhancing nuclear export and proteasomal degradation. Once a cell experiences a stress, the concentration of p53 rapidly rises to stimulate the appropriate response: e.g. cell cycle arrest or apoptosis.

As in vitro experiments have shown, the binding affinity of the P53 TAD–Mdm2 complex can be reduced 5–25-fold solely by phosphorylation of Thr18 [8–10].

In contrast, the interactions of p53 with its positive cofactors generally start out weak and increase in proportion with increasing phosphorylation. This allows for a nuanced response in which the interactions of p53 with different subgroups of cofactors change over time. The strength of the effect depends on the location of the phosphorylation within the TAD sequence and varies for the different domains of CBP/p300. Single phosphorylation of Ser15, Thr18, Ser20, Ser33, Ser37 or Ser46 generally increases the binding affinities to the Taz1, Taz2 and KIX domains by 2- to 7-fold [11, 12].

The structure of the NMR of the Taz2 complex of the p300 protein and the N-terminal transactivation domain of P53 was reported in [3].

In the complex, p53 forms a short alpha helix and interacts with the Taz2 region through the expanded surface. Mutational analyses demonstrate the importance of hydrophobic residues for complex stabilization. In addition, they suggest that the increased affinity is partly due to electrostatic interactions of phosphate with a neighboring arginine residue in the Taz2 domain of the p53 protein. Thermodynamic experiments have shown the importance of hydrophobic interactions in the complex of Taz2 with p53 in the phosphorylation of Ser (15) and Thr (18).

In [3, 13], Ser20 phosphorylation reduces the binding between p53 and Mdm2, and hence p53 is activated and stabilized. Thus, one mechanism by which p53 is protected against Mdm2 in response to DNA damage involves Ser20 phosphorylation.

The functions and role of checkpoint kinase 2 (Chk2), which is a key mediator of various cellular responses to genotoxic stress protecting the integrity of the eukaryotic genome were reviewed in [14].

In particular, Chk2 takes part in the phosphorylation of the tumor suppressor p53, which results in the stabilization of p53 and transactivation of the p53 target genes. Specific ubiquitin ligase Mdm2 can also be a substrate for the Chk2 kinase.

6.3 Description of the Physical Model

To account for the phosphorylation effect, we made the following assumptions:

- the selected amino acid residues of serine20 and threonine 18 of the P53 protein were replaced by negatively charged phosphoric acid residues OPO_3H_2, which we represented in the form of spheres with a radius equal to 0.3×10^{-9} m;
- residues of phosphoric acid interact with five charged amino acids (aspartic acid, glutamic acid, arginine, histidine, lysine) with a charge of 0.9^{-19} C;
- residues of phosphoric acid interact with selected hydrophobic amino acids (methionine, asparagine, leucine, tyrosine, valine) with a charge of 0.1^{-19} C;
- the distance between the centers of the phosphoric acid residue and the amino acid residue is 1.76^{-9} m.

To analyze the biochemical processes we use the notion of condition number matrix of the potential energy of the pair electrostatic interaction between peptides. In this physical formulation of the problem, it will characterize the degree of stability of the configuration of the biological complex. In order to choose a more stable biochemical compound between proteins, we select the matrix of potential energy of electrostatic interaction with the **smallest** value of the condition number (see Chap. 2).

6.4 Results of a Numerical Calculation of the Formation of Biological Complexes by Different Sites of the P53, Mdm2 and P300 Proteins, Taking into Account the Effect of Phosphorylation of the Flexible N-Terminus of the P53 Protein

We selected next sites of the proteins $P53_{(1-22)}$, $P53_{(10-51)}$, $Mdm2_{(25-104)}$, $Mdm2_{(50-14)}$, $P300_{(1726-1806)}$.

The list of involved amino acid sequences is shown below.

$P53_{(1-22)}$

MEEPQSDPSVEPPLSQEXFXDL

$P53_{(10-51)}$

EPPLSQETFSDLWKLLPENNVLSPLPSQAMDDLMLSPDDIE

Table 6.1 Results of mathematical modelling of the effects of phosphorylation and dephosphorylation on the stability of complexes formed by different sites of P53, Mdm2 and P300 proteins

$N^{\underline{0}}$	Proteins	Phosphorylation, lg(cond(W))	Dephosphorylation, lg(cond(W))
1	$P53_{(1-22)}$–$Mdm2_{(25-104)}$	18.298	17.831
2	$P53_{(1-22)}$–$Mdm2_{(51-104)}$	18.689	18.155
3	$P53_{(10-51)}$–$Mdm2_{(51-104)}$	19.012	18.647
4	$P53_{(1-22)}$–$P300_{(1726-1806)}$	18.314	18.370
5	$P53_{(10-51)}$–$P300_{(1726-1806)}$	18.632	18.771

lg(cond(W)) is common logarithm of condition number

P300 $TAZ2_{(1726-1806)}$
SPGDSRRLSIQRCIQSLVHACQCRNANCSLPSCQKMKRVVQHTKGC
KRKTNGGCPICKQLIALCCYHAKHCQENKCPVPFC

$Mdm2_{(25-104)}$
ETLVRPKPLLLKLLKSVGAQKDTYTMKEVLFYLGQYIMTKRLYDEKQ
QHIVYCSNDLLGDLFGVPSFSVKEHRKIYTMIY

$Mdm2_{(51-104)}$
KEVLFYLGQYIMTKRLYDEKQQHIVYCSNDLLGDLFGVPSFSVKEHR
KIYTMIY

We performed a numerical calculation of the potential energy matrix of electrostatic interaction of different pairs of protein sequences: $P53_{(1-22)}-Mdm2_{(25-104)}$, $P53_{(10-51)}-Mdm2_{(51-104)}$, $P53_{(1-22)}-Mdm2_{(50-14)}$, $P53_{(1-22)}-P300_{(1726-1806)}$, $P53_{(10-51)}-P300_{(1726-1806)}$ before and after phosphorylation of 18 aa. and 20 a.a. protein P53.

The results of the numerical calculation are shown in Table 6.1.

As can be seen from the table, upon interaction of the phosphorylated flexible N-terminus of the P53 protein with the Mdm2 protein, an increase in the lg(cond (W)) value is observed for the $P53_{(1-22)}$–$Mdm2_{(25-104)}$ dimers, $P53_{(1-22)}$–$Mdm2_{(51-104)}$, and $P53_{(10-51)}$–$Mdm2_{(51-104)}$, compared with the interaction of the non-phosphorylated region of the P53 protein, which suggests a decrease in the stability of protein complexes formed by different regions of the P53 and Mdm2 proteins and phosphorylation of two amino acid residues from the N-terminus of the P53 protein.

Table 6.2 Results of numerical calculations before and after the phosphorylation of two amino acid residues of the protein P53

$N^{\underline{0}}$	Proteins	Phosphorylation, lg(condW)	Dephosphorylation, lg(condW)
1	$Mdm2_{(54-100)}$–$P53_{(17-30)}$	20.139	17.870

lg(cond(W)) is common logarithm of condition number

Analysis of the data in the table indicates a decrease in the values of lg(cond(W)) during the interaction of the phosphorylated sites of $P53_{(1-22)}$ and $P53_{(10-51)}$ with the protein region $P300_{(1726-1806)}$ as compared to the interaction of $P300_{(1726-1806)}$ with the non-phosphorylated N-termini of the $P53_{(1-22)}$ and $P53_{(10-51)}$ proteins.

Thus, we come to the conclusion that the phosphorylation of the flexible N-region of the protein $P53_{(1-22)}$ and $P53_{(10-51)}$ positively affects the interaction with the Taz2 domain of the protein P300.

If researchers know the exact areas of interaction of two proteins, then numerical calculations should be performed of the effect of phosphorylation of the amino acid residues of one of the proteins, taking into account the precisely defined regions of the two interacting proteins. According to previous experiments [15], active parts of the interaction between proteins P53 and Mdm2 were identified. Active amino acid residues of protein P53 are from 18 a.a. to 27 a.a., arranged consecutively one after another, and active amino acids of Mdm2: L54, L57, G58, I61, M62, Y67, Q72, V75, F91, V93, H96, I99, Y100 [15].

Our next numerical calculation will be performed between the interacting regions of the two proteins before and after phosphorylation of the two amino acid residues of the protein P53.

$P53_{(17-30)}$:17-E**<u>T</u>**F **<u>S</u>**DLWKLLPENN-30

$Mdm2_{(54-100)}$: L54, L57, G58, I61, M62, Y67, Q72, I74, V75, F91, V93, H96, I99, Y100.

Two amino acid residues of the protein P53 are phosphorylated, while in the program complex phosphorylation is performed by changing the letter designations of the amino acid residues to the letter designation ≪X≫, which in this formulation of the problem denotes a negatively charged phosphoric acid residue. Then, taking into account the amino acid substitutions performed for the phosphoric acid residue, the studding amino acid sequence of the P53 protein will look like this:

$P53_{(17-30)}$:17-E**<u>X</u>**F **<u>X</u>**DLWKLLPENN-30

The results of the numerical calculation are shown in Table 6.2.

As we can see from the table above, the phosphorylation processes of two amino acid residues of the protein P53 leads to destabilization of the protein complex $Mdm2_{(54-100)}$–$P53_{(17-30)}$, which may affect the stability of the protein complex formed by whole protein sequences of Mdm2 and P53. Analysis of the numerical calculations is in good agreement with articles [8–10], which pointed to the significant influence of phosphorylation processes on the interaction of proteins Mdm2 and P53.

In this section, we developed a mathematical model of the phosphorylation of the amino acid residues of the polypeptide chain using the protein P53 as an example. We performed a numerical calculation of the potential energy matrixes of the electrostatic interaction of various amino acid sequences of the P53, P300, and Mdm2 proteins, taking into account the phosphorylation of the polypeptide sequence of the P53B protein.

We found that the phosphorylation of two amino acid residues of Thr18 and Ser20 from the N-terminus of the P53 protein leads to an increase in the stability of the biological complexes formed by different regions of the proteins: $P53_{(1-22)}$–$P300_{(1726-1806)}$, $P53_{(10-51)}$–$P300_{(1726-1806)}$, and as well as to a decrease in the stability of biological complexes formed by protein sites:
$P53_{(1-22)}$–$Mdm2_{(25-104)}$, $P53_{(1-22)}$–$Mdm2_{(51-104)}$, $P53_{(10-51)}$–$Mdm2_{(51-104)}$.

The authors suggest that such a model of accounting for phosphorylation of amino a acid residues in the interaction of small regions of proteins will help in future to determine the effect of phosphorylation processes on the stability of whole protein complexes.

6.5 Matlab Script for Mathematical Modelling of the Effect of Phosphorylation on the Stability of the Formation of Biological Complexes P53–Mdm2 and P53–P300

Input parameters:

1. S_1, S_{20} are amino acid sequences of biological complexes ($S_1 \geq S_{20}$)
2. epsilon is the dielectric constant of the medium.

Output parameters:

lg(cond(W) is the common logarithm of the condition number of the matrix W, where its elements are composed of the electrostatic potential energy which is created based on the interaction between pair of amino acid residues of biological complexes.

Compute:

lg(cond(W)) is the common logarithm of the condition number of the matrix W, which will allow a prediction the reactivity of the studied biological complexes.

```
clear all
clc
format long e
%p53 10-50
S_20=[ 'E'  'P'  'P'  'L'  'S'  'Q'  'E'  'T'  'F'  'S' ...
'D'  'L'  'W'  'K'  'L'  'L'  'P'  'E'  'N' 'N'     'V'  ...
'L'  'S'  'P'  'L'  'P'  'S'  'Q'  'A'  'M'    ...
'D' 'D'  'L'  'M'  'L'  'S'  'P'  'D'  'D'  'I'  'E' ]
%MDM2 50-104
S_1=['K' 'E' 'V' 'L' 'F' 'Y' 'L' 'G' 'Q' 'Y'  'I' 'M' ...
'T' 'K' 'R' 'L' 'Y' 'D' 'E' 'K'  'Q' 'Q' 'H' 'I' 'V' 'Y' ...
'C' 'S' 'N' 'D'  'L' 'L' 'G' 'D' 'L' 'F' 'G' 'V' 'P' 'S' ...
'F' 'S' 'V' 'K' 'E' 'H' 'R' 'K' 'I' 'Y'  'T' 'M' 'I' 'Y']
epsilon=1;
len_S1=length(S_1);
len_S20=length(S_20);
N1=10*len_S20;
[S_1,S_20,Q1,Q2,R1,R2,h,M,N]=potential__phospho(S_1,S_20);
[A]=electrostatic(Q1,Q2, R1,R2,h,M,N,N1,epsilon);
[R1]=condmy(A)
%-----------------------------------------------
function [S_1,S_2,Q1,Q2,R1,R2,h,M,N]=...
potential__phospho(S_1,S_20);
N=length(S_1);
M=length(S_20);
S_2=S_20;
Q1=[];
Q2=[];
R1=[];
R2=[];
for i=1:length(S_1);
for j=1:length(S_2);
%D
if (S_1(i)=='D' & S_2(j)=='E')| (S_1(i)=='E' & S_2(j)=='D');
Q1(i,j)= 0.16e-19;
Q2(i,j)= 0.16e-19;
else
if (S_1(i)=='D' & S_2(j)=='D');
Q1(i,j)= 0.07e-19;
Q2(i,j)= 0.07e-19;
else
if (S_1(i)=='D' & S_2(j) =='C')|(S_1(i)=='C' & S_2(j) =='D');
Q1(i,j)= 0.05e-19;
Q2(i,j)= 0.05e-19;
else
if (S_1(i)=='D' &S_2(j)=='N')|(S_1(i)=='N' &S_2(j)=='D')|...
(S_1(i)=='D' &S_2(j)=='F')|(S_1(i)=='D' &S_2(j)=='Y')|...
(S_1(i)=='D' &S_2(j)=='Q')|(S_1(i)=='D' &S_2(j)=='S')|...
(S_1(i)=='F' &S_2(j)=='D')|(S_1(i)=='Y' &S_2(j)=='D')|...
(S_1(i)=='Q' &S_2(j)=='D')|(S_1(i)=='S' &S_2(j)=='D');
Q1(i,j)= 0.57e-19;
Q2(i,j)= 0.57e-19;
```

```
53 else
54 if ((S_1(i)=='D' & S_2(j)=='M')|(S_1(i)=='D' & S_2(j)=='T')|...
55 (S_1(i)=='D' & S_2(j)=='I')|(S_1(i)=='D' & S_2(j)=='G')|...
56 (S_1(i)=='D' & S_2(j)=='V')|(S_1(i)=='D' & S_2(j)=='W')|...
57 (S_1(i)=='D' & S_2(j)=='L')|(S_1(i)=='D' & S_2(j)=='A')|...
58 (S_1(i)=='M' & S_2(j)=='D')|(S_1(i)=='T' & S_2(j)=='D')|...
59 (S_1(i)=='I' & S_2(j)=='D')|(S_1(i)=='G' & S_2(j)=='D')|...
60 (S_1(i)=='V' & S_2(j)=='D')|(S_1(i)=='W' & S_2(j)=='D')|...
61 (S_1(i)=='L' & S_2(j)=='D')|(S_1(i)=='A' & S_2(j)=='D'));
62 Q1(i,j)= 0.64e-19;
63 Q2(i,j)= 0.64e-19;
64 else
65 if ((S_1(i)=='D'& S_2(j)=='P')|(S_1(i)=='P'& S_2(j)=='D'));
66 Q1(i,j)= 0.78e-19;
67 Q2(i,j)= 0.78e-19;
68 else
69 if ((S_1(i)=='D' & S_2(j)=='H')|(S_1(i)=='H'& S_2(j)=='D'));
70 Q1(i,j)= 0.99e-19;
71 Q2(i,j)= 0.99e-19;
72 else
73 if ((S_1(i)=='D'& S_2(j)=='K')|(S_1(i)=='K'& S_2(j)=='D'));
74 Q1(i,j)= 1.4e-19;
75 Q2(i,j)= 1.4e-19;
76 else
77 if ((S_1(i)=='D' & S_2(j)=='R')|(S_1(i)=='R'& S_2(j)=='D'));
78 Q1(i,j)= 1.59e-19;
79 Q2(i,j)= 1.59e-19;
80 else
81 if ((S_1(i)=='E'&S_2(j)=='E'));
82 Q1(i,j)= 0.16e-19;
83 Q2(i,j)= 0.16e-19;
84 else
85 if ((S_1(i)=='E'  & S_2(j)=='C')|(S_1(i)=='E' & S_2(j)=='F')|...
86 (S_1(i)=='E' & S_2(j)=='N')|(S_1(i)=='C'  & S_2(j)=='E')|...
87 (S_1(i)=='F' & S_2(j)=='E')|(S_1(i)=='N' & S_2(j)=='E'));
88 Q1(i,j)= 0.55e-19;
89 Q2(i,j)= 0.55e-19;
90 else
91 if((S_1(i)=='E' & S_2(j)=='Q')|(S_1(i)=='E' & S_2(j)=='Y')|...
92  (S_1(i)=='E' & S_2(j)=='S')|(S_1(i)=='E' & S_2(j)=='M')|...
93  (S_1(i)=='E' & S_2(j)=='T')|(S_1(i)=='E' & S_2(j)=='I')|...
94  (S_1(i)=='E' & S_2(j)=='G')|(S_1(i)=='E' & S_2(j)=='V')|...
95  (S_1(i)=='E' & S_2(j)=='W')|(S_1(i)=='E' & S_2(j)=='L')|...
96  (S_1(i)=='E' & S_2(j)=='A')|(S_1(i)=='Q' & S_2(j)=='E')|...
97  (S_1(i)=='Y' & S_2(j)=='E')| (S_1(i)=='S' & S_2(j)=='E')|...
98  (S_1(i)=='M' & S_2(j)=='E')|(S_1(i)=='T' & S_2(j)=='E')|...
99  (S_1(i)=='I' & S_2(j)=='E')| (S_1(i)=='G' & S_2(j)=='E')|...
100  (S_1(i)=='V' & S_2(j)=='E')|(S_1(i)=='W' & S_2(j)=='E')|...
101  (S_1(i)=='L' & S_2(j)=='E')|(S_1(i)=='A' & S_2(j)=='E'));
102 Q1(i,j)= 0.64e-19;
103 Q2(i,j)= 0.64e-19;
104 else
105 if ((S_1(i)=='E' & S_2(j)=='P' )|(S_1(i)=='P' & S_2(j)=='E'));
```

```
106 Q1(i,j)= 0.78e-19;
107 Q2(i,j)= 0.78e-19;
108 else
109 if ((S_1(i)=='E' & S_2(j)=='H')|(S_1(i)=='H' &S_2(j)=='E'));
110 Q1(i,j)= 0.99e-19;
111 Q2(i,j)= 0.99e-19;
112 else
113 if (S_1(i)=='E'& S_2(j)=='K')| (S_1(i)=='K'& S_2(j)=='E');
114 Q1(i,j)= 1.34e-19;
115 Q2(i,j)= 1.34e-19;
116 else
117 if (S_1(i)=='E' & S_2(j)=='R')|(S_1(i)=='R' & S_2(j)=='E');
118 Q1(i,j)= 1.58e-19;
119 Q2(i,j)= 1.58e-19;
120 else
121 if (S_1(i)=='C' & S_2(j)=='C')|(S_1(i)=='C' & S_2(j)=='F')|...
122 (S_1(i)=='C' & S_2(j)=='Q')|(S_1(i)=='C'& S_2(j)=='Y')|...
123 (S_1(i)=='C' & S_2(j)=='S')|(S_1(i)=='C' & S_2(j)=='M')|...
124 (S_1(i)=='C' & S_2(j)=='T')|(S_1(i)=='C' & S_2(j)=='I')|...
125 (S_1(i)=='C' & S_2(j)=='G')|(S_1(i)=='C' & S_2(j)=='V')|...
126 (S_1(i)=='C' & S_2(j)=='W')|(S_1(i)=='C' & S_2(j)=='L')|...
127 (S_1(i)=='C' & S_2(j)=='L')|(S_1(i)=='C' & S_2(j)=='A')|...
128 (S_1(i)=='F' & S_2(j)=='C')|(S_1(i)=='Q' & S_2(j)=='C')|...
129 (S_1(i)=='Y'& S_2(j)=='C')|(S_1(i)=='S' & S_2(j)=='C')|...
130 (S_1(i)=='M' & S_2(j)=='C')|(S_1(i)=='T' & S_2(j)=='C')|...
131 (S_1(i)=='I' & S_2(j)=='C')|(S_1(i)=='G' & S_2(j)=='C')|...
132 (S_1(i)=='V' & S_2(j)=='C')|(S_1(i)=='W' & S_2(j)=='C')|...
133 (S_1(i)=='L' & S_2(j)=='C')|( S_1(i)=='A' & S_2(j)=='C');
134 Q1(i,j)=0.74e-19;
135 Q2(i,j)=0.74e-19;
136 else
137 if (S_1(i)=='C' & S_2(j)=='H')| (S_1(i)=='H' & S_2(j)=='C');
138 Q1(i,j)= 0.99e-19;
139 Q2(i,j)= 0.99e-19;
140 else
141 if (S_1(i)=='C' & S_2(j)=='K')|(S_1(i)=='K' & S_2(j)=='C');
142 Q1(i,j)= 1.34e-19;
143 Q2(i,j)= 1.34e-19;
144 else
145 if (S_1(i)=='C' & S_2(j)=='R')|(S_1(i)=='R' & S_2(j)=='C');
146 Q1(i,j)= 1.59e-19;
147 Q2(i,j)= 1.59e-19;
148 else
149 if (S_1(i)=='N' & S_2(j)=='N')|(S_1(i)=='N' & S_2(j)=='F')|...
150 (S_1(i)=='N' & S_2(j)=='Q')|(S_1(i)=='N' & S_2(j)=='Y')|...
151 (S_1(i)=='N' & S_2(j)=='S')|(S_1(i)=='N'& S_2(j)=='M')|...
152 (S_1(i)=='F' & S_2(j)=='N')|(S_1(i)=='Q' & S_2(j)=='N')|...
153 (S_1(i)=='Y' & S_2(j)=='N')|(S_1(i)=='S' & S_2(j)=='N')|...
154 (S_1(i)=='M'& S_2(j)=='N');
155 Q1(i,j)=0.74e-19;
156 Q2(i,j)=0.74e-19;
157 else
```

```
158 if (S_1(i)=='N' & S_2(j)=='H')|(S_1(i)=='H' & S_2(j)=='N')
159 Q1(i,j)= 0.99e-19;
160 Q2(i,j)= 0.99e-19;
161 else
162 if(S_1(i)=='N' & S_2(j)=='K')|(S_1(i)=='K' & S_2(j)=='N');
163 Q1(i,j)= 1.05e-19;
164 Q2(i,j)= 1.05e-19;
165 else
166 if (S_1(i)=='N' & S_2(j)=='R')|(S_1(i)=='R' & S_2(j)=='N');
167 Q1(i,j)= 1.1e-19;
168 Q2(i,j)= 1.1e-19;
169 else
170 if ((S_1(i)=='F' & S_2(j)=='F')|(S_1(i)=='F' & S_2(j)=='Q'));
171   Q1(i,j)=0.74e-19;
172 Q2(i,j)=0.74e-19;
173 else
174  if((S_1(i)=='F' & S_2(j)=='Y')|(S_1(i)=='F' & S_2(j)=='S')|...
175  (S_1(i)=='F' & S_2(j)=='M')|(S_1(i)=='Q' & S_2(j)=='F')|...
176  (S_1(i)=='Y' & S_2(j)=='F'));
177 Q1(i,j)=0.74e-19;
178 Q2(i,j)=0.74e-19;
179  else
180     if (S_1(i)=='S' & S_2(j)=='F')|(S_1(i)=='M' & S_2(j)=='F');
181 Q1(i,j)=0.74e-19;
182 Q2(i,j)=0.74e-19;
183     else
184 if (S_1(i)=='F' & S_2(j)=='H')|(S_1(i)=='H' & S_2(j)=='F');
185 Q1(i,j)= 0.99e-19;
186 Q2(i,j)= 0.99e-19;
187 else
188 if (S_1(i)=='F' & S_2(j)=='K')|(S_1(i)=='K' & S_2(j)=='F');
189 Q1(i,j)= 1.05e-19;
190 Q2(i,j)= 1.05e-19;
191 else
192 if (S_1(i)=='F' & S_2(j)=='R')|(S_1(i)=='R' & S_2(j)=='F');
193 Q1(i,j)= 1.1e-19;
194 Q2(i,j)= 1.1e-19;
195 else
196 if (S_1(i)=='Q' & S_2(j)=='H')|(S_1(i)=='H' & S_2(j)=='Q');
197 Q1(i,j)= 0.99e-19;
198 Q2(i,j)= 0.99e-19;
199 else
200 if (S_1(i)=='Q' & S_2(j)=='K')|(S_1(i)=='K' & S_2(j)=='Q');
201 Q1(i,j)= 1.05e-19;
202 Q2(i,j)= 1.05e-19;
203 else
204 if (S_1(i)=='Q' & S_2(j)=='R')|(S_1(i)=='R' & S_2(j)=='Q');
205 Q1(i,j)= 1.1e-19;
206 Q2(i,j)= 1.1e-19;
207 else
208 if (S_1(i)=='Q' & S_2(j)=='H')|(S_1(i)=='H' & S_2(j)=='Q');
209 Q1(i,j)= 0.99e-19;
```

```
210  Q2(i,j)= 0.99e-19;
211  else
212  if (S_1(i)=='Y' & S_2(j)=='K')|(S_1(i)=='K' & S_2(j)=='Y')
213  Q1(i,j)= 1.05e-19;
214  Q2(i,j)= 1.05e-19;
215  else
216  if (S_1(i)=='Y' & S_2(j)=='R')|(S_1(i)=='R' & S_2(j)=='Y');
217  Q1(i,j)= 1.1e-19;
218  Q2(i,j)= 1.1e-19;
219  else
220  if (S_1(i)=='S' & S_2(j)=='H')|(S_1(i)=='H' & S_2(j)=='S');
221  Q1(i,j)= 0.99e-19;
222  Q2(i,j)= 0.99e-19;
223  else
224  if (S_1(i)=='S' & S_2(j)=='K')|(S_1(i)=='K' & S_2(j)=='S');
225  Q1(i,j)= 1e-19;
226  Q2(i,j)= 1e-19;
227  else
228  if (S_1(i)=='S' & S_2(j)=='R')|(S_1(i)=='R' & S_2(j)=='S');
229  Q1(i,j)= 1.1e-19;
230  Q2(i,j)= 1.1e-19;
231  else
232  if (S_1(i)=='M' & S_2(j)=='H')|(S_1(i)=='H' & S_2(j)=='M');
233  Q1(i,j)= 0.99e-19;
234  Q2(i,j)= 0.99e-19;
235  else
236  if (S_1(i)=='M' & S_2(j)=='K')|(S_1(i)=='K' & S_2(j)=='M');
237  Q1(i,j)= 1e-19;
238  Q2(i,j)= 1e-19;
239  else
240  if (S_1(i)=='M' & S_2(j)=='R')|(S_1(i)=='R' & S_2(j)=='M');
241  Q1(i,j)= 1.1e-19;
242  Q2(i,j)= 1.1e-19;
243  else
244  if (S_1(i)=='T' & S_2(j)=='H')|(S_1(i)=='H' & S_2(j)=='T');
245  Q1(i,j)= 0.99e-19;
246  Q2(i,j)= 0.99e-19;
247  else
248  if (S_1(i)=='T' & S_2(j)=='K')|(S_1(i)=='K' & S_2(j)=='T');
249  Q1(i,j)= 1e-19;
250  Q2(i,j)= 1e-19;
251  else
252  if (S_1(i)=='T' & S_2(j)=='R')|(S_1(i)=='R' & S_2(j)=='T');
253  Q1(i,j)= 1.05e-19;
254  Q2(i,j)= 1.05e-19;
255  else
256  if (S_1(i)=='I' & S_2(j)=='H')|(S_1(i)=='H' & S_2(j)=='I');
257  Q1(i,j)= 0.99e-19;
258  Q2(i,j)= 0.99e-19;
259  else
260  if (S_1(i)=='I' & S_2(j)=='K')|(S_1(i)=='K' & S_2(j)=='I');
261  Q1(i,j)= 1e-19;
```

```
262 Q2(i,j)= 1e-19;
263 else
264 if (S_1(i)=='I' & S_2(j)=='R')|(S_1(i)=='R' & S_2(j)=='I');
265 Q1(i,j)= 1.05e-19;
266 Q2(i,j)= 1.05e-19;
267 else
268 if (S_1(i)=='G' & S_2(j)=='H')|(S_1(i)=='H' & S_2(j)=='G');
269 Q1(i,j)= 0.99e-19;
270 Q2(i,j)= 0.99e-19;
271 else
272 if (S_1(i)=='G' & S_2(j)=='K')|(S_1(i)=='K' & S_2(j)=='G');
273 Q1(i,j)= 1e-19;
274 Q2(i,j)= 1e-19;
275 else
276 if (S_1(i)=='G' & S_2(j)=='R')|(S_1(i)=='R' & S_2(j)=='G');
277 Q1(i,j)= 1.05e-19;
278 Q2(i,j)= 1.05e-19;
279 else
280 if (S_1(i)=='V' & S_2(j)=='H')|(S_1(i)=='H' & S_2(j)=='V');
281 Q1(i,j)= 0.99e-19;
282 Q2(i,j)= 0.99e-19;
283 else
284 if (S_1(i)=='V' & S_2(j)=='K')|(S_1(i)=='K' & S_2(j)=='V');
285 Q1(i,j)= 1e-19;
286 Q2(i,j)= 1e-19;
287 else
288 if (S_1(i)=='V' & S_2(j)=='R')|(S_1(i)=='R' & S_2(j)=='V');
289 Q1(i,j)= 1.05e-19;
290 Q2(i,j)= 1.05e-19;
291 else
292 if (S_1(i)=='W' & S_2(j)=='H')|(S_1(i)=='H' & S_2(j)=='W');
293 Q1(i,j)= 0.99e-19;
294 Q2(i,j)= 0.99e-19;
295 else
296 if (S_1(i)=='W' & S_2(j)=='K')|(S_1(i)=='K' & S_2(j)=='W');
297 Q1(i,j)= 1e-19;
298 Q2(i,j)= 1e-19;
299 else
300 if (S_1(i)=='W' & S_2(j)=='R')|(S_1(i)=='R' & S_2(j)=='W');
301 Q1(i,j)= 1.05e-19;
302 Q2(i,j)= 1.05e-19;
303 else
304 if (S_1(i)=='L' & S_2(j)=='H')|(S_1(i)=='H' & S_2(j)=='L');
305 Q1(i,j)= 0.99e-19;
306 Q2(i,j)= 0.99e-19;
307 else
308 if (S_1(i)=='L' & S_2(j)=='K')|(S_1(i)=='K' & S_2(j)=='L');
309 Q1(i,j)= 1e-19;
310 Q2(i,j)= 1e-19;
311 else
312 if (S_1(i)=='L' & S_2(j)=='R')|(S_1(i)=='R' & S_2(j)=='L');
313 Q1(i,j)= 1.05e-19;
```

```
314 Q2(i,j)= 1.05e-19;
315 else
316 if (S_1(i)=='A' & S_2(j)=='H')|(S_1(i)=='H' & S_2(j)=='A');
317 Q1(i,j)= 0.99e-19;
318 Q2(i,j)= 0.99e-19;
319 else
320 if (S_1(i)=='A' & S_2(j)=='K')|(S_1(i)=='K' & S_2(j)=='A');
321 Q1(i,j)= 1e-19;
322 Q2(i,j)= 1e-19;
323 else
324 if (S_1(i)=='A' & S_2(j)=='R')|(S_1(i)=='R' & S_2(j)=='A');
325 Q1(i,j)= 1.05e-19;
326 Q2(i,j)= 1.05e-19;
327 else
328 if (S_1(i)=='P' & S_2(j)=='H')|(S_1(i)=='H' & S_2(j)=='P');
329 Q1(i,j)= 0.99e-19;
330 Q2(i,j)= 0.99e-19;
331 else
332 if (S_1(i)=='P' & S_2(j)=='K')|(S_1(i)=='K' & S_2(j)=='P');
333 Q1(i,j)= 0.82e-19;
334 Q2(i,j)= 0.82e-19;
335 else
336 if (S_1(i)=='P' & S_2(j)=='R')|(S_1(i)=='R' & S_2(j)=='P');
337 Q1(i,j)= 0.96e-19;
338 Q2(i,j)= 0.96e-19;
339 else
340 if (S_1(i)=='H' & S_2(j)=='H');
341 Q1(i,j)= 0.82e-19;
342 Q2(i,j)= 0.82e-19;
343 else
344 if (S_1(i)=='H' & S_2(j)=='K')|(S_1(i)=='K' & S_2(j)=='H');
345 Q1(i,j)= 0.82e-19;
346 Q2(i,j)= 0.82e-19;
347 else
348 if (S_1(i)=='H' & S_2(j)=='R')|(S_1(i)=='R' & S_2(j)=='H');
349 Q1(i,j)= 0.74e-19;
350 Q2(i,j)= 0.74e-19;
351 else
352 if (S_1(i)=='K' & S_2(j)=='K');
353 Q1(i,j)= 0.54e-19;
354 Q2(i,j)= 0.54e-19;
355 else
356 if (S_1(i)=='K' & S_2(j)=='R')|(S_1(i)=='R' & S_2(j)=='K');
357 Q1(i,j)= 0.41e-19;
358 Q2(i,j)= 0.41e-19;
359 else
360 if (S_1(i)=='R' & S_2(j)=='R');
361 Q1(i,j)= 0.16e-19;
362 Q2(i,j)= 0.16e-19;
363 else
364 if ((S_1(i)=='X'& S_2(j)=='D')|(S_1(i)=='X'& S_2(j)=='K')|...
365 (S_1(i)=='X'& S_2(j)=='E')|(S_1(i)=='X'& S_2(j)=='H')|...
```

```
366 (S_1(i)=='X'& S_2(j)=='R')|(S_2(j)=='X'& S_1(i)=='D')|...
367 (S_2(j)=='X'& S_1(i)=='K')|(S_2(j)=='X'& S_1(j)=='E')|...
368 (S_2(j)=='X'& S_1(j)=='H')|(S_2(j)=='X'& S_1(j)=='R'))
369     Q1(i,j)=0.9*10^(-19);
370     Q2(i,j)=0.9*10^(-19);
371     else
372 if(S_1(i)=='X'& S_2 (j)=='M')|(S_1(i)=='X' & S_2(j)=='N')|...
373 (S_1(i)=='X' & S_2(j)=='L')|(S_1(i)=='X' & S_2(j)=='Y')|...
374 (S_1(i)=='X' & S_2(j)=='V')|(S_1(i)=='M'& S_2 (j)=='X')|...
375 (S_1(i)=='N' & S_2 (j)=='X')|(S_1(i)=='L' & S_2(j)=='X')|...
376 (S_1(i)=='Y' & S_2(j)=='X')|(S_1(i)=='V' & S_2(j)=='X')
377 Q1(i,j)= 0.1e-19;
378 Q2(i,j)= 0.1e-19;
379 else
380 Q1(i,j)= 0.824e-19;
381 Q2(i,j)= 0.824e-19;
382 end
383 end
384 end
385 end
386 end
387 end
388 end
389 end
390 end
391 end
392 end
393 end
394 end
395 end
396 end
397 end
398 end
399 end
400 end
401 end
402 end
403 end
404 end
405 end
406 end
407 end
408 end
409 end
410 end
411 end
412 end
413 end
414 end
415 end
416 end
417 end
```

```
418 end
419 end
420 end
421 end
422 end
423 end
424 end
425 end
426 end
427 end
428 end
429 end
430 end
431 end
432 end
433 end
434 end
435 end
436 end
437 end
438 end
439 end
440 end
441 end
442 end
443 end
444 end
445 end
446 end
447 end
448 end
449 end
450 end
451 end
452 end
453 end
454 end
455 end
456 end
457 end
458  Q3=[];
459  Q4=[];
460  R1=[];
461  R2=[];
462 for i=1:length(S_1);
463 if (S_1(i)=='A');
464 R1(i)=0.6e-9;
465 else
466 if (S_1(i)=='R');
467 R1(i)=0.809e-9;
468 else
469 if (S_1(i)=='N');
```

```
470 R1(i)=0.682e-9;
471 else
472 if (S_1(i)=='D');
473 R1(i)=0.665e-9;
474 else
475 if (S_1(i)=='C');
476 R1(i)=0.629e-9;
477 else
478 if (S_1(i)=='Q');
479 R1(i)=0.725e-9;
480 else
481 if (S_1(i)=='E');
482 R1(i)=0.714e-9;
483 else
484 if (S_1(i)=='G');
485 R1(i)=0.537e-9;
486 else
487 if (S_1(i)=='H');
488 R1(i)=0.732e-9;
489 else
490 if (S_1(i)=='I');
491 R1(i)=0.735e-9;
492 else
493 if (S_1(i)=='L');
494 R1(i)=0.734e-9;
495 else
496 if (S_1(i)=='K');
497 R1(i)=0.737e-9;
498 else
499 if (S_1(i)=='M');
500 R1(i)=0.741e-9;
501 else
502 if (S_1(i)=='F');
503 R1(i)=0.781e-9;
504 else
505 if (S_1(i)=='P');
506 R1(i)=0.672e-9;
507 else
508 if (S_1(i)=='S');
509 R1(i)=0.615e-9;
510 else
511 if (S_1(i)=='T');
512 R1(i)=0.659e-9;
513 else
514 if (S_1(i)=='W');
515 R1(i)=0.826e-9;
516 else
517 if (S_1(i)=='Y');
518 R1(i)=0.781e-9;
519 else
520 if (S_1(i)=='V');
521 R1(i)=0.694e-9;
```

```
522                 else
523                if (S_1(i)=='X')
524                 R1(i)=0.3E-9;
525
526  end
527  end
528  end
529  end
530  end
531  end
532  end
533  end
534  end
535  end
536  end
537  end
538  end
539  end
540  end
541  end
542  end
543  end
544  end
545  end
546  end
547  end
548  for j=1:length(S_2);
549  if (S_2(j)=='A');
550  R2(j)=0.6e-9;
551  else
552  if (S_2(j)=='R');
553  R2(j)= 0.809e-9;
554  else
555  if (S_2(j)=='N');
556  R2(j)=0.682e-9;
557  else
558  if (S_2(j)=='D');
559  R2(j)=0.665e-9;
560  else
561  if (S_2(j)=='C');
562  R2(j)=0.629e-9;
563  else
564  if (S_2(j)=='Q');
565  R2(j)=0.725e-9;
566  else
567  if (S_2(j)=='E');
568  R2(j)=0.714e-9;
569  else
570  if (S_2(j)=='G');
571  R2(j)=0.537e-9;
572  else
573  if (S_2(j)=='H');
```

```
574 R2(j)=0.732e-9;
575 else
576 if (S_2(j)=='I');
577 R2(j)=0.735e-9;
578 else
579 if(S_2(j)=='L');
580 R2(j)=0.734e-9;
581 else
582 if (S_2(j)=='K')
583 R2(j)=0.737e-9;
584 else
585 if (S_2(j)=='M')
586 R2(j)=0.741e-9;
587 else
588 if (S_2(j)=='F')
589 R2(j)=0.781e-9;
590 else
591 if (S_2(j)=='P');
592 R2(j)=0.672e-9;
593 else
594 if (S_2(j)=='S');
595 R2(j)=0.615e-9;
596 else
597 if (S_2(j)=='T');
598 R2(j)=0.659e-9;
599 else
600 if (S_2(j)=='W');
601 R2(j)=0.826e-9;
602 else
603 if (S_2(j)=='Y');
604 R2(j)=0.781e-9;
605 else
606 if (S_2(j)=='V');
607 R2(j)=0.694e-9;
608              else
609             if (S_2(j)=='X')
610 R2(j)=0.3E-9;
611 end
612 end
613 end
614 end
615 end
616 end
617 end
618 end
619 end
620 end
621 end
622 end
623 end
624 end
625 end
```

```
626 end
627 end
628 end
629 end
630 end
631 end
632 end
633  Ra=0.6e-9;
634  Rr=0.809e-9;
635  Rn=0.682e-9;
636  Rd=0.665e-9;
637  Rc=0.629e-9;
638  Rq=0.725e-9;
639  Re=0.714e-9;
640  Rg=0.725e-9;
641  Rh=0.732e-9;
642  Ri=0.735e-9;
643  Rl=0.734e-9;
644  Rk=0.737e-9;
645  Rm=0.741e-9;
646  Rf=0.781e-9;
647  Rp=0.672e-9;
648  Rs=0.615e-9;
649  Rt=0.659e-9;
650  Rw=0.826e-9;
651  Ry=0.781e-9;
652  Rv=0.694e-9;
653  Rx=0.3E-9;
654 for i=1:length(S_1);
655 for j=1:length(S_2);
656 if (S_1(i)=='R'& S_2(j)=='D');
657    h(i,j)=.15*10^(-9)+Rr+Rd;
658 else
659 if (S_1(i)=='R'& S_2(j)=='E');
660     h(i,j)=.15*10^(-9)+Rr+Re;
661 else
662 if (S_1(i)=='D'& S_2(j)=='R');
663     h(i,j)=.15*10^(-9)+Rd+Rr;
664 else
665 if (S_1(i)=='D'& S_2(j)=='H');
666    h(i,j)=.15*10^(-9)+Rd+Rh;
667 else
668 if (S_1(i)=='D'& S_2(j)=='R');
669     h(i,j)=.15*10^(-9)+Rd+Rr;
670 else
671 if (S_1(i)=='D'& S_2(j)=='H');
672     h(i,j)=.15*10^(-9)+Rd+Rh;
673 else
674 if (S_1(i)=='D'& S_2(j)=='K');
675     h(i,j)=.15*10^(-9)+Rd+Rk;
676 else
677 if (S_1(i)=='E')& (S_2(j)=='R');
```

```
678      h(i,j)=.15*10^(-9)+Re+Rr;
679  else
680  if (S_1(i)=='E'& S_2(j)=='H');
681     h(i,j)=.15*10^(-9)+Re+Rh;
682  else
683  if (S_1(i)=='E'& S_2(j)=='K');
684     h(i,j)=.15*10^(-9)+Re+Rk;
685  else
686  if (S_1(i)=='H'& S_2(j)=='D')
687      h(i,j)=.15*10^(-9)+Rh+Rd;
688  else
689  if (S_1(i)=='H'& S_2(j)=='E')
690      h(i,j)=.15*10^(-9)+Rh+Re;
691  else
692  if (S_1(i)=='R'& S_2(j)=='R')
693      h(i,j)=.4*10^(-9)+Rr+Rr;
694  else
695  if (S_1(i)=='R'& S_2(j)=='H')
696     h(i,j)=.4*10^(-9)+Rr+Rh;
697  else
698  if (S_1(i)=='R'& S_2(j)=='H')
699      h(i,j)=.4*10^(-9)+Rr+Rh;
700  else
701  if (S_1(i)=='R'& S_2(j)=='K')
702      h(i,j)=.4*10^(-9)+Rr+Rk;
703  else
704  if (S_1(i)=='D'& S_2(j)=='E');
705      h(i,j)=.4*10^(-9)+Rd+Re;
706  else
707  if (S_1(i)=='D'& S_2(j)=='D');
708      h(i,j)=.4*10^(-9)+Rd+Rd;
709  else
710  if (S_1(i)=='H'& S_2(j)=='R')
711      h(i,j)=.4*10^(-9)+Rh+Rr;
712  else
713  if (S_1(i)=='H'& S_2(j)=='H')
714      h(i,j)=.4*10^(-9)+Rh+Rh;
715  else
716  if (S_1(i)=='H'& S_2(j)=='K')
717       h(i,j)=.4*10^(-9)+Rh+Rk;
718  else
719  if (S_1(i)=='K'& S_2(j)=='R')
720     h(i,j)=.4*10^(-9)+Rk+Rr;
721  else
722  if (S_1(i)=='K'& S_2(j)=='H')
723      h(i,j)=.4*10^(-9)+Rk+Rh;
724  else
725  if (S_1(i)=='K'& S_2(j)=='K')
726      h(i,j)=.4*10^(-9)+Rk+Rk;
727  else
728  if (S_1(i)=='N'& S_2(j)=='Q')
729     h(i,j)=.25*10^(-9)+Rn+Rq;
```

```
730 else
731 if (S_1(i)=='N'& S_2(j)=='S')
732    h(i,j)=.25*10^(-9)+Rn+Rs;
733 else
734 if (S_1(i)=='N'& S_2(j)=='Y')
735     h(i,j)=.25*10^(-9)+Rn+Ry;
736 else
737 if (S_1(i)=='Q'& S_2(j)=='S')|(S_1(i)=='Q')& (S_2(j)=='Y');
738     h(i,j)=.25*10^(-9)+Rq+Rs;
739 else
740 if (S_1(i)=='Q')& (S_2(j)=='Y');
741      h(i,j)=.25*10^(-9)+Rq+Ry;
742 else
743 if (S_1(i)=='S'& S_2(j)=='Y');
744     h(i,j)=.25*10^(-9)+Rs+Ry;
745 else
746     h(i,j)=1.76*10^(-9);
747
748
749 end
750 end
751 end
752 end
753 end
754 end
755 end
756 end
757 end
758 end
759 end
760 end
761 end
762 end
763 end
764 end
765 end
766 end
767 end
768 end
769 end
770 end
771 end
772 end
773 end
774 end
775 end
776 end
777 end
778 end
779 end
780 end
781
```

```
782 function[A]=electrostatic(Q1,Q2, R1,R2,h,M,N,N1,epsilon)
783 for i=1:N
784     for j=1:M
785         if R1(i)>R2(j)
786             gamma(i,j)=R1(i)/R2(j);
787         else
788             if  R1(i)<R2(j)
789                 gamma(i,j)=R2(j)/R1(i);
790               else if R1(i)==R2(j);
791      gamma(i,j)=R2(j)/R1(i);
792          end
793             end
794         end
795         if h(i,j)>(R1(i)+R2(j))
796             r(i,j)=h(i,j)/(R1(i)+R2(j));
797         else if  h(i,j)<=(R1(i)+R2(j))
798             r(i,j)=(R1(i)+R2(j))/h(i,j);
799         end
800         end
801     y(i,j)=(((r(i,j)^2*...
802     (1+gamma(i,j))^2)-(1+(gamma(i,j))^2))/(2*gamma(i,j)));
803     beta(i,j)=acosh(y(i,j));
804     z(i,j)=exp(-beta(i,j));
805     S12=0;
806     S22=0;
807     S11=0;
808     for k=1:N1
809 
810         S_1(k)=(z(i,j)^k)/(((1-z(i,j)^(2*k)))*...
811         ((gamma(i,j)+y(i,j))-(y(i,j)^2-1)^(1/2)*...
812         (1+z(i,j)^(2*k))/(1-z(i,j)^(2*k))));
813         S11=S11+S_1(k);
814         S_2(k)=(z(i,j)^(2*k))/(1-(z(i,j)^(2*k)));
815         S12=S12+S_2(k);
816         S_3(k)=(z(i,j)^k)/(((1-z(i,j)^(2*k)))*...
817 ((1-gamma(i,j)*y(i,j))-gamma(i,j)*(y(i,j)^2-1)^(1/2)*...
818 (1+z(i,j)^(2*k))/(1-z(i,j)^(2*k))));
819         S22=S22+S_3(k);
820     end
821     epsilon0=8.85418781762*10^(-12);
822     c11(i,j)=(2*gamma(i,j)*((y(i,j)^2-1)^(1/2))).*S11;
823     c22(i,j)=(2*gamma(i,j)*((y(i,j)^2-1)^(1/2))).*S22;
824     c12(i,j)=-((2*gamma(i,j)*...
825     ((y(i,j)^2-1))^(1/2))/(r(i,j)*(1+gamma(i,j)))).*S12;
826     delta(i,j)=((c11(i,j)*c22(i,j)-c12(i,j)^2));
827      k=1/(4*pi*epsilon0);
828     k1=1/(4*pi*epsilon* epsilon0);
829         alpha(i,j)=Q2(i,j)/Q1(i,j);
830     if R1(i)>R2(j)
831         gamma(i,j)=R1(i)/R2(j);
832   W1(i,j)=((1/k1)*R2(j)*gamma(i,j))*...
833    ((1+gamma(i,j))/(2*alpha(i,j)))*...
```

```
834     ((alpha(i,j)^2*c11(i,j)-2*...
835    alpha(i,j)*c12(i,j)+c22(i,j))/delta(i,j));
836           else if (R1(i)<R2(j))
837               gamma(i,j)=R2(j)/R1(i);
838  W1(i,j)=((1/k1)*R1(i)*gamma(i,j))*...
839  ((1+gamma(i,j))/(2*alpha(i,j)))*...
840  ((alpha(i,j)^2*c11(i,j)-2*...
841  alpha(i,j)*c12(i,j)+c22(i,j))/delta(i,j));
842        else if R1(i)==R2(j);
843  W1(i,j)=((1/k1)*R1(i)*gamma(i,j))*...
844  ((1+gamma(i,j))/(2*alpha(i,j)))*...
845  ((alpha(i,j)^2*c11(i,j)-2*...
846  alpha(i,j)*c12(i,j)+c22(i,j))/delta(i,j));
847               end
848               end
849      end
850      W2(i,j)=(k*(Q1(i,j)*Q2(i,j)))/(R1(i)+R2(j));
851      A1(i,j)=W1(i,j);
852      A2(i,j)=W2(i,j);
853      A(i,j)=A1(i,j)/A2(i,j);
854       end
855  end
856  return
857
858  function[cond2]=condmy(A)
859  [U,S,V]=SVD_2(A);
860  lambda_max=max(diag(S));
861  lambda_min=min(diag(S));
862  cond_1=(((lambda_max)/(lambda_min)));
863  cond2=(log(cond_1))/(log(10));
864  return
865
866  function [Uout,Sout,Vout] = SVD_2(A)
867            m = size(A,1);
868        n = size(A,2);
869        U = eye(m);
870        V = eye(n);
871        e = eps*fro(A);
872        while (sum(abs(A(~eye(m,n)))) > e)
873            for i = 1:n
874                for j = i+1:n
875                    [J1,J2] = jacobi(A,m,n,i,j);
876                    A = mtimes(J1,mtimes(A,J2));
877                    U = mtimes(U,J1');
878                    V = mtimes(J2',V);
879                end
880                for j = n+1:m
881                    J1 = jacobi2(A,m,n,i,j);
882                    A = mtimes(J1,A);
883                    U = mtimes(U,J1');
884                end
885            end
```

```
886         end
887         S = A;
888 
889         if (nargout < 3)
890             Uout = diag(S);
891         else
892             Uout = U; Sout = times(S,eye(m,n)); Vout = V;
893         end
894       end
895 
896 
897       function [J1,J2] = jacobi(A,m,n,i,j)
898          B = [A(i,i), A(i,j); A(j,i), A(j,j)];
899          [U,S,V] = tinySVD(B); %
900 
901          J1 = eye(m);
902          J1(i,i) = U(1,1);
903          J1(j,j) = U(2,2);
904          J1(i,j) = U(2,1);
905          J1(j,i) = U(1,2);
906 
907          J2 = eye(n);
908          J2(i,i) = V(1,1);
909          J2(j,j) = V(2,2);
910          J2(i,j) = V(2,1);
911          J2(j,i) = V(1,2);
912       end
913 
914       function J1 = jacobi2(A,m,n,i,j)
915          B = [A(i,i), 0; A(j,i), 0];
916          [U,S,V] = tinySVD(B);
917 
918          J1 = eye(m);
919          J1(i,i) = U(1,1);
920          J1(j,j) = U(2,2);
921          J1(i,j) = U(2,1);
922          J1(j,i) = U(1,2);
923       end
924 
925       function [Uout,Sout,Vout] = tinySVD(A)
926  t=rdivide((minus(A(1,2),A(2,1))),(plus(A(1,1),A(2,2))));
927         c = rdivide(1,sqrt(1+t^2));
928         s = times(t,c);
929         R = [c,-s;s,c];
930         M = mtimes(R,A);
931         [U,S,V] = tinySymmetricSVD(M);
932         U = mtimes(R',U);
933 
934         if (nargout < 3)
935             Uout = diag(S);
936         else
937             Uout = U; Sout = S; Vout = V;
```

```
938         end
939       end
940
941       function [Uout,Sout,Vout] = tinySymmetricSVD(A)
942         if (A(2,1) == 0)
943            S = A;
944            U = eye(2);
945            V = U;
946         else
947
948            w = A(1,1);
949            y = A(2,1);
950            z = A(2,2);
951            ro = rdivide(minus(z,w),times(2,y));
952 t2 = rdivide(sign(ro),plus(abs(ro),sqrt(plus(times(ro,ro),1))));
953            t = t2;
954            c = rdivide(1,sqrt(plus(1,times(t,t))));
955            s = times(t,c);
956            U = [c, -s; s, c];
957            V = [c,  s;-s, c];
958            S = mtimes(U,mtimes(A,V));
959            U = U';
960            V = V';
961         end
962
963         [U,S,V] = fixSVD(U,S,V);
964
965         if (nargout < 3)
966             Uout = diag(S);
967         else
968            Uout = U; Sout = S; Vout = V;
969         end
970       end
971
972       function [U,S,V] = fixSVD(U,S,V)
973         Z = [sign(S(1,1)),0; 0,sign(S(2,2))];
974         U = mtimes(U,Z);
975         S = mtimes(Z,S);
976         if (S(1,1) < S(2,2))
977            P = [0,1;1,0];
978            U = mtimes(U,P);
979            S = mtimes(P,mtimes(S,P));
980            V = mtimes(P,V);
981         end
982       end
983
984       function f = fro(M)
985         f = sqrt(sum(sum(times(M,M))));
986       end
987
988       function s = sign(x)
989          if (x > 0)
```

```
990             s = 1;
991         else
992             s = -1;
993         end
994     end
```

References

1. K.M. ElSawy, A. Sim, D.P. Lane, C.S. Verma, L.S. Caves, A spatiotemporal characterization of the effect of p53 phosphorylation on its interaction with MDM2. Cell Cycle **14**(2), 179–188 (2015)
2. J. Chen, Intra molecular interactions in the regulation of p53 pathway. Transl. Cancer Res. **5**(6) (2016)
3. H. Feng, L.M. Jenkins, S.R. Durell, R. Hayashi, S.J. Mazur, S. Cherry, J.E. Tropea, M. Miller, A. Wlodawer, E. Appella, Y. Bai, Structural basis for p300 Taz2-p53 TAD1 binding and modulation by phosphorylation. Structure **17**(2), 202–210 (2009)
4. L.M. Miller Jenkins, H. Feng, S.R. Durell, H.D. Tagad, S.J. Mazur, J.E. Tropea, Y. Bai, E. Appella, Characterization of the p300 Taz2-p53 TAD2 complex and comparison with the p300 Taz2–P53 TAD1 complex. Biochemistry **54**(11), 2001–2010 (2015)
5. L.M. Jenkins, S.R. Durell, S.J. Mazur, E. Appella, P53 N-terminal phosphorylation: a defining layer of complex regulation. Carcinogenesis **33**(8), 1441–1449 (2012)
6. G. Plopper, D. Sharp, E. Sikorski, *Lewin's CELLS* (Jones & Bartlett Learning, Burlington, 2015)
7. P. Vise et al., Identifying long-range structure in the intrinsically unstructured transactivation domain of p53. Proteins **67**, 526–530 (2007)
8. C.W. Lee et al., Graded enhancement of p53 binding to CREB-binding protein (CBP) by multisite phosphorylation. Proc. Natl. Acad. Sci. USA **107**, 19290–19295 (2010)
9. J.C. Ferreon et al., Cooperative regulation of p53 by modulation of ternary complex formation with CBP/p300 and HDM2. Proc. Natl. Acad. Sci. USA **106**, 6591–6596 (2009)
10. D.P. Teufel et al., Regulation by phosphorylation of the relative affinities of the N-terminal transactivation domains of p53 for p300 domains and Mdm2. Oncogene **28**, 2112–2118 (2009)
11. L.M. Jenkins et al., Two distinct motifs within the p53 transactivation domain bind to the Taz2 domain of p300 and are differentially affected by phosphorylation. Biochemistry **48**, 1244–1255 (2009)
12. S. Polley et al., Differential recognition of phosphorylated transactivation domains of p53 by different p300 domains. J. Mol. Biol. **376**, 8–12 (2008)
13. S.P. Deb, S. Deb (eds.), *Mutant p53 and MDM2 in Cancer* (Springer, Berlin, 2014)
14. J. Bartek, J. Falck, J. Lukas, CHK2 kinase-a busy messenger. Nat. Rev. Mol. Cell Biol. **12**, 877–886 (2001)
15. P.H. Kussie, S. Gorina, V. Marechal, B. Elenbaas, J. Moreau, A.J. Levine, N.P. Pavletich, Structure of the MDM2 oncoprotein bound to the p53 tumor suppressor transactivation domain. Science **274**(5289), 948–953 (1996)

Chapter 7
Mathematical Modelling of the Interaction of BH3-Peptides with Full-Length Proteins, and Account of the Influence of Point Mutations on the Stability of the Formed Biological Complex on the Example of the Bcl-2 Family Proteins

Abstract This chapter presents a new method that allows one to qualitatively determine the effect of point mutations in peptides on the stability of the formed complex with full-length proteins. On the basis of the developed approach, a qualitative correlation of the obtained results with the dissociation constant was revealed using the example of the formation of the BH3 peptide biological complex of Bmf, Puma, Bad, Hrk, Bax, Bik, Noxa, Bid, Bim, and Bak proteins with the Bcl-xl protein and the BH3 peptides protein Bax with the Bcl-2 protein, taking into account the replacement of amino acid residues.

7.1 Introduction

This chapter is devoted to the investigation of the interaction of BH3-peptides with anti-apoptotic proteins of the Bcl-2 family, which are regulators of mitochondrial apoptosis pathways. Note that disturbances in the process of apoptosis are a sign of a large number of diseases, such as cancer, sarcomas, carcinomas, lymphomas, and leukemias.

Studies results [1–3] indicate that peptides have a pronounced protective effect in various diseases and have a modulating effect on various body systems. Unlike chemotherapy drugs, peptides are selective and effective signaling molecules that bind to certain receptors or ion channels, where they cause intracellular effects. Because peptides are highly selective and effective and at the same time relatively safe and well tolerated, they represent an excellent starting point for the development of new therapeutic agents, and their specificity demonstrates excellent safety, tolerability, and efficacy profiles in people with various pathologies.

We believe that the further development of peptide drugs will be based on the use of computer and mathematical design to find the optimal peptides, taking into account the affinity for their targets, and also to improve their chemical and physical properties.

T. Koshlan and K. Kulikov, *Mathematical Modeling of Protein Complexes*, Biological and Medical Physics, Biomedical Engineering,
https://doi.org/10.1007/978-3-319-98304-2_7

In this chapter, the effect of point mutations in BH3 peptides on the stability of the biological complexes with Bcl-2 will be determined, as well as the qualitative determination of the dissociation constant for binding different BH3 peptides to Bcl-xl proteins.

Let us examine some of the works devoted to the family of Bcl-2 proteins, as well as the study of the affinity of BH3 peptides for Bcl-2 family proteins. In [4] review recent advances in understanding how BCL2 family proteins control MOMP (mitochondrial outer membrane permeabilization) as well as new nonapoptotic functions for these proteins.

In [5] it was found that the Bcl-2 protein binds to the Bax protein through two interdependent interfaces, which leads to the inhibition of the proapoptotic oligomerization of Bax. Studies of various interfaces with a large number of involved amino acid residues bring additional clarity to the nature of the interaction of proteins of the Bcl-2 family.

In [6] the molecular basis of the binding specificity of proapoptotic BH3 peptides, which contain different motifs, with a pro-apoptotic Bcl-xl protein, was investigated. Various motifs were identified in the BH3 domains of proteins which influenced the binding affinity of the Bcl-xl protein.

In this chapter, in contrast to the above, we propose a mathematical model for determining the affinity of different BH3 peptides for Bcl-2 family proteins, as well as taking into account the effect of point mutations in peptides on the stability of the biological complex formed by them in dependence at amino acid sequence of protein.

The first part of the chapter contains information about the structure and functions of the studied proteins of the Bcl-2 family. The second part is devoted to the numerical calculation of the interaction of peptides of the Bcl-2 family proteins containing the BH3 region with the Bcl-xl protein. In this part, the numerical calculations of the interaction of protein peptides Bmf, Puma, Bad, Hrk, Bax, Bik, Noxa, Bid, Bim, and Bak with the Bcl-xl protein were analyzed and compared with the dissociation constant (K_d). The effect of point mutations in BH3 peptides of the Bax protein on the stability of the biological complexes formed with the Bcl-2 protein was studied. A qualitative comparison of lg(cond(W)) and logarithmic constants of dissociation of K_d is carried out on the example of the interplay of BH3 peptides of Puma, Bad, Hrk, Bax, Bik, Noxa proteins with the Bcl-xl protein.

To analyze the biochemical processes we use the notion of condition number matrix of the potential energy of the pair electrostatic interaction between peptides. In this physical formulation of the problem, it will characterize the degree of stability of the configuration of the biological complex. In order to choose a more stable biochemical compound between proteins, we select the matrix of potential energy of electrostatic interaction with the **smallest** value of the condition number (see Chap. 2).

7.2 Structure and Functions of Bcl-2 Family Proteins

Proteins of the B-cell lymphoma-2 family (Bcl-2) control their own apoptotic path, regulating the process of permeabilization of the outer membrane of the mitochondria through protein-protein interactions.

Structural and biochemical studies have shown the dual role of anti-apoptotic proteins of the Bcl-2 family in inhibiting BH3-only proteins and the activated proteins Bax and Bak. Details of the interactions between the Bcl-2 family proteins are presented in [4].

The proteins of the Bcl-2 family can be divided into 3 groups based on their structure and intracellular functions:

(1) One group includes Bcl-2 antogonist/killer (Bak) and Bcl-2 associated X protein (Bax), which are known as apoptosis effectors. Also called multidomain pro-apoptotic BCL2 family proteins, BAX and BAK contain BCL2 homology (BH) domains 1–3 and can directly permeabilize MOM when activated. Whether Bcl-2-related ovarian killer (Bok) belongs to this same subfamily is not clear. Structurally it is similar to Bak and Bax [7]; however, functionally it does not have the ability to permeabilize the MOM by itself but instead induces apoptosis only in the presence of Bak or Bax [8].

Structural studies have demonstrated that Bak and Bax monomers are globular structures consisting of a central hydrophobic core helix(alfa5) surrounded by eight alpha helices [9–11].

In the Bak monomer, four of these helices($\alpha 1$, $\alpha 3$, $\alpha 4$, and $\alpha 6$) are long helices that form a circle around the central helix $\alpha 5$ while the others ($\alpha 2$, $\alpha 7$, and $\alpha 8$) are shorter and link either the longer helices or the main structure to the transmembrane (TM) domain which consists of helix $\alpha 9$ [9].

The major structural difference between monomeric Bak and Bax is the orientation of helix $\alpha 9$. In Bax, this helix is buried in a hydrophobic groove formed by helices $\alpha 3$, $\alpha 4$, and $\alpha 5$ [11].

In contrast, the hydrophobic groove of Bak is empty; and the $\alpha 9$ helix of Bak extends away from the remainder of the globular protein.

(2) The second group, called anti-apoptotic or pro-survival Bcl-2 family members, includes Bcl-2, Bcl-x large(Bcl-xl), Bcl-2-like protein 2(Bcl-W), Bcl-2-like protein 10(Bcl-B), myeloid cell leukemia 1(Mcl1), and Bcl-2-related protein A1(Bfl-1) (A1 in mouse). These proteins, which contain four BH domains (BH1-BH4), inhibit apoptosis by binding and sequestering their pro-apoptotic counterparts.

The anti-apoptotic Bcl-2 proteins possess a remarkably similar globular structure containing a so-called ≪Bcl-2 core≫ [12].

This core consists of a bundle of eight alfa-helices that form a hydrophobic groove flanked by the BH1 and BH3 domains. In Bcl-W, the core also includes a short C-terminal helix α-8 attached to the BH2 domain. The hydrophobic groove made by $\alpha 3-\alpha 5$ is termed the ≪BC groove≫ [12] because it binds the BH3 region of binding partners [15] (Scheme of structure Bcl-2, see Fig. 7.1).

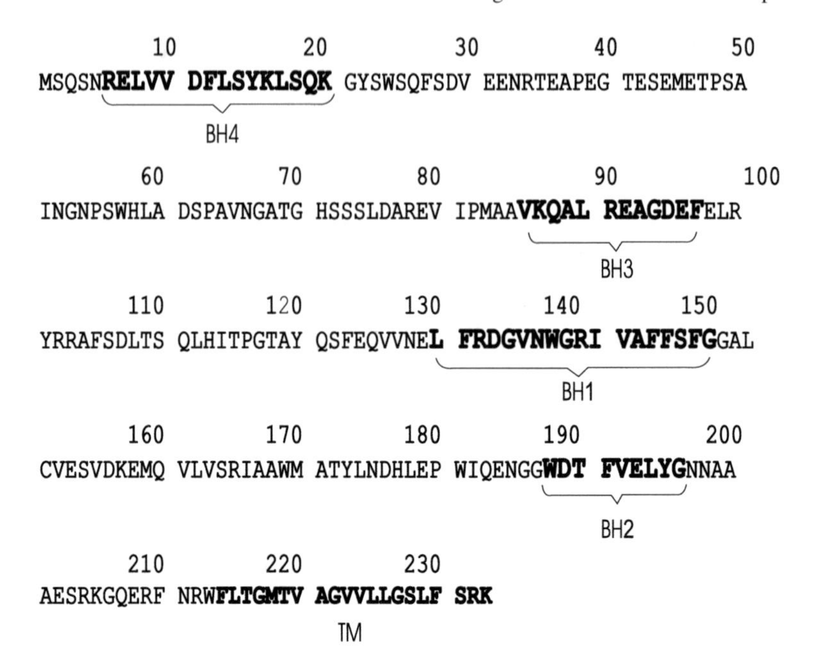

Fig. 7.1 Schema of a Bcl-2 protein. The amino acid sequence Q071817 [13]. Specifying BH domains [14]

The BC groove is crucial for the biology of anti-apoptotic BCL2 family proteins, as it provides an interface for the interaction with the BH3 domain of BH3-only proteins and the apoptotic effectors BAK and BAX. Based on the structure of the BC groove, several BH3 mimics that can occupy this groove and thus inactivate these proteins anti-apoptotic function have been developed and are currently being tested in the clinic [16–18].

(3) The final group, termed BH3-only proteins, includes Bim, Puma, Bid, Bad, Noxa, Bik, Bmf, and Hrk. These polypeptides share only a 15–25 residue BH3 domain in common with other Bcl-2 family proteins. This BH3 domain, however, is critical for the interactions of these proteins with other BCL2 family proteins to regulate MOMP.

The most important role of BH3-only proteins such as Bim, Puma, Bid, and Noxa is to act as integrators of various signals to initiate MOMP. The BH3-only proteins are activated by distinct cytotoxic stimuli in various ways, including enhanced transcription and posttranslational modifications [19].

BH3-only proteins can be divided into direct activators and sensitizers [20, 21].

As pro-apoptotic signals are received, for example, DNA damage or cellular stress, proteins such as Bid or Bad stimulate and compete with effectors for binding to repressors, and not only neutralize the antiapoptical actions of repressors, but also lead to a pro-apoptotic effect of the effectors.

The effectors subsequently initiate apoptotic cell death by their ability to integrate into the outer membrane of the mitochondria, which causes the formation of pores in the membranes. This results in the release of apoptogenic factors, such as cytochrome c and Smac/Diablo from the mitochondria into the cytosol [22].

Thus, the concerted action of different Bcl-2 proteins allows one to keep apoptosis under control in a healthy cell, while a disorder in the regulation leads to serious pathological consequences.

The formation of heterodimers between different proteins of the Bcl-2 family determines whether the cell survives or not [6]. The Bcl-2 family is an important therapeutic target due to over-expression in some cancer cells where these proteins contribute to oncogenesis and resistance to chemotherapy.

In particular, overexpression of the Bcl-xl and Bcl-2 proteins of apoptosis repressors is associated with the development of various oncological diseases. The Bcl-xl and Bcl-2 proteins are the most suitable targets for anticancer therapy.

Thus, in this paper, interactions of various BH3 peptides with the anti-apoptotic proteins Bcl-xl and Bcl-2 will be investigated.

To analyze the biochemical processes we use the notion of condition number matrix of the potential energy of the pair electrostatic interaction between peptides. In this physical formulation of the problem, it will characterize the degree of stability of the configuration of the biological complex. In order to choose a more stable biochemical compound between proteins, we select the matrix of potential energy of electrostatic interaction with the **smallest** value of the condition number (see Chap. 2).

7.3 Numerical Calculation Results. Conclusion

This part of the chapter presents the results of the numerical calculations of the interaction of BH3 peptides with Bcl-xl and Bcl-2 proteins, as well as an analysis of the effect of point mutations in BH3 peptides on their ability to form stable biological complexes with the pro-apoptotic Bcl-xl protein.

7.3.1 *Results of Numerical Simulation of the Interaction of BH3-Peptides of Bmf, Puma, Bad, Hrk, Bax, Bik, Noxa, Bid, Bim, and Bak Proteins with the Bcl-xl$_{(1-200)}$ Protein*

In this section, the interaction of short BH3-peptide proteins: Bmf, Puma, Bad, Hrk, Bax, Bik, Noxa, Bid, Bim, and Bak with the protein Bcl-xl$_{(1-200)}$ is investigated.

Each of these peptides has its own affinity for the Bcl-xl protein since each peptide has its own unique amino acid sequence.

In this section, we compared the affinity involved in the study of BH3-peptides Bmf, Puma, Bad, Hrk, Bax, Bik, Noxa, Bid, Bim, and Bak during their successive interactions with the protein Bcl-xl$_{(1-200)}$ according to the previously developed algorithm 1.

Note that BH3-peptides take the conformation of a α-helix and are associated with the $\alpha 2$–$\alpha 5$ helixes protein Bcl-xl. This interaction is stabilized through a large number of intramolecular contacts.

According to previous studies [6], the above short peptides related to the full-size Bcl-xl protein can be divided into 2 groups.

The first group (I) includes BH3 peptides (Hrk, Bax, Bik, Noxa) with a higher dissociation constant for Bcl-xl, and the second group (II) (Bmf, Puma, Bad, Bid, Bim, Bak) includes BH3-peptides characterized by a lower Kd value for the Bcl-xl protein. The list of amino acid sequences is given in Table 7.1.

Dissociation constants (K_d) for each peptide group interaction with Bcl-xl$_{(1-200)}$ are given in Table 7.2 according to [6].

As we see from Table 7.2, group I consists of BH3 peptides with a lower affinity for Bcl-xl, the Kd of which varies from 4.69 μmol to 26.01 μmol. Group II comprised BH3 peptides with a higher affinity level; their Kd varied from 0.2 μmol to 0.65 μmol.

Algorithm 1 (see Chap. 5) was used in the implementation of the numerical calculations of the interaction of each BH3 peptide with Bcl-xl$_{(1-200)}$. As the short BH3 peptide moves along the Bcl-xl$_{(1-200)}$ protein, we get the value of lg(cond(W)) for each step of the shift along the long Bcl-xl protein.

Table 7.1 Amino acid sequences of BH3 peptides [6]

$N^{\underline{0}}$	Protein name	Amino acid sequences
1	Bmf	QAEVQIARKLQCIADQFHRL
2	Puma	QWAREIGAQLRRMADDLNAQ
3	Bad	WAAQRYGRELRRMSDEFVDS
4	Hrk	SAAQLTAARLKALGDELHQR
5	Bax	ASTKKLSESLKRIGDELDSN
6	Bik	EGSDALALRLACIGDEMDVS
7	Noxa	ELEVECATQLRRFGDKLNFR
8	Bid	DIIRNIARHLAQVGDSMDRS
9	Bim	RPEIWIAQELRRIGDEFNAY
10	Bak	STMGQVGRQLAIIGDDINRR

Table 7.2 Groups of BH3-peptides according to the degree of affinity for the protein Bcl-xl$_{(1-200)}$

Groups	3-peptides	Kd, μmol
I	Hrk, Bax, Bik, Noxa	4.69–26.01
II	Bmf, Puma, Bad, Bid, Bim, Bak	0.2–0.65

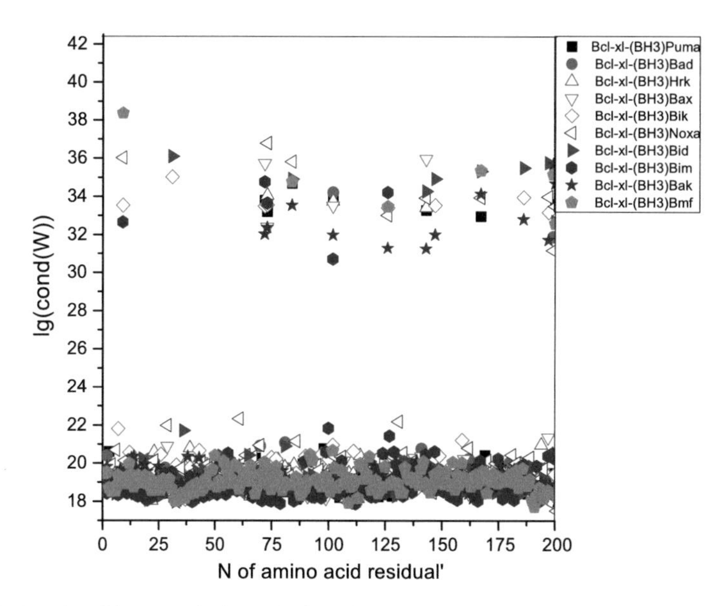

Fig. 7.2 Results of the numerical calculation of the interaction of the BH3-peptides of Bmf, Puma, Bad, Hrk, Bax, Bik, Noxa, Bid, Bim, and Bak proteins with the Bcl-xl$_{(1-200)}$ protein. The colored figures indicate the results obtained during the interaction of BH3 peptides of group II with Bcl-xl$_{(1-200)}$; uncolored figures indicate the results obtained by the interaction of BH3 peptides of group I with Bcl-xl$_{(1-200)}$, $\varepsilon = 1$

In Fig. 7.2 is a graph of the numerical calculations obtained for all ten BH3 peptides upon interaction with the protein Bcl-xl$_{(1-200)}$. We note that the Bcl-xl protein region was involved in the calculation from 1 a.a. to 200 a.a., in a way similar to the experimental article [6].

To analyze the obtained data, we divided the region of the smallest values of lg(cond(W)) in several gradations, starting with the value of 17.65. In this case, we believe that the most stable biological complex is characterized by the smallest value of lg(cond(W)).

In Fig. 7.3 are represented the regions of the smallest values of lg(cond(W)) obtained by the interaction of 10 short peptides with the protein Bcl-xl$_{(1-200)}$. The least-significant part is represented by four areas: I, II, III, and IV, with each successive region including the previous one. Thus, region I is in the range of values of lg(cond(W)) 17.65–17.9; region II−17.65–18.0; region III−17.65–18.1; region IV−17.65–18.2. For 100%, we take the total number of points that fall into each particular area under consideration: I, II, III, IV. The percentage of values for each group of peptides for different regions is summarized in Table 7.3.

As we see from Table 7.3, the regions with the lowest values of lg(cond(W)) (regions I–IV) are represented by the majority of values obtained for peptides of group II, which are characterized by a high affinity for the Bcl-xl$_{(1-200)}$. Thus, region I contains 100% of the values obtained when the BH3 peptides of group II interact

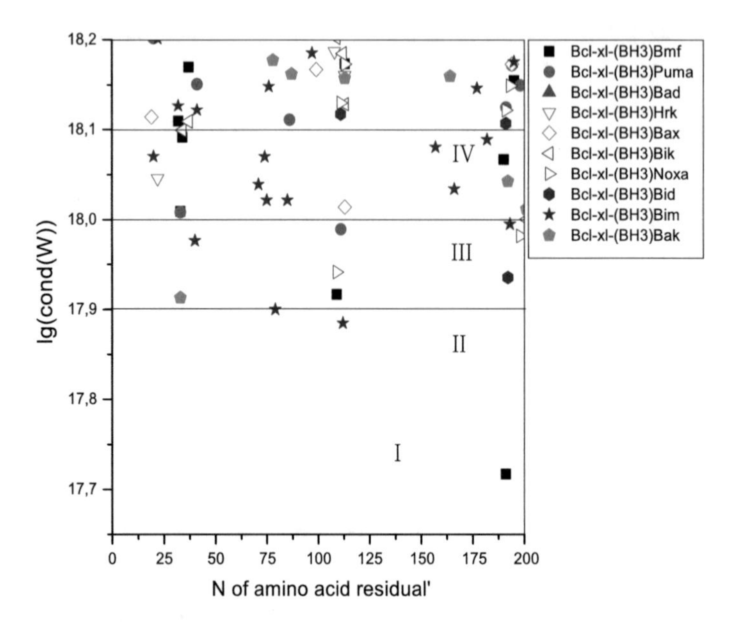

Fig. 7.3 Results of the numerical calculation of the interaction of the BH3-peptides of Bmf, Puma, Bad, Hrk, Bax, Bik, Noxa, Bid, Bim, and Bak proteins with the Bcl-xl$_{(1-200)}$ protein in the region of the smallest values of lg(cond(W)). The colored figures indicate the results obtained during the interaction of BH3 peptides of group II with Bcl-xl$_{(1-200)}$; uncolored figures indicate the results obtained by the interaction of BH3 peptides of group I with Bcl-xl$_{(1-200)}$, $\varepsilon = 1$

Table 7.3 Values of lg(cond(W)) for each group of BH3-peptides

Region	Range of values lg(cond(W))	Hrk, Bax, Bik, Noxa	Bmf, Puma, Bad, Bid, Bim, Bak
I	17.65–17.9	0.0	100.0
II	17.65–18	18.18	81.82
III	17.65–18.1	16	84
IV	17.65–18.2	28.81	71.19
V	17.65–18.3	23.1	76.9

with the protein Bcl-xl$_{(1-200)}$. Region II contains 18.18% of the values of lg(cond(W)) obtained from the interaction of BH3 peptides of the group I with the protein Bcl-xl$_{(1-200)}$ and 81.82% of the values of lg(cond(W)), obtained during the interaction of BH3 peptides of group II with the protein Bcl-xl$_{(1-200)}$. The regions containing higher values of lg(cond(W)) (regions III and IV) contain 16 and 28.81% of the values of lg(cond(W)) obtained from the interaction of group I peptides with the Bcl-xl$_{(1-200)}$ and 84 and 71.19% of the values of lg(cond(W)) for the interaction of BH3 peptides of group II with the protein Bcl-xl$_{(1-200)}$, respectively.

As we see from the presented table, as the values of lg(cond(W)) increase, a gradual uneven decrease in the values of lg(cond(W)) in the range of values of

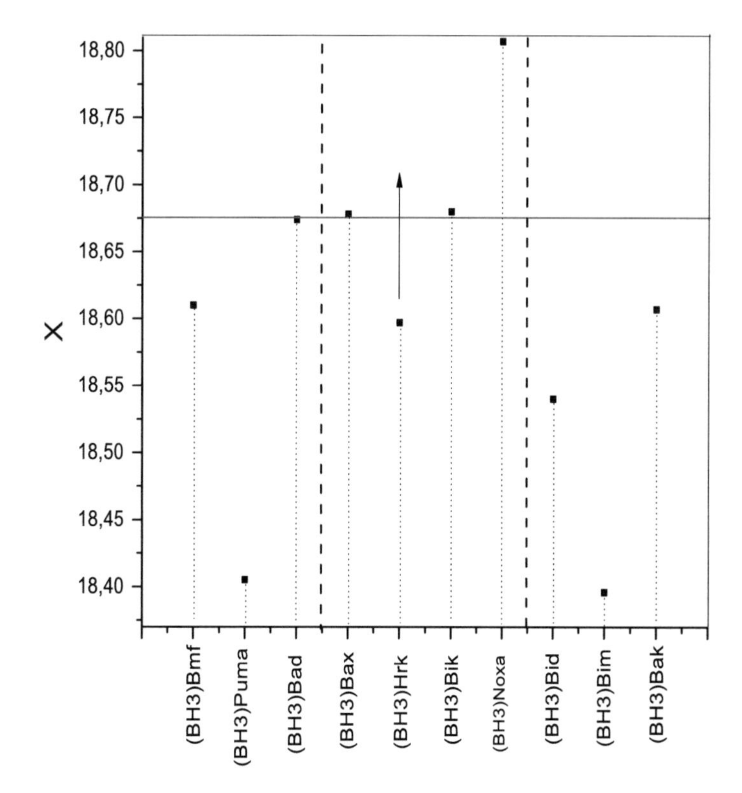

Fig. 7.4 The value of X obtained at interaction of each BH3-peptide protein Bmf, Puma, Bad, Hrk, Bax, Bik, Noxa, Bid, Bim, Bak with the Bcl-xl$_{(1-200)}$, $\varepsilon = 1$

lg(cond(W)) occurs for the second group of proteins and an increase in the values of lg(cond(W)) occurs for the first group of proteins.

Thus, the minimum values of lg(cond(W)) are in higher values for group I BH3-peptides, and in lower values for BH3-peptides of group II, which indicates a higher affinity of BH3 -peptides of group II for the protein Bcl-xl$_{(1-200)}$ than of BH3-peptides of group I for the protein Bcl-xl$_{(1-200)}$.

To verify this assumption, we calculated the mean value of the 100 minimum values, which we will mark as X, for each BH3 peptide in interaction with the Bcl-xl protein. Next, we presented graphically the values obtained for each interacting BH3 peptide with the Bcl-xl$_{(1-200)}$protein (see Fig. 7.4). On this histogram, a red baseline was drawn, which conditionally separates the BH3 peptides of the first and second groups. Above the baseline lie the values of the 100 minimum average values of lg(cond(W)) obtained during the interaction of the BH3 peptides of the group I proteins with the Bcl-xl$_{(1-200)}$ protein. Below the baseline lie the values obtained in the interaction BH3 peptides of group II proteins with the protein Bcl-xl$_{(1-200)}$.

As can be seen from the histogram, six values of X corresponding to the interactions of BH3 peptides of the protein group II with Bcl-xl$_{(1-200)}$ (Bmf, Puma, Bad,

Bid, Bim, Bak) are in the range of values below the baseline. Their values amounted to: 18.61, 18.405, 18.674, 18.54, 18.396, and 18.607. Three values of X corresponding to the interactions of group I BH3 peptides (Hrk, Bik, Noxa) with Bcl-xl$_{(1-200)}$ are in the range of higher values of X and are located above the baseline. However, one value of X corresponding to the interaction of the BH3 peptide of group I of the Bax protein with Bcl-xl$_{(1-200)}$ is below the baseline and its value amounted to 18.597. Thus, an analysis of the given criterion (X) demonstrated that 9 out of 10 BH3 peptides satisfy the given criterion.

Note that in this calculation and subsequent analysis, the 100 minimum values of lg(cond(W)) were chosen for the interaction of each BH3 peptide with the whole amino acid sequence of the Bcl-xl protein.

To achieve more accurate results in the future, it is proposed to take into account the folding of proteins, the structure of the formed dimeric complex, and to analyze the minimum values of lg(cond(W)) with the participation of exclusively interacting regions of whole proteins.

Thus, in the presented section, a qualitative analysis of the determination of the affinity of short BH3 peptides of Bmf, Puma, Bad, Hrk, Bax, Bik, Noxa, Bid, Bim, Bak proteins with the Bcl-xl$_{(1-200)}$ protein was performed and a qualitative agreement with [6] was revealed.

7.3.2 Interaction of Modified BH3-Peptides of Bax Protein with Bcl-2 Protein Taking into Account the Replacement of Amino Acid Residues

In this section, the binding of BH3 peptides of the Bax protein was numerically simulated, taking into account various changes in amino acid residues with the Bcl-2 protein. The obtained result will allow us to determine the influence of point mutations in the BH3 peptides of the Bax protein on the stability of the complexes formed with the Bcl-2 protein.

In [23], the binding structure of the Bcl-2 protein to the Bax protein region was determined. The Bax protein peptide forms an amphiphilic α–helix and binds to the BH3-binding hydrophobic groove of the Bcl-2 protein. The intramolecular interaction between the Bcl-2 protein and the Bax protein peptide is mediated by hydrophobic and ionic interactions. Some of the main amino acid residues from the protein Bcl-2 are the amino acid residues in the region from 107a.a. to 146 a.a. as well as the amino acid residue in the 200 a.a. region. The amino acid sequences of the BH3-peptide of the Bax protein with the performed amino acid residue substitutions, as well as the dissociation constants for binding each peptide to the Bcl-2 protein are shown in Table 7.4.

Interaction of the BH3 peptide of the $Bax_{(wt)}$ protein with the Bcl-2 protein was taken as the **main** interaction of the BH3 peptide with the Bcl-2 protein. When point exchanges of amino acid residues in the BH3 peptide of the Bax protein are performed, we assume that the main interactions of these modified BH3 peptides fall on the same sections of the Bcl-2 protein as the $Bax_{(wt)}$ BH3-peptide of the

Table 7.4 List of amino acid sequences of BH3 peptides of the Bax protein with amino acid substitutions and dissociation constant for each peptide when interacting with the Bcl-2 protein [23]

Location of point mutations in the region $Bax_{(49-84)}$	Amino acid sequence	K_d, nmol
$Bax_{(49-84)(wt)}$	QPPQDASTKKLSECLRRIGDELDSNMELQRMIADVD	15.1
$mBax_{(61A,R64A,R78A)}$	QPPQDASTKKLS**A**CL**A**RIGDELDSNMELQ**A**MIADVD	787
$mBax_{(E61A)}$	QPPQDASTKKLS**A**CLRRIGDELDSNMELQRMIADVD	95.2
$mBax_{(R64A)}$	QPPQDASTKKLSECL**A**RIGDELDSNMELQRMIADVD	129
$mBax_{(D68A)}$	QPPQDASTKKLSECLRRIG**A**ELDSNMELQRMIADVD	1040
$mBax_{(E69A)}$	QPPQDASTKKLSECLRRIGD**A**LDSNMELQRMIADVD	476
$mBax_{(R78A)}$	QPPQDASTKKLSECLRRIGDELDSNMELQ**A**MIADVD	57.1

wild type when interacting with the Bcl- 2 and the point replacements of the amino acid residues do not essentially change the binding site with the Bcl-2 protein, but have a significant effect on the affinity of complex formation. When analyzing the interaction of modified BH3 peptides with Bcl-2, the sites identified as the main sites in the interaction of the wild-type BH3 peptide with Bcl-2 will be analyzed.

Figure 7.5 shows the numerical results obtained for the interaction of all $Bax_{(wt)}$ and modified BH3 peptides of the Bax protein with Bcl-2 at low values of lg(cond(w)). The results of numerical simulation of the interaction of the BH3-peptide $Bax_{(wt)}$ with Bcl-2 on the graph are presented by a black square, while the results of the interaction of all other modified BH3 peptides of the Bax protein with Bcl-2 are represented by empty figures.

For further analysis of the data, we identified three significant areas with the lowest values of lg(cond(W)) in the interaction of the BH3 peptide $Bax_{(wt)}$ with the Bcl-2 protein in the range of lg(cond(w)) from 18.75 to 19.15. Recall that each point on the graph represents the first a.a. when binding two amino acid sequences.

The first area lies in the interval from 60 a.a. to 70 a.a., the second region lies in the region from 105 a.a. to 130 a.a., and the third area from 160 a.a. to 180 a.a.

For each of these areas, we calculated the number of hits for each interaction of the BH3 peptide with Bcl-2 and plotted the result in the form of a histogram, see Fig. 7.6.

As can be seen from the histogram, the greatest number of hits-8, is typical for the interaction of the BH3-peptide $Bax_{(wt)}$ with Bcl-2. The number of all other hits for modified BH3 peptides Bax c Bcl-2 corresponds to a smaller number of hits of the minimum values of lg(cond(W)) in these regions. Also, from the given histogram 7.6, the interaction of the BH3-peptide $Bax_{(49-85)}$ with the Bcl-2 protein is characterized by the most frequent hit of the lg(cond(W)) values in the selected regions in comparison with the other modified BH3 peptides of the Bax protein, in which the amino acid residues were substituted.

Thus, six modified BH3-peptides of the protein $Bax_{(49-85)}$ in which the amino acid residue substitutions were performed bind to the Bcl-2 protein worse than the

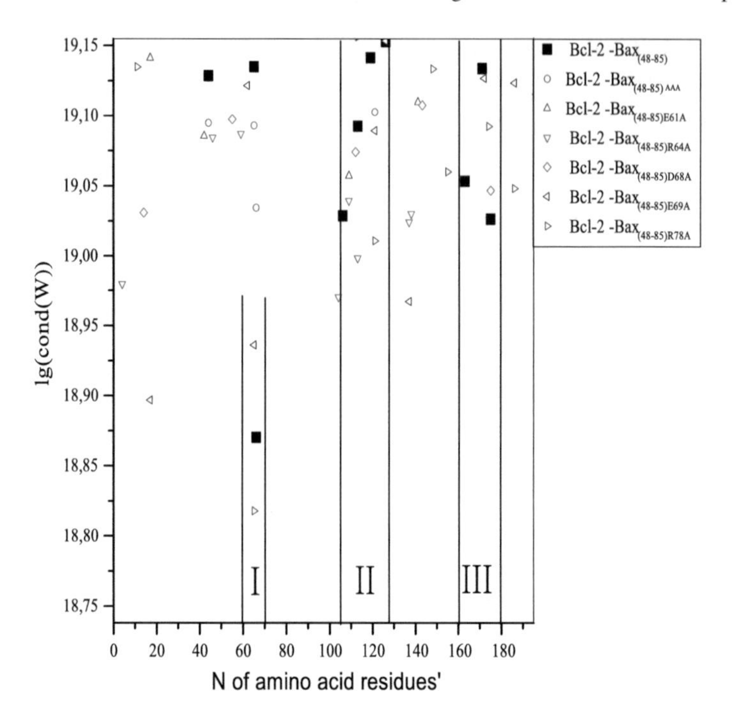

Fig. 7.5 Results of the numerical calculation of the interaction of the Bax3 peptides of the Bax protein with the Bcl-2 protein (fill figures) and the Bax3 peptides of the Bax protein, in which the amino acid residue substitutions (empty figures) were performed, with the Bcl-2 protein in the region of the smallest values of lg(cond(W)), $\varepsilon = 1$

BH3-peptide $Bax_{(wt)}$, while the site $\mathrm{Bax}_{(49-85)}$ is prone to form the most stable biological complexes with the Bcl-2 protein compared to all other $\mathrm{Bax}_{(49-85)}$ peptides in which one or more amino acid residues were replaced by the amino acid residue of alanine (A).

This result is in good agreement with the previously performed experimental article, which indicates that the dissociation constant of mutant peptides upon binding to the Bcl-2 protein is higher than when bounding to the natural site of $\mathrm{Bax}_{(49-85)}$) [23].

7.3.3 *Qualitative Definition of the Logarithm of the Dissociation Constant K_d in the Interaction of BH3 Peptides with the Bcl-2 Protein*

The task was to find the correspondence and correlation between lg(cond(W)) and the logarithm of the dissociation constant K_d for the interaction of protein molecules. To do this, we compared the available data of the value of $\lg(K_d)$ with the calculated value of lg(cond(W)) for binding of the BH3 peptides with Bcl-xl [23] protein.

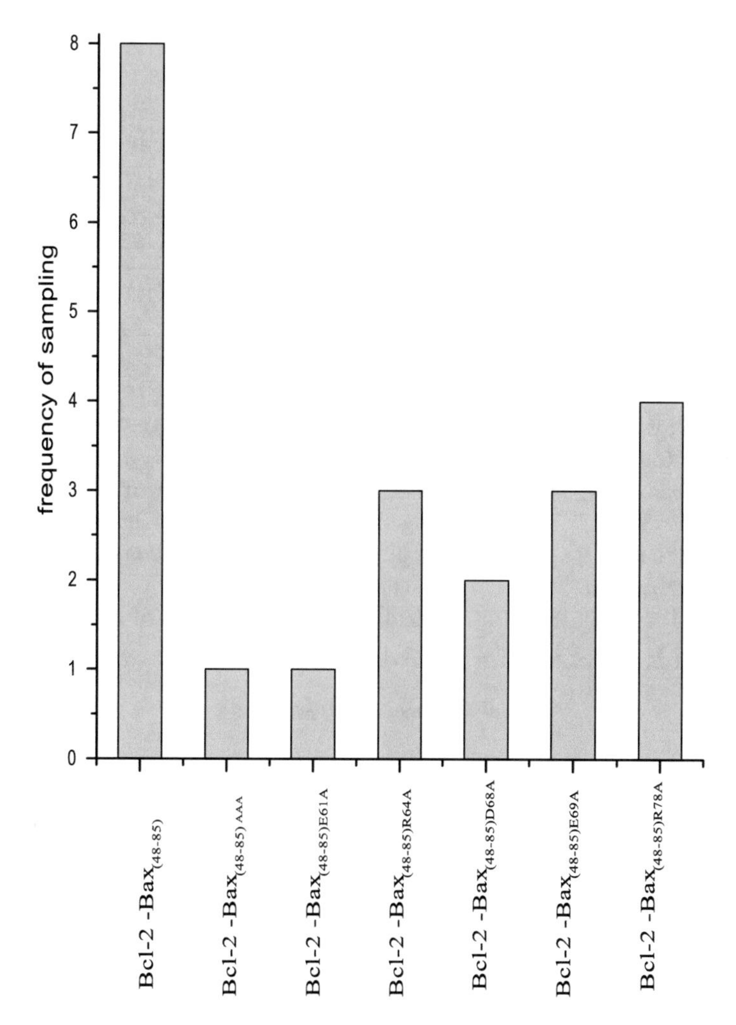

Fig. 7.6 Frequency of BH3 peptides Bax in the three designated regions of the Bcl-2 protein, $\varepsilon = 1$

We performed the calculation of the interaction of the BH3 peptides of proteins $\text{Puma}_{(127-162)}$, $\text{Hrk}_{(23-58)}$, $\text{Bad}_{(137-172)}$, $\text{Bik}_{(41-76)}$, $\text{Bax}_{(48-85)}$, $\text{Noxa}_{(65-100)}$ with the entire amino acid sequence Bcl-xl, as well as with the truncated amino acid sequence of the protein $\text{Bcl-xl}_{(86-233)}$, assuming that not the entire amino acid sequence of the Bcl-xl protein participates in the direct formation of the biological complex with BH3 peptides, in order to determine the correspondence between the theoretical values of lg(cond(W)) and to compare with the experimental values of K_d.

The list of amino acid sequences of BH3-peptides is given in the Table 7.5, as well as their numbers from [13].

Table 7.5 List of amino acid sequences of BH3-peptides [13]

Protein	Amino acid sequence	Number in uniprot
$\text{Puma}_{(127-162)}$	RVEEEEWAREIGAQLRRMADDLNAQYERRRQEEQHR	Q99ML1
$\text{Bad}_{(137-172)}$	APPNLWAAQRYGRELRRMSDEFEGSFKGLPRPKSAG	Q61337
$\text{Hrk}_{(23-58)}$	PGLRWAAAQVTALRLQALGDELHRRAMRRRARPRDP	P62817
$\text{Bik}_{(41-76)}$	LMECVEGRNQVALRLACIGDEMDLCLRSPRLVQLPG	070337
$\text{Bax}_{(49-84)}$	QPPQDASTKKLSECLRRIGDELDSNMELQRMIADVD	Q07813
$\text{Noxa}_{(65-100)}$	TRVPADLKDECAQLRRIGDKVNLRQKLLNLISKLFN	Q9JM54

The mean values for the 100 minimal values of lg(cond(W)) were calculated for each interacting BH3 peptide of the proteins $\text{Puma}_{(127-162)}$, $\text{Hrk}_{(23-58)}$, $\text{Bad}_{(137-172)}$, $\text{Bik}_{(41-76)}$, $\text{Bax}_{(48-85)}$, $\text{Noxa}_{(65-100)}$ with the Bcl-xl protein. The results obtained are shown in the Fig. 7.7. The graphs Fig. 7.8 show the mean values for 100 minimal values of lg(cond(W)) obtained by the interaction of BH3 peptides with the truncated protein $\text{Bcl-xl}_{(86-233)}$.

The specific values of K_d corresponding to the interactions of each BH3-peptide with the Bcl-xl protein are given below [23].

$$BH3\text{-}peptide\ \ PUMA\ 4.65$$

$$BH3\text{-}peptide\ \ BAD\ 18.4$$

$$BH3\text{-}peptide\ \ HRK\ 17.9$$

$$BH3\text{-}peptide\ \ BIK\ 23.6$$

$$BH3\text{-}peptide\ \ Bax > 100$$

$$BH3\text{-}peptide\ \ Noxa > 1000$$

The values of K_d corresponding to the interaction of $\text{Bax}_{(48-85)}$, $\text{Noxa}_{(65-100)}$ with the Bcl-xl protein are expressed as $\lg(K_d)$.

In the presented graphs Figs. 7.7 and 7.8, BH3 peptides are listed in order with increasing $\lg(K_d)$ when interacting with the Bcl-xl protein. The interaction of BH3 peptides of the proteins $\text{Bax}_{(48-85)}$ and $\text{Noxa}_{(65-100)}$ with Bcl-xl is characterized by the greatest K_d. Thus, the graphs should show an increase in the values of the lg(cond(W)) value, starting from the BH3 peptide of $\text{Puma}_{(127-162)}$ to the BH3 peptide of $\text{Noxa}_{(65-100)}$.

Let us now analyze the results presented in each graphs.

In Fig. 7.7, a baseline was made separating the last two BH3 peptides of $\text{Bax}_{(48-85)}$ and $\text{Noxa}_{(65-100)}$ from the remaining BH3 peptides of $\text{Puma}_{(127-162)}$, $\text{Hrk}_{(23-58)}$,

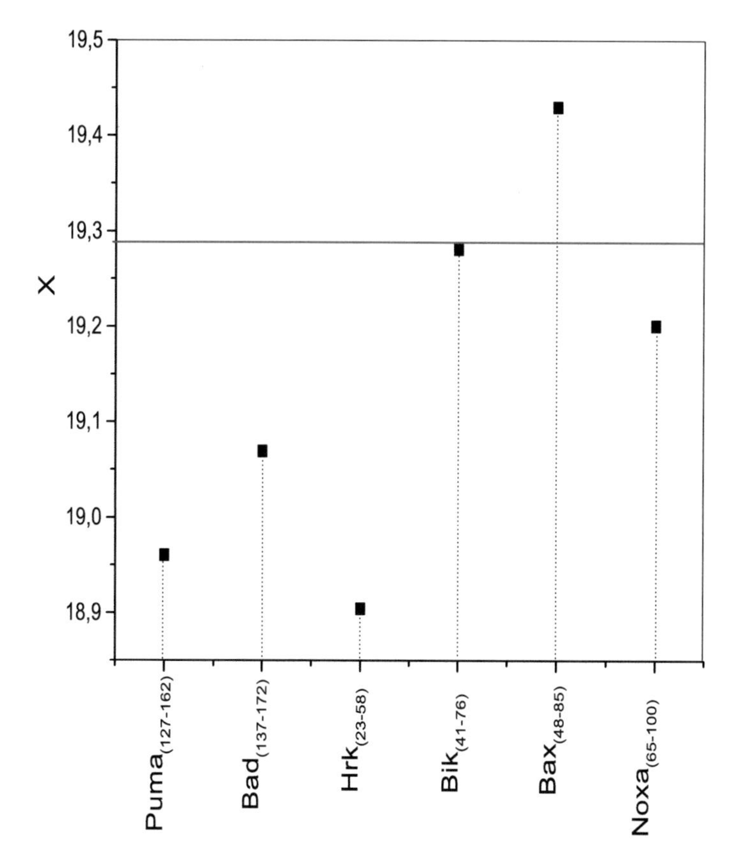

Fig. 7.7 Average value of the one hundred minimum values of lg(cond(W)) for each BH3 peptide upon interaction with the Bcl-xl, $\varepsilon = 1$

$\mathrm{Bad}_{(137-172)}$, $\mathrm{Bik}_{(41-76)}$ with the Bcl-2 protein depending on K_d interacting proteins, since the K_d of the last two BH3 peptides is significantly larger compared to the K_d of previous BH3-peptides.

As seen from the graph, the interaction of the BH3 peptide $\mathrm{Bax}_{(48-85)}$ with Bcl-xl is characterized by a higher K_d and value of lg(cond(W)) compared to the interaction of BH3 peptides $\mathrm{Puma}_{(127-162)}$, $\mathrm{Hrk}_{(23-58)}$, $\mathrm{Bad}_{(137-172)}$, $\mathrm{Bik}_{(41-76)}$ with Bcl-xl. However, the calculated value of lg(cond(W)) in the interaction of the BN3 peptide of $\mathrm{Noxa}_{(65-100)}$ with Bcl-xl lies in a lower range of lg(cond(W)) than the results of the interaction of BH3-peptides, $\mathrm{Bik}_{(41-76)}$ and $\mathrm{Bax}_{(48-85)}$ with Bcl-xl.

One of the reasons that the latter value lies below the baseline is that not all the polypeptide chain of the Bcl-xl protein takes a direct part in the formation of the biological complex with BH3-peptides.

We assume that the main participation in the formation of a biological complex involving Bcl-2 and BH3-peptide proteins occurs in the B1–BH3 domains of the

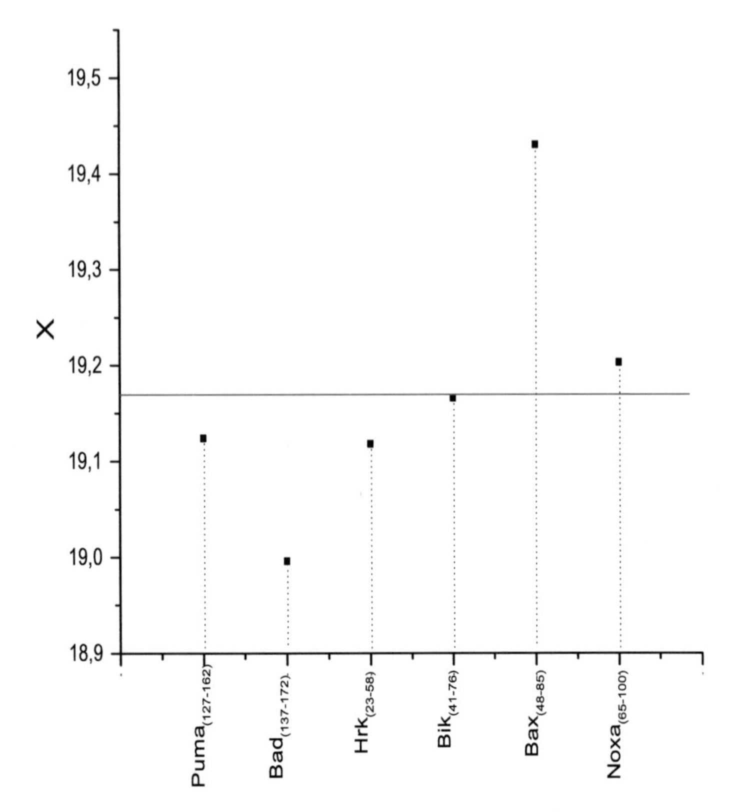

Fig. 7.8 Mean value of the fifty minimum values of lg(cond(W)) for each BH3 peptide when interacting with the truncated Bcl-xl$_{(86-233)}$, $\varepsilon = 80$

Bcl-xl protein. Therefore, an additional calculation of the interaction between BH3-peptides and Bcl-xl$_{(86-233)}$ was carried out.

In Fig. 7.8, a baseline was also made that separates the interactions of BH3 peptides Puma$_{(127-162)}$, Hrk$_{(23-58)}$, Bad$_{(137-172)}$, Bik$_{(41-76)}$ with Bcl-xl$_{(100-233)}$ from the interaction of BH3-peptides Bax, Noxa with Bcl-xl$_{(100-233)}$.

As can be seen from the figure, the values of lg(cond(w)) of the BH3-peptide interaction of Bax$_{(48-85)}$ and Noxa$_{(65-100)}$ with Bcl-xl are above the baseline, as K_d is found for interactions of Bax$_{(48-85)}$–Bcl-xl and Noxa$_{(65-100)}$–Bcl-xl [23] are at higher values than K_d, characterizing the interactions of Puma$_{(127-162)}$, Hrk$_{(23-58)}$, Bad$_{(137-172)}$, and Bik$_{(41-76)}$ with Bcl-xl. We believe that in order to obtain better data, it is necessary to perform a calculation between the contacting regions of the polypeptide chains of the interacting proteins.

Thus in this chapter, a new method has been developed that allows us:

- to qualitatively determine the lg(K_d) of peptides for full-length proteins;
- to determine the effect of point mutations in peptides on the stability of the formed complex with whole proteins.

A qualitative agreement of the results with Kd on the example of the formation of the biological complex of BH3 peptides of Bmf, Puma, Bad, Hrk, Bax, Bik, Noxa, Bid, Bim, and Bak proteins with $Bcl\text{-}xl_{(1-200)}$ protein was determined.

The influence of point mutations on the stability of the formed biological complexes was also studied on the example of the interaction of the BH3 peptides of the Bax protein in which point replacements of amino acid residues were made with the whole Bcl-2 protein. The formation of the biological complex of BH3-peptide of the protein $Bax_{(49-85)}$ with Bcl-2 was taken as the main interaction. The numerical results of the interaction of the protein $Bax_{(49-85)}$ with Bcl-2 were compared with the remaining results of the interaction of BH3 peptides Bax with the Bcl-2 protein taking into account the substitution of amino acid residues. As a result, the Bcl-2 protein regions with the largest number of minimum values of lg(cond(W)) were found in the interaction with $Bax_{(49-85)}$. The subsequent analysis of these regions revealed that the other modified BH3 peptides contain much less than the minimum values of lg(cond(W)) in the previously designated regions.

Thus, it is possible to use the obtained result to determine the binding site of the peptide with the whole protein in order to determine the stability of the formation of the biological complex by any modified BH3 peptide of the Bax protein in which the amino acid residues have been replaced with the Bcl-2 protein.

The third stage of the theoretical studies of the interaction of BH3 peptides with proteins of the Bcl-2 family was devoted to finding a qualitative correlation between the values of lg(cond(W)) and $\lg(K_d)$. To perform this comparison, we used the results of the values of $\lg(K_d)$ [23] obtained by the interaction of BH3 peptides Puma, Hrk, Bad, Bik, Bax, Noxa with the whole Bcl-xl protein. The result was a qualitative determination of the value of $\lg(K_d)$ by analyzing lg(cond(W)).

Application of the developed mathematical algorithms will allow us to find the optimal peptides taking into account the affinity for their target proteins and to develop inhibitors or activators of proteins in the future.

References

1. B. Santiago, I. Gutierrez-Canas, J. Dotor, G. Palao, J.J. Lasarte, J. Ruiz, J. Prieto, F. Borras-Cuesta, J.L. Pablos, Topical application of a peptide inhibitor of transforming growth factor-$\beta 1$ ameliorates bleomycin-induced skin fibrosis. J. Investig. Dermatol. **125**(3), 450–455 (2005)
2. N. Suzuki, S. Hazama, H. Iguchi, K. Uesugi, H. Tanaka, K. Hirakawa, A. Aruga, T. Hatori, H. Ishizaki, Y. Umeda, T. Fujiwara, T. Ikemoto, M. Shimada, K. Yoshimatsu, R. Shimizu, H. Hayashi, K. Sakata, H. Takenouchi, H. Matsui, Y. Shindo, M. Iida, Y. Koki, H. Arima, H. Furukawa, T. Ueno, S. Yoshino, Y. Nakamura, M. Oka, H. Nagano, Phase II clinical trial of peptide cocktail therapy for patients with advanced pancreatic cancer: VENUS-PC study. Cancer Sci. **108**(1), 73–80 (2017)
3. R. Arafeh, K. Flores, A. Keren-Paz, G. Maik-Rachline, N. Gutkind, S. Rosenberg, R. Seger, Y. Samuels, Combined inhibition of MEK and nuclear ERK translocation has synergistic antitumor activity in melanoma cells. Sci. Rep. **7**(1), 16345 (2017)
4. H. Dai, X.W. Meng, S.H. Kaufmann, BCL2 family, mitochondrial apoptosis, and beyond. Cancer Transl. Med. **2**(1), 7–20 (2016)

5. J. Ding, Z. Zhang, G.J. Roberts, M. Falcone, Y. Miao, Y. Shao, X.C. Zhang, D.W. Andrews, J. Lin, Bcl-2 and Bax interact via the BH1-3 groove-BH3 motif interface and a novel interface involving the BH4 motif. J. Biol. Chem. **285**(37), 28749–28763 (2010)
6. V. Bhat, M.B. Olenick, B.J. Schuchardt, D.C. Mikles, C.B. McDonald, A. Farooq, Molecular determinants of the binding specificity of BH3 ligands to BclXL apoptotic repressor. Biopolymers **101**(6), 573–582 (2014)
7. P.E. Czabotar, G. Lessene, A. Strasser, J.M. Adams, Control of apoptosis by the Bcl-2 protein family: implications for physiology and therapy. Nat. Rev. Mol. Cell Biol. **15**(1), 49–63 (2014)
8. N. Echeverry, D. Bachmann, F. Ke, A. Strasser, H.U. Simon, T. Kaufmann, Intracellular localization of the BCL-2 family member BOK and functional implications. Cell Death Differ. **20**(6), 785–799 (2013)
9. T. Moldoveanu, Q. Liu, A. Tocilj, M. Watson, G. Shore, K. Gehring, The X-ray structure of a BAK homodimer reveals an inhibitory zinc binding site. Mol. Cell **24**(5), 677–688 (2006)
10. H. Wang, C. Takemoto, R. Akasaka, T. Uchikubo-Kamo, S. Kishishita, K. Murayama, T. Terada, L. Chen, Z.J. Liu, B.C. Wang, S. Sugano, A. Tanaka, M. Inoue, T. Kigawa, M. Shirouzu, S. Yokoyama, Novel dimerization mode of the human Bcl-2 family protein Bak, a mitochondrial apoptosis regulator. J. Struct. Biol. **166**(1), 32–37 (2009)
11. M. Suzuki, R.J. Youle, N. Tjandra, Structure of Bax: coregulation of dimer formation and intracellular localization. Cell **103**(4), 645–654 (2000)
12. T. Moldoveanu, A.V. Follis, R.W. Kriwacki, D.R. Green, Many players in BCL-2 family affairs. Trends Biochem. Sci. **39**(3), 101–111 (2014)
13. http://www.uniprot.org/
14. A.M. Petros, A. Medek, D.G. Nettesheim, D.H. Kim, H.S. Yoon, K. Swift, E.D. Matayoshi, T. Oltersdorf, S.W. Fesik, Solution structure of the antiapoptotic protein bcl-2. Proc. Natl. Acad. Sci. USA **98**(6), 3012–3017 (2001)
15. M.G. Hinds, M. Lackmann, G.L. Skea, P.J. Harrison, D.C. Huang, C.L. Day, The structure of Bcl-w reveals a role for the C-terminal residues in modulating biological activity. EMBO J. **22**(7), 1497–1507 (2003)
16. C. Correia, S.H. Lee, X.W. Meng, N.D. Vincelette, K.L. Knorr, H. Ding, G.S. Nowakowski, H. Dai, S.H. Kaufmann, Emerging understanding of Bcl-2 biology: implications for neoplastic progression and treatment. Biochem. Biophys. Acta. **1853**(7), 1658–1671 (2015)
17. G. Lessene, P.E. Czabotar, P.M. Colman, BCL-2 family antagonists for cancer therapy. Nat. Rev. Drug Discov. **7**(12), 989–1000 (2008)
18. C. Billard, BH3 mimetics: status of the field and new developments. Mol. Cancer Ther. **12**(9), 1691–1700 (2013)
19. H. Puthalakath, A. Strasser, Keeping killers on a tight leash: transcriptional and post-translational control of the pro-apoptotic activity of BH3-only proteins. Cell Death Differ. **9**(5), 505–512 (2002)
20. M.C. Wei, T. Lindsten, V.K. Mootha, S. Weiler, A. Gross, M. Ashiya, C.B. Thompson, S.J. Korsmeyer, tBID, a membrane-targeted death ligand, oligomerizes BAK to release cytochrome c. Genes Dev. **14**(16), 2060–2071 (2000)
21. A. Letai, M.C. Bassik, L.D. Walensky, M.D. Sorcinelli, S. Weiler, S.J. Korsmeyer, Distinct BH3 domains either sensitize or activate mitochondrial apoptosis, serving as prototype cancer therapeutics. Cancer Cell **2**(3), 183–192 (2002)
22. G. Dewson, Interplay of Bcl-2 proteins decides the life or death fate. Open Cell Signal. J. **3**, 3–8 (2011)
23. B. Ku, C. Liang, J.U. Jung, B.H. Oh, Evidence that inhibition of BAX activation by BCL-2 involves its tight and preferential interaction with the BH3 domain of BAX. Cell Res. **21**, 627–641 (2011)

Chapter 8
Mathematical Algorithms for Finding the Optimal Composition of the Amino Acid Composition of Peptides Used as a Therapy

Abstract In this chapter, two algorithms have been developed, one of which (Algorithm 3) was developed specifically for the selection of amino acid residues in peptides to improve their affinity in the interaction of peptides with full-length proteins, and Algorithm 4 was developed to search for "scattered" active region of the protein when bound to the peptide.

8.1 Introduction

In this part of the chapter, we will discuss how to improved peptide vaccines by developed algorithms.

Global concern is the rise in the incidence of cancer, recent data released revealed 12.7 million new cases and 7.6 million deaths, just in 2008 [1], in Europe alone, 3.45 million new cases were diagnosed and 1.75 million deaths occurred during 2012 [2].

Nowadays, cancer is the second most common cause of death worldwide [3], caused by an abnormal cellular growth, in a uncontrolled manner, with the ability to invade other tissues, leading to the formation of tumor masses, neo-vascularization (angiogenesis), and metastasis [4] . Lung, colorectal, prostate, and breast cancer are the most diagnosed forms of this disease [5].

Considering the numbers revealed, it is urgent to find new anticancer drugs able to control tumor growth with minimal side effects [6–8].

Potential clinical approaches using ACPs (Anti Cancer peptides).

Although a wide variety of drugs are commercially available, treatments for cancer have one thing in common: the emergence of resistance against multiple drugs [9].

Another associated problem is the lack of selectivity of the available drugs, and their consequent undesirable side effects for the patients [10].

Thus, there is a need for the development of new antineoplastic, with higher selectivity, leading to fewer side effects than current ones. It is desirable that these new compounds present different mechanisms of action, without dependence on activity toward a single specific molecule in the target cells. The main goal is resistance prevention, overcoming the existing mechanisms that cancer cells use, being active and diminishing the side effects [11–13].

T. Koshlan and K. Kulikov, *Mathematical Modeling of Protein Complexes*,
Biological and Medical Physics, Biomedical Engineering,
https://doi.org/10.1007/978-3-319-98304-2_8

Peptide-based cancer vaccines represent the most specific approach to polarize the immune system against malignant cells, since they are preparations made of single epitopes, the minimal immunogenic region of an antigen [14].

In addition, peptides also play an important role in cancer, including early diagnosis, prognostic predictors, and the treatment of cancer patients. Unlike other therapies, peptides show superiority due to their specificity. Recently, peptide-based therapy against cancer, such as peptide vaccines, has attracted increased attention [15, 16].

Here are some results of vaccine trials in Phase II clinical trials.

The complex effect of peptides derived from cancerous tissue of proteins such as LY6K, CDCA1, IMP3, whose peptides were tested in head and neck squamous cell cancer (HNSCC) immunotherapy, was studied [17].

In clinical trials of Phase II, vaccines were used based on the following peptides

$LY6K_{(177-186)}$ RYCNLEGPPI

$CDCA_{(156-164)}$ VYGIRLEHF

$IMP3_{(508-518)}$ KTVNELQNL

HIV is specific peptide ILKEPVHGV

CMV is specific peptide RYLRDQQLL

Phase II clinical trials demonstrate that antigenic vaccination, based on the five above-mentioned peptides, induces an immune response. Positive CTL responses (cytotoxic T-lymphocytes) specific for LY6K peptide after vaccination were observed in 85.7% of patients.

Patients with a positive CTL response showed significantly longer survival periods (overall survival OC) than those who did not have a CTL immune response.

Also, the MST (Median survival time) was 8.1 months for patients with an immune response and 1.4 months for patients who did not have an induced immune response to the vaccination. Also, in some patients, there was a stable remission and an increase in the duration of the period without progression of the disease.

It should also be noted that the result of phase II clinical trials is the complete recovery of the patient from the fourth stage of the disease. After 16 cycles of vaccination, recurrent and metastatic tumors disappeared. Thus, such a combinatorial vaccine with multi-epitope peptides, as monotherapy, can help circumvent the heterogeneity of cancer cells and avoid a peptide-specific immune response due to loss of antigen expression.

This chapter proposes the search for optimal solutions for future improvement and strategies in immunotherapy for various types of tumors.

Vaccines will be developed by improving existing ways of creating peptide vaccines based on the created therapeutic molecular approaches:

1. Increased affinity of existing ligand-receptor interactions

2. Regulation of protein functions by synthesizing new highly selective peptides that will bind to the active site of the target protein

3. Development of highly selective peptides for target proteins, which will activate or inhibit cascade pathways of chemical reactions in cells.

So, based on the method we developed, existing peptide vaccines can be improved by increasing the affinity for the target protein or increasing the affinity of the ligand for the receptor by selecting the amino acid composition of the peptide vaccines. We

Fig. 8.1 Peptide $Bcl2_{(185-200)}$ with the amino acid residues to be replaced

propose an algorithm developed by us that will allow the automatic setting of point replacements of amino acid residues in peptides and obtain numerical results.

To analyze the biochemical processes we use the notion of condition number matrix of the potential energy of the pair electrostatic interaction between peptides. In this physical formulation of the problem, it will characterize the degree of stability of the configuration of the biological complex. In order to choose a more stable biochemical compound between proteins, we select the matrix of potential energy of electrostatic interaction with the **smallest** value of the condition number (see Chap. 2).

8.2 Description of the Algorithm 3

Let the Bcl2 protein peptide be given from 185 a.a. to 200 a.a., in which it is necessary to determine the effects of amino acid substitutions with order numbers 4, 6, 8, 10, 12 on the stability of the formation of the biological complex of each such modified peptide with the Bax protein. In this case, it is necessary to evaluate the binding of each modified peptide to a specific region of the Bax protein, say, from 100 a.a. to 140 a.a. For example, let us analyze the interaction of the peptide Bcl-$2_{(185-200)}$ with the protein Bax. Red color indicates the variable parameters that the researcher can set independently, depending on the task. In the program, a short array is specified as:

```
1  S_20=['W'  'I'  'Q' 'D'  'N'  'G'  'G'  'W' ...
2   'D'  'A'  'F' 'V'  'E' 'L'  'Y'  'G' ]
```

The amino acid sequence of the whole protein Bax is given by the sequence:

```
1  S_100=['M'  'D'  'G'  'S'  'G'  'E'  'Q'  'P'  'R' ...
2  'G'  'G'  'G' 'P'  'T'  'S'  'S'  'E'  'Q'  'I' ...
3  'M'  'K'  'T'  'G'  'A'  'L'  'L'  'L' 'I'  'F' ...
4  'V'  'A'  'G'  'V'  'L'  'T'  'A'  'S'  'L'  'T' ...
5  'I'  'W'  'K' 'K'  'M'  'G']
```

In Fig. 8.1 such a peptide is represented. The red numbers denote the serial numbers of the amino acid residues that will be changed. In the algorithm, it is possible to specify up to 5 replacements of the amino acid residues in one peptide. When one runs the program, one can specify the number of required changes:

```
1  "Enter the number of replacements"
2  You enter the required number of replacements: 5
```

After the number of substitutions of a.a. from 1 to 5 is chosen, you should specify the sequence numbers for each a.a. substitution:

```
1 You must enter the sequence number of a.a. in the peptide: 4
2 "Enter the sequence number for replacement number 2"
3 You must enter the sequence number of a.a. in the peptide: 6
4 "Enter the sequence number for replacement number 3"
5 You must enter the sequence number of a.a. in the peptide: 8
6 "Enter the sequence number for replacement number 4"
7 You must enter the sequence number of a.a. in the peptide: 10
8 "Enter the sequence number for replacement number 5"
9 You must enter the sequence number of a.a. in the peptide: 12
```

In the body of the program, one-dimensional arrays consisting of a list of amino acid residues for rotation with point replacements a.a. are preset manually:

```
1 Sub1=['A' 'T']; %replacement matrix №1
2 Sub2=['H'];     %replacement matrix №2
3 Sub3=['K'];     %replacement matrix №3
4 Sub4=['Y'];     %replacement matrix №4
5 Sub5=['T'];     %replacement matrix №5
```

The red color denotes a.a., which the researcher sets himself. In this case, such one-dimensional arrays is 5. Next, the program asks you to specify the boundaries of protein 2 (in this case, the full-length protein Bax):

```
1 "Enter the LEFT boundary of the vector S_100="
2 Enter the value: 100
3 "Enter the RIGHT boundary of the vector S_100="
4 Enter the value: 140
```

The steps and the number of minimum values are set in the program body manually:

```
1 sh=1;          % step shift
2 n_el=10;       %amount of minimal elements.
```

All data will be written to the Excel file at the end of the program.

In Fig. 8.2 the average values $(\overline{X})$ of the 10 minimal values of lg(cond(W)) obtained from the interaction of each modified peptide with a region of the whole protein are presented. The minimum average value of the 10 minimal lg(cond(W)) in this case was 17.647. On the graph, this value is seen opposite the sequence number of the fifth amino acid substitution. Such modified peptide, the mean value of 10 minimal lg(cond(W)) of which was the minimum value, is the peptide WIQDNGGKDAFVELYG. In this methodological example, a modified Bcl-2 protein peptide was identified which, when interacting with the protein region $Bax_{(100-140)}$, showed the lowest average value of 10 minimal values of lg(cond(W)). Red color indicates the changed amino acid residues. Recall that the researcher can put any number of the smallest values of lg(cond (W)), which correspond to the interaction of the modified peptide with a protein. Thus, to obtain modified peptides with the lowest dissociation constant, it is proposed to analyze peptides that fall within the lower range of the minimum mean values of lg(cond(W)) and not be limited to one modified peptide.

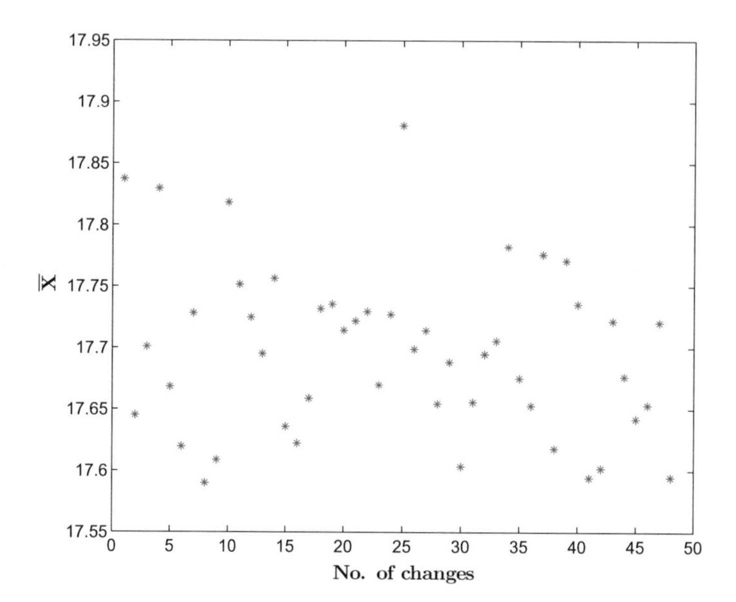

Fig. 8.2 Results of numerical calculation according to Algorithm 3

8.3 Description of the Algorithm 4

Algorithm 4 was developed for cases where the catalytic center of a protein, when bound to a peptide, is formed by different sections of a polypeptide chain located close to three-dimensional space by folding the protein into a native conformation, for example, the Bcl-2 family proteins.

So if BH3-only proteins form complexes due to the participation of the BH3 domain with proteins containing BH1-BH3 domains, such as Bcl-2, Bcl-xl, Bcl-w, then one assumes that the formation of the protein complex is due to the involvement of different sites of a polypeptide chain of a protein containing BH1-BH4 domains. Algorithm 4 allows providing such partial ≪scattered≫ of amino acid residues involved in the formation of a biological complex with another protein. A part of the peptide is supposed to be permanently installed in one of the active regions of the protein polypeptide sequence, the rest of the peptide begins its progress along the amino acid sequence of the protein, and when the second part of the peptide enters the second part of the ≪scattered≫ catalytic center, it is assumed that the value of lg(cond(W)) will decrease. Thus, we assume that it is possible to ≪feel≫ the regions of the ≪scattered≫ catalytic center along the polypeptide chain of the whole protein.

Input data:

The segment protein $Bax_{(59-74)}$

```
1  S_20=[ 'L'  'S'  'E'  'C'  'L'  'K'  'R' 'I'  'G'  'D'  'E'...
2  'L'  'D'  'S'  'N'  'M' ]
```

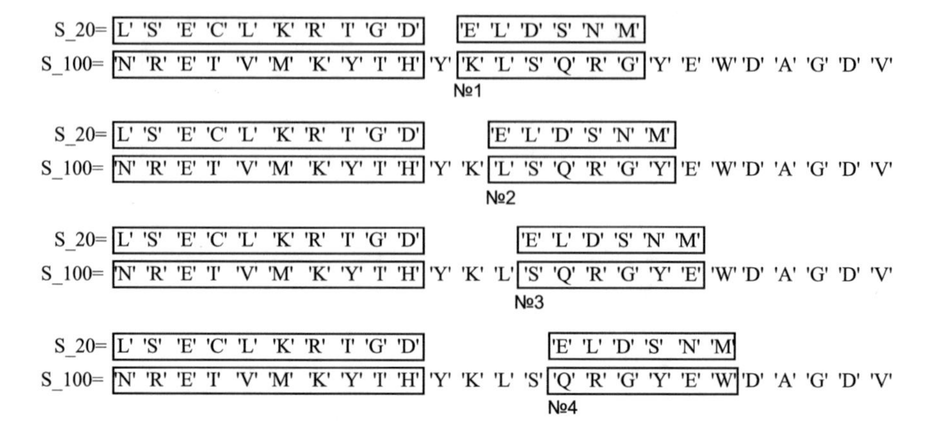

Fig. 8.3 The scheme of Algorithm 4

Table 8.1 The four lowest values of lg(cond(w)) and their corresponding amino acid sequences

$N^{\underline{0}}$	Amino acid sequence of Bcl-2	Amino acid sequence $Bax_{(59-74)}$	lg(cond(W))
7	**NREIVMK**YIHYEWDAG	LSECLKRIGDELDSNM	17.471
3	**NREIVMK**YIHSQRGYE	LSECLKRIGDELDSNM	17.479
1	**NREIVMK**YIHKLSQRG	LSECLKRIGDELDSNM	17.539
8	**NREIVMK**YIHEWDAGD	LSECLKRIGDELDSNM	17.549

The segment protein Bcl-2:

```
1 S_100=[ 'N'  'R'  'E'  'I' 'V'  'M'  'K'  'Y'  'I'...
2 'H'  'Y'  'K' 'L'  'S'  'Q'  'R'  'G'  'Y'  'E'  'W' 'D'...
3 'A'  'G'  'D'  'V'  ]
4 n_el=4    %amount of minimal elements:
```

Thus, the first matrix will be formed when calculating the interactions between a.a. of the following two one-dimensional arrays Fig. 8.3:

Bold characters denotes an invariable part of the formed one-dimensional array (Table 8.1). The result is shown in Fig. 8.4.

8.4 Schematic Representation of the Increased Affinity of Peptides to Proteins Targets

The above algorithms for analyzing the interaction of short peptides with full-length proteins will allow the selection of higher affinity peptides for receptors or for protein targets. Let us consider several options for the application of mathematical modeling to improve existing peptides and the development of new biologically active peptides.

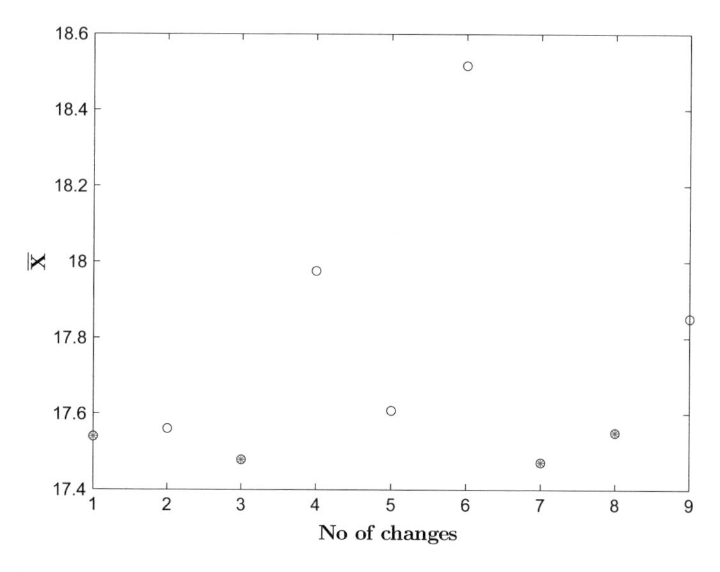

Fig. 8.4 Results of numerical calculation according to Algorithm 4

8.4.1 *Increase of the Affinity of the Existing Peptides to the Active Center of the Target Protein*

In Figs. 8.5 and 8.6 a model for improving the existing L1 ligand by the method of selecting amino acid residues in the peptide using Algorithm 3 is presented. After the mathematical calculation of the substitution of one amino acid residue for other amino-acid residues, the L1 ligand is converted to the ligand L2, which has an increased affinity for the receptor, compared to the L1 ligand.

The dissociation constants K1 and K2 characterize the affinity of the original ligand L1 to the receptor. The constants K3 and K4 are new constants characterizing the affinity of the new L2 ligand to the same receptor. Ligand L2 as a result of our mathematical algorithm has an increased affinity for the receptor.

Thus, the association constant K1 is smaller than the new K3 association constant. In turn, the dissociation constant K2 is greater than the new dissociation constant K4.

8.4.2 *Creation of a New Peptide that Binds to a Given Active Protein Center*

One can also synthesize special peptides that will bind to a given site on the receptor (as shown in the Fig. 8.7). Thus, in addition to improving existing ligands, it will be possible to develop new peptides that will selectively bind to a selected region of other proteins.

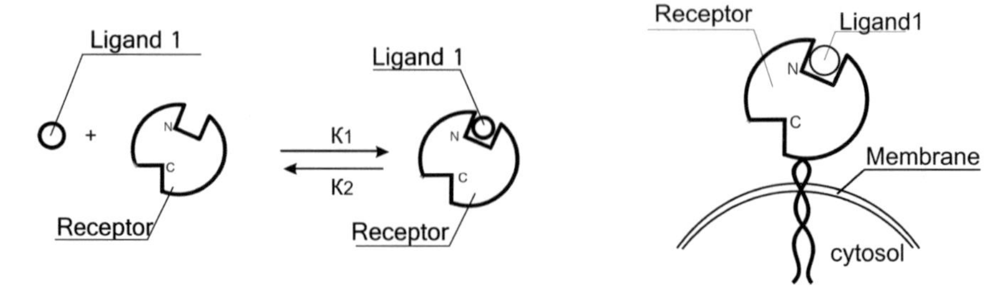

Fig. 8.5 The previously existing peptide ligand L1

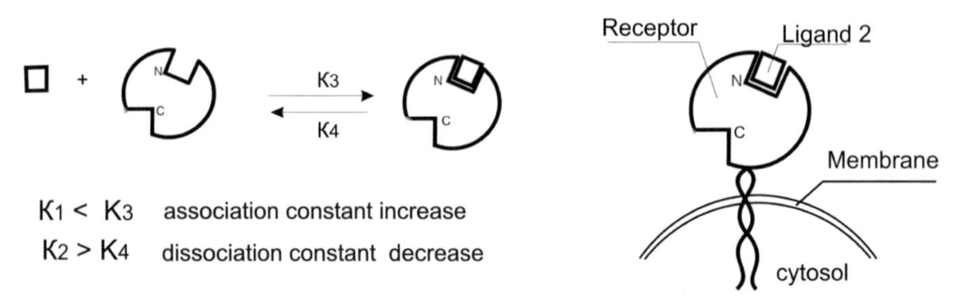

Fig. 8.6 Modified peptide ligand L2

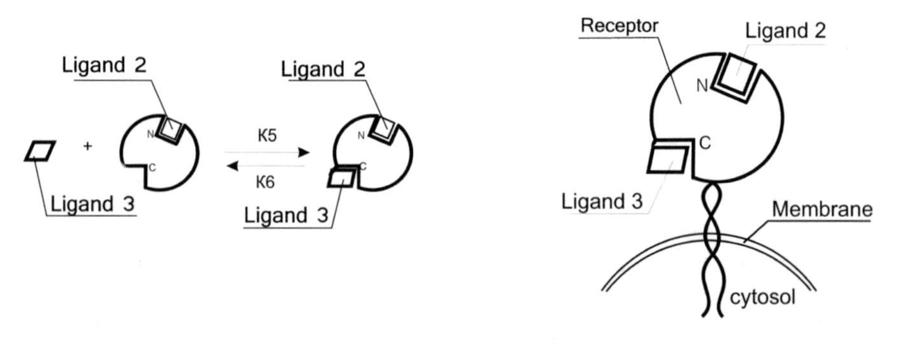

Fig. 8.7 Schema of joining the synthesized peptide (Ligand 3) to a given active site of the target protein (Receptor)

8.4.3 Creation of Peptides That Will Interfere with the Formation of Homo- and Heterodimers

The developed mathematical algorithms will also allow calculating amino acid sequences of peptides that will interfere with the formation of homo and heterodimers, as shown in Fig. 8.8.

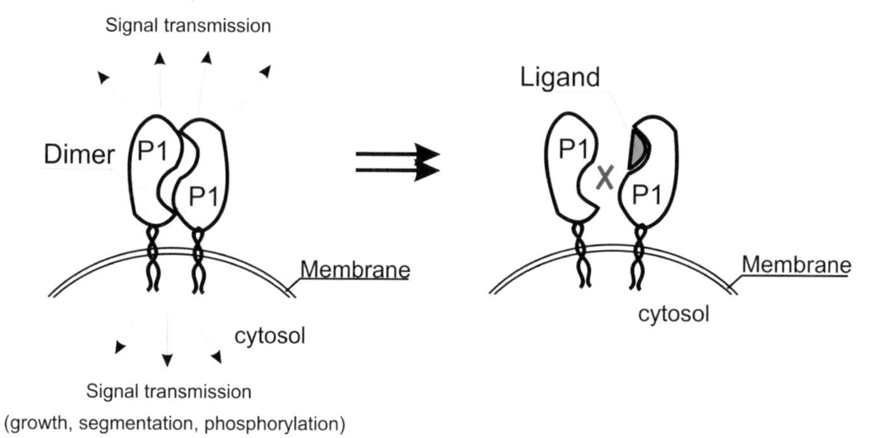

Fig. 8.8 Schema of inhibiting the active site of the protein (P1) by the synthesized peptide (L3)

Thus, the Algorithms 3–4 and methods presented in this chapter will allow to determine of point changes in amino acid residues in peptides. The second proposed algorithm makes it possible to determine the active region of a protein that is ≪scattered≫ along the entire length of the amino acid sequence of the protein when bound to the peptide.

8.5 Matlab Script Algorithm 3 for Finding the Optimal Composition of the Amino Acid Composition of Peptides Used as a Therapy

Input parameters:

1. S_{100}, S_{20} are amino acid sequences of biological complexes ($S_{100} \geq S_{20}$)
2. epsilon is the dielectric constant of the medium
3. sb is the number of replacements
4. a is left boundary of the vector S_{100}
5. b is right boundary of the vector S_{100}
6. sh is step shift

Output parameters:

lg(cond(W) is the common logarithm of the condition number of the matrix W, where its elements are composed of the electrostatic potential energy which is created based on the interaction between pair of amino acid residues of biological complexes.

Compute:

lg(cond(W) is the common logarithm of the condition number of the matrix W, which will allow a prediction the reactivity of the studied biological complexes.

```
1  clc,clear all
2  close all
3  format long e
4  %Bcl-xl
5  S_100=['M'  'S'  'Q'  'S'  'N'  'R'  'E'  'L'  'V'  'V' 'D'...
6  'F'  'L'  'S'  'Y'  'K'  'L'  'S'  'Q'  'K' 'G'  'Y'  'S' ...
7  'W' 'S'  'Q'  'F'  'S'  'D'  'V' 'E'  'E'  'N'  'R'  'T' ...
8  'E'  'A' 'P'  'E'  'G' 'T'  'E'  'S'  'E'  'M'  'E'  'T'...
9  'P'  'S'  'A'  'I'  'N'  'G'  'N'  'P'  'S'  'W' 'H' ...
10 'L'  'A' 'D'  'S'  'P' 'A'  'V'  'N'  'G'  'A'  'T' ...
11 'G' 'H'  'S'  'S'  'S'  'L'  'D'  'A'  'R'  'E'  'V' ...
12 'I'  'P'  'M'  'A'  'A'  'V'  'K'  'Q'  'A'  'L' 'R' ...
13 'E'  'A'  'G'  'D'  'E'  'F'  'E'  'L'  'R'  'Y' 'R'  ...
14 'R'  'A'  'F'  'S'  'D'  'L'  'T'  'S' 'Q'  'L'  'H'...
15 'I'  'T' 'P'  'G'  'T'  'A'  'Y' 'Q'  'S'  'F'  'E' ...
16 'Q'  'V'  'V'  'N'  'E'  'L' 'F'  'R'  'D'  'G'  'V'...
17 'N'  'W'  'G'  'R'  'I' 'V'  'A'  'F'  'F'  'S'  'F' ...
18 'G'  'G'  'A'  'L'  'C' 'V'  'E'  'S'  'V'  'D'  'K' ...
19 'E'  'M'  'Q' 'V'  'L'  'V'  'S'  'R'  'I'  'A' ...
20 'A'  'W'  'M' 'A'  'T'  'Y'  'L'  'N'  'D'  'H'...
21 'L'  'E'  'P'  'W'  'I'  'Q'  'E'  'N'  'G'  'G' ...
22 'W'  'D'  'T' 'F'  'V'  'E'  'L'  'Y'  'G'  'N' ...
23 'N'  'A'  'A'  'A'  'E'  'S'  'R'  'K'  'G'...
24 'Q'  'E'  'R'  'F' 'N'  'R'  'W'  'F'  'L' ...
25 'T'  'G'  'M'  'T'  'V' 'A'  'G'  'V'  'V'  'L'...
26 'L'  'G'  'S'  'L'  'F' 'S'  'R'  'K'] ;
27 %Bcl2 185-200
28 S_20=[  'W'  'I'  'Q' 'D'  'N'  'G'  'G'  'W'  'D' ...
29 'A'  'F' 'V' 'E' 'L'  'Y'  'G' ]
30 epsilon=1;
31 %----------------------------------------------------------
32 MEANS=[];
33 nomer=0;
34 Sub1=['A' 'T']; %replacement matrix №1
35 Sub2=['H'];     %replacement matrix №2
36 Sub3=['K'];     %replacement matrix №3
37 Sub4=['Y'];     %replacement matrix №4
38 Sub5=['T'];     %replacement matrix №5
39 len_Sub1=length(Sub1);
40 len_Sub2=length(Sub2);
41 len_Sub3=length(Sub3);
42 len_Sub4=length(Sub4);
43 len_Sub5=length(Sub5);
44 disp ('--------------------------------')
45 disp ('DIMENSIONS OF VECTORS:')
46 old_len_S20=length(S_20)
47 old_len_S100=length(S_100)
48 disp ('--------------------------------')
49 sb=input('Enter the number of replacements  = ');
50 while sb~=1 && sb~=2 && sb~=3 && sb~=4 && sb~=5
51 sb=input('Enter the number of replacements   = ');
52 end
```

```
53 nums_sb(1)=input('Enter the sequence number for...
54 replacement number 1  = ');
55 buf_S_20(1)=S_20(nums_sb(1));
56     if sb==2
57 nums_sb(2)=input('Enter the sequence number for...
58 replacement number 2  = ');
59          while nums_sb(2)<=nums_sb(1)
60 nums_sb(2)=input('Enter the sequence number for...
61 replacement number 2  = ');
62          end
63          buf_S_20(2)=S_20(nums_sb(2));
64     end
65     if sb==3
66 nums_sb(2)=input('Enter the sequence number for...
67 replacement number 2  = ');
68          while nums_sb(2)<=nums_sb(1)
69 nums_sb(2)=input('Enter the sequence number for...
70 replacement number 2  = ');
71          end
72          buf_S_20(2)=S_20(nums_sb(2));
73 nums_sb(3)=input('Enter the sequence number for...
74 replacement number 3  = ');
75          while nums_sb(3)<=nums_sb(2)
76 nums_sb(3)=input('Enter the sequence number for ...
77 replacement number 3  = ');
78          end
79          buf_S_20(3)=S_20(nums_sb(3));
80     end
81     if sb==4
82 nums_sb(2)=input('Enter the sequence number for ...
83 replacement number 2  = ');
84          while nums_sb(2)<=nums_sb(1)
85 nums_sb(2)=input('Enter the sequence number for...
86  replacement number 2  = ');
87          end
88          buf_S_20(2)=S_20(nums_sb(2));
89 nums_sb(3)=input('Enter the sequence number for...
90  replacement number 3  = ');
91          while nums_sb(3)<=nums_sb(2)
92 nums_sb(3)=input('Enter the sequence number for...
93  replacement number 3  = ');
94          end
95          buf_S_20(3)=S_20(nums_sb(3));
96 nums_sb(4)=input('Enter the sequence number for...
97 replacement number 4  = ');
98          while nums_sb(4)<=nums_sb(3)
99 nums_sb(4)=input('Enter the sequence number for...
100 replacement number 4  = ');
101          end
102          buf_S_20(4)=S_20(nums_sb(4));
103     end
104     if sb==5
```

```
105 nums_sb(2)=input('Enter the sequence number for...
106 replacement number 2  = ');
107         while nums_sb(2)<=nums_sb(1)
108 nums_sb(2)=input('Enter the sequence number for...
109 replacement number 2  = ');
110         end
111         buf_S_20(2)=S_20(nums_sb(2));
112 nums_sb(3)=input('Enter the sequence number for...
113 replacement number 3  = ');
114         while nums_sb(3)<=nums_sb(2)
115 nums_sb(3)=input('Enter the sequence number for...
116 replacement number 3  = ');
117         end
118         buf_S_20(3)=S_20(nums_sb(3));
119 nums_sb(4)=input('Enter the sequence number for...
120 replacement number 4  = ');
121         while nums_sb(4)<=nums_sb(3)
122 nums_sb(4)=input('Enter the sequence number for...
123 replacement number 4  = ');
124         end
125         buf_S_20(4)=S_20(nums_sb(4));
126 nums_sb(5)=input('Enter the sequence number for...
127 replacement number 5  = ');
128         while nums_sb(5)<=nums_sb(4)
129 nums_sb(5)=input('Enter the sequence number for...
130 replacement number 5  = ');
131         end
132         buf_S_20(5)=S_20(nums_sb(5));
133     end
134 nums_sb;
135 a=input('Enter the LEFT boundary of the vector S_100= ');
136 b=input('Enter the RIGHT boundary of the vector S_100 = ');
137 while a>b
138 b=input('Reentry. ...
139 Enter the RIGHT boundary of the vector S_100 = ');
140 end
141 old_S_100=S_100
142 AB_S_100=S_100(a:b)
143 S_100=S_100(a:b)
144 len_S20=length(S_20);
145 len_S100=length(S_100);
146 N1=1*len_S100;
147 sh=1;
148 %-----------------------------------------------------
149 del_len=len_S100-len_S20;
150 br=ceil(del_len/sh)-1;
151 for nsub1=0:len_Sub1
152       if nsub1~=0
153           S_20(nums_sb(1))=Sub1(nsub1)
154      end
155     if sb>=2
156      for nsub2=0:len_Sub2
```

```
157             if nsub2~=0
158 S_20(nums_sb(2))=Sub2(nsub2)
159           end
160                   if sb>=3
161               for nsub3=0:len_Sub3
162                   if nsub3~=0
163 S_20(nums_sb(3))=Sub3(nsub3)
164                   end
165                    if sb>=4
166                     for nsub4=0:len_Sub4
167                       if nsub4~=0
168 S_20(nums_sb(4))=Sub4(nsub4)
169                           end
170                           if sb==5
171                               for nsub5=0:len_Sub5
172                            if nsub5~=0
173 S_20(nums_sb(5))=Sub5(nsub5)
174                                   end
175                               S_20_change=S_20
176                               X=[];
177                               Out=[];
178                               V=[];
179                               Z=[];
180                               F=[];
181                               for ii=0:br+1
182                                   if ii~=br+1
183 X=[S_100(ii*sh+1:ii*sh+1+len_S20-1)];
184                                   else
185 X=[S_100(del_len+1:len_S100)];
186                                   end
187                                   S_1=X;
188                                   num=ii;
189                                   N=length(S_1);
190                                   M=length(S_20);
191                                   S_2=S_20;
192                                   Q1=[];
193                                   Q2=[];
194                                   R1=[];
195                                   R2=[];
196 [S_1,S_2,Q1,Q2,R1,R2,h]=potential(S_1,S_2,N1,N,M);
197 [A]=electrostatic(Q1,Q2, R1,R2,h,M,N,N1,epsilon);
198 [cond2]=condmy(A)
199  Out=[Out; X];
200 F=[F {num, S_1,S_2,(real(cond2))}'];
201                              end
202 SortF = sortrows(F',4);
203 minelem=[SortF(1:n_el,1) SortF(1:n_el,2) SortF(1:n_el,3)...
204  SortF(1:n_el,4)]
205 mean_minelem=sum(cell2mat(minelem(:,4)))/n_el;
206                         nomer=nomer+1;
207 MEANS=[MEANS; {nomer, S_20_change,S_100, mean_minelem,F}];
208                                   end
```

```
209         S_20(nums_sb(5))=buf_S_20(5);
210                             else
211                             S_20_change=S_20
212                             X=[];
213                             Out=[];
214                             V=[];
215                             Z=[];
216                             F=[];
217                             for ii=0:br+1
218                                 if ii~=br+1
219 X=[S_100(ii*sh+1:ii*sh+1+len_S20-1)];
220                                 else
221 X=[S_100(del_len+1:len_S100)];
222                                 end
223                                 S_1=X;
224                                 num=ii;
225                                 N=length(S_1);
226                                 M=length(S_20);
227                                 S_2=S_20;
228                                 Q1=[];
229                                 Q2=[];
230                                 R1=[];
231                                 R2=[];
232 [S_1,S_2,Q1,Q2,R1,R2,h]=potential(S_1,S_2,N1,N,M);
233 [A]=electrostatic(Q1,Q2, R1,R2,h,M,N,N1,epsilon);
234                               [cond2]=condmy(A)
235                                 Out=[Out; X];
236 F=[F {num, S_1,S_2,(real(cond2))}'];
237                             end
238                   SortF = sortrows(F',4);
239 minelem=[SortF(1:n_el,1) SortF(1:n_el,2) SortF(1:n_el,3)...
240 SortF(1:n_el,4)]
241 mean_minelem=sum(cell2mat(minelem(:,4)))/n_el;
242                   nomer=nomer+1;
243 MEANS=[MEANS; {nomer, S_20_change,S_100, mean_minelem,F}];
244                         end
245                         end
246 S_20(nums_sb(4))=buf_S_20(4);
247                       else
248                       S_20_change=S_20
249                       X=[];
250                       Out=[];
251                       V=[];
252                       Z=[];
253                       F=[];
254                       for ii=0:br+1
255                           if ii~=br+1
256 X=[S_100(ii*sh+1:ii*sh+1+len_S20-1)];
257                           else
258                               X=[S_100(del_len+1:len_S100)];
259                           end
260                           S_1=X;
```

```
261                         num=ii;
262                         N=length(S_1);
263                         M=length(S_20);
264                         S_2=S_20;
265                         Q1=[];
266                         Q2=[];
267                         R1=[];
268                         R2=[];
269 [S_1,S_2,Q1,Q2,R1,R2,h]=potential(S_1,S_2,N1,N,M);
270 [A]=electrostatic(Q1,Q2, R1,R2,h,M,N,N1,epsilon);
271 [cond2]=condmy(A)
272 Out=[Out; X];
273 F=[F {num, S_1,S_2,(real(cond2))}'];
274                     end
275 SortF = sortrows(F',4);
276 minelem=[SortF(1:n_el,1) SortF(1:n_el,2) SortF(1:n_el,3)...
277 SortF(1:n_el,4)]
278 mean_minelem=sum(cell2mat(minelem(:,4)))/n_el;
279                         nomer=nomer+1;
280 MEANS=[MEANS; {nomer, S_20_change,S_100, mean_minelem,F}];
281                   end
282                 end
283 S_20(nums_sb(3))=buf_S_20(3);
284         else
285             S_20_change=S_20
286             X=[];
287             Out=[];
288             V=[];
289             Z=[];
290             F=[];
291             for ii=0:br+1
292                 if ii~=br+1
293 X=[S_100(ii*sh+1:ii*sh+1+len_S20-1)];
294                 else
295 X=[S_100(del_len+1:len_S100)];
296                 end
297                 S_1=X;
298                 num=ii;
299                 N=length(S_1);
300                 M=length(S_20);
301                 S_2=S_20;
302                 Q1=[];
303                 Q2=[];
304                 R1=[];
305                 R2=[];
306 [S_1,S_2,Q1,Q2,R1,R2,h]=potential(S_1,S_2,N1,N,M);
307 [A]=electrostatic(Q1,Q2, R1,R2,h,M,N,N1,epsilon);
308 [cond2]=condmy(A)
309 Out=[Out; X];
310 F=[F {num, S_1,S_2,(real(cond2))}'];
311             end
312 SortF = sortrows(F',4);
```

```
313 minelem=[SortF(1:n_el,1) SortF(1:n_el,2) SortF(1:n_el,3)...
314 SortF(1:n_el,4)]
315 mean_minelem=sum(cell2mat(minelem(:,4)))/n_el;
316 nomer=nomer+1;
317 MEANS=[MEANS; {nomer, S_20_change,S_100, mean_minelem,F}];
318         end
319     end
320     S_20(nums_sb(2))=buf_S_20(2);
321  else
322 S_20_change=S_20
323       X=[];
324       Out=[];
325       V=[];
326       Z=[];
327       F=[];
328       for ii=0:br+1
329            if ii~=br+1
330 X=[S_100(ii*sh+1:ii*sh+1+len_S20-1)];
331            else
332 X=[S_100(del_len+1:len_S100)];
333            end
334            S_1=X;
335            num=ii;
336            N=length(S_1);
337            M=length(S_20);
338            S_2=S_20;
339            Q1=[];
340            Q2=[];
341            R1=[];
342            R2=[];
343 [S_1,S_2,Q1,Q2,R1,R2,h]=potential(S_1,S_2,N1,N,M);
344 [A]=electrostatic(Q1,Q2, R1,R2,h,M,N,N1,epsilon);
345           [cond2]=condmy(A)
346            Out=[Out; X];
347 F=[F {num, S_1,S_2,(real(cond2))}'];
348       end
349       SortF = sortrows(F',4);
350 minelem=[SortF(1:n_el,1) SortF(1:n_el,2) SortF(1:n_el,3)...
351 SortF(1:n_el,4)]
352 mean_minelem=sum(cell2mat(minelem(:,4)))/n_el;
353 nomer=nomer+1;
354 MEANS=[MEANS; {nomer, S_20_change,S_100, mean_minelem,F}];
355  end
356 end
357 S_20(nums_sb(1))=buf_S_20(1);
358 figure();
359 bar(cell2mat(MEANS(:,1)),cell2mat(MEANS(:,4)))
360 hold on
361 set(0,'DefaultTextInterpreter', 'latex');
362 set(0,'DefaultTextFontSize',14,...
363 'DefaultTextFontName','Arial Cyr');
364 xlabel('\bf No. of changes');
```

```
365 set(0,'DefaultTextFontSize',14,...
366 'DefaultTextFontName','Arial Cyr');
367 ylabel('MEANS');
368 figure();
369 plot(cell2mat(MEANS(:,1)),cell2mat(MEANS(:,4)),'*r')
370 hold on
371 set(0,'DefaultTextInterpreter', 'latex');
372 set(0,'DefaultTextFontSize',14,...
373 'DefaultTextFontName','Arial Cyr');
374 xlabel('\bf No. of changes');
375 set(0,'DefaultTextFontSize',14,...
376 'DefaultTextFontName','Arial Cyr');
377 ylabel('MEANS');
378 %-------------------------------------------
379 function [S_1,S_2,Q1,Q2,R1,R2,h]=potential(S_1,S_2,N1,N,M);
380 N=length(S_1);
381 M=length(S_2);
382 Q1=[];
383 Q2=[];
384 R1=[];
385 R2=[];
386 for i=1:length(S_1);
387 for j=1:length(S_2);
388 if (S_1(i)=='D' & S_2(j)=='E')| (S_1(i)=='E' & S_2(j)=='D');
389 Q1(i,j)= 0.16e-19;
390 Q2(i,j)= 0.16e-19;
391 else
392 if (S_1(i)=='D' & S_2(j)=='D');
393 Q1(i,j)= 0.07e-19;
394 Q2(i,j)= 0.07e-19;
395 else
396 if (S_1(i)=='D' & S_2(j) =='C')|(S_1(i)=='C' & S_2(j) =='D');
397 Q1(i,j)= 0.05e-19;
398 Q2(i,j)= 0.05e-19;
399 else
400 if (S_1(i)=='D' &S_2(j)=='N')|(S_1(i)=='N' &S_2(j)=='D')|...
401 (S_1(i)=='D' &S_2(j)=='F')|(S_1(i)=='D' &S_2(j)=='Y')|...
402 (S_1(i)=='D' &S_2(j)=='Q')|(S_1(i)=='D' &S_2(j)=='S')|...
403 (S_1(i)=='F' &S_2(j)=='D')|(S_1(i)=='Y' &S_2(j)=='D')|...
404 (S_1(i)=='Q' &S_2(j)=='D')|(S_1(i)=='S' &S_2(j)=='D');
405 Q1(i,j)= 0.57e-19;
406 Q2(i,j)= 0.57e-19;
407 else
408 if ((S_1(i)=='D' & S_2(j)=='M')|(S_1(i)=='D' & S_2(j)=='T')|...
409 (S_1(i)=='D' & S_2(j)=='I')|(S_1(i)=='D' & S_2(j)=='G')|...
410 (S_1(i)=='D' & S_2(j)=='V')|(S_1(i)=='D' & S_2(j)=='W')|...
411 (S_1(i)=='D' & S_2(j)=='L')|(S_1(i)=='D' & S_2(j)=='A')|...
412 (S_1(i)=='M' & S_2(j)=='D')|(S_1(i)=='T' & S_2(j)=='D')|...
413 (S_1(i)=='I' & S_2(j)=='D')|(S_1(i)=='G' & S_2(j)=='D')|...
414 (S_1(i)=='V' & S_2(j)=='D')|(S_1(i)=='W' & S_2(j)=='D')|...
415 (S_1(i)=='L' & S_2(j)=='D')|(S_1(i)=='A' & S_2(j)=='D'));
416 Q1(i,j)= 0.64e-19;
```

```
417 Q2(i,j)= 0.64e-19;
418 else
419 if ((S_1(i)=='D'& S_2(j)=='P')|(S_1(i)=='P'& S_2(j)=='D'));
420 Q1(i,j)= 0.78e-19;
421 Q2(i,j)= 0.78e-19;
422 else
423 if ((S_1(i)=='D' & S_2(j)=='H')|(S_1(i)=='H'& S_2(j)=='D'));
424 Q1(i,j)= 0.99e-19;
425 Q2(i,j)= 0.99e-19;
426 else
427 if ((S_1(i)=='D'& S_2(j)=='K')|(S_1(i)=='K'& S_2(j)=='D'));
428 Q1(i,j)= 1.4e-19;
429 Q2(i,j)= 1.4e-19;
430 else
431 if ((S_1(i)=='D' & S_2(j)=='R')|(S_1(i)=='R'& S_2(j)=='D'));
432 Q1(i,j)= 1.59e-19;
433 Q2(i,j)= 1.59e-19;
434 else
435 if ((S_1(i)=='E'&S_2(j)=='E'));
436 Q1(i,j)= 0.16e-19;
437 Q2(i,j)= 0.16e-19;
438 else
439 if ((S_1(i)=='E'  & S_2(j)=='C')|(S_1(i)=='E' & S_2(j)=='F')|...
440 (S_1(i)=='E' & S_2(j)=='N')|(S_1(i)=='C'  & S_2(j)=='E')|...
441 (S_1(i)=='F' & S_2(j)=='E')|(S_1(i)=='N' & S_2(j)=='E'));
442 Q1(i,j)= 0.55e-19;
443 Q2(i,j)= 0.55e-19;
444 else
445 if  ((S_1(i)=='E' & S_2(j)=='Q')|(S_1(i)=='E' & S_2(j)=='Y')|...
446 (S_1(i)=='E' & S_2(j)=='S')|(S_1(i)=='E' & S_2(j)=='M')|...
447 (S_1(i)=='E' & S_2(j)=='T')|(S_1(i)=='E' & S_2(j)=='I')|...
448 (S_1(i)=='E' & S_2(j)=='G')|(S_1(i)=='E' & S_2(j)=='V')|...
449 (S_1(i)=='E' & S_2(j)=='W')|(S_1(i)=='E' & S_2(j)=='L')|...
450 (S_1(i)=='E' & S_2(j)=='A')|(S_1(i)=='Q' & S_2(j)=='E')|...
451 (S_1(i)=='Y' & S_2(j)=='E')| (S_1(i)=='S' & S_2(j)=='E')|...
452 (S_1(i)=='M' & S_2(j)=='E')|(S_1(i)=='T' & S_2(j)=='E')|...
453 (S_1(i)=='I' & S_2(j)=='E')| (S_1(i)=='G' & S_2(j)=='E')|...
454 (S_1(i)=='V' & S_2(j)=='E')|(S_1(i)=='W' & S_2(j)=='E')|...
455 (S_1(i)=='L' & S_2(j)=='E')|(S_1(i)=='A' & S_2(j)=='E'));
456 Q1(i,j)= 0.64e-19;
457 Q2(i,j)= 0.64e-19;
458 else
459 if ((S_1(i)=='E' & S_2(j)=='P' )|(S_1(i)=='P' & S_2(j)=='E'));
460 Q1(i,j)= 0.78e-19;
461 Q2(i,j)= 0.78e-19;
462 else
463 if ((S_1(i)=='E' & S_2(j)=='H')|(S_1(i)=='H' &S_2(j)=='E'));
464 Q1(i,j)= 0.99e-19;
465 Q2(i,j)= 0.99e-19;
466 else
467 if (S_1(i)=='E'& S_2(j)=='K')| (S_1(i)=='K'& S_2(j)=='E');
468 Q1(i,j)= 1.34e-19;
```

```
469 Q2(i,j)= 1.34e-19;
470 else
471 if (S_1(i)=='E' & S_2(j)=='R')|(S_1(i)=='R' & S_2(j)=='E');
472 Q1(i,j)= 1.58e-19;
473 Q2(i,j)= 1.58e-19;
474 else
475 if (S_1(i)=='C' & S_2(j)=='C')|(S_1(i)=='C' & S_2(j)=='F')|...
476 (S_1(i)=='C' & S_2(j)=='Q')|(S_1(i)=='C'& S_2(j)=='Y')|...
477 (S_1(i)=='C' & S_2(j)=='S')|(S_1(i)=='C' & S_2(j)=='M')|...
478 (S_1(i)=='C' & S_2(j)=='T')|(S_1(i)=='C' & S_2(j)=='I')|...
479 (S_1(i)=='C' & S_2(j)=='G')|(S_1(i)=='C' & S_2(j)=='V')|...
480 (S_1(i)=='C' & S_2(j)=='W')|(S_1(i)=='C' & S_2(j)=='L')|...
481 (S_1(i)=='C' & S_2(j)=='L')|(S_1(i)=='C' & S_2(j)=='A')|...
482 (S_1(i)=='F' & S_2(j)=='C')|(S_1(i)=='Q' & S_2(j)=='C')|...
483 (S_1(i)=='Y'& S_2(j)=='C')|(S_1(i)=='S' & S_2(j)=='C')|...
484 (S_1(i)=='M' & S_2(j)=='C')|(S_1(i)=='T' & S_2(j)=='C')|...
485 (S_1(i)=='I' & S_2(j)=='C')|(S_1(i)=='G' & S_2(j)=='C')|...
486 (S_1(i)=='V' & S_2(j)=='C')|(S_1(i)=='W' & S_2(j)=='C')|...
487 (S_1(i)=='L' & S_2(j)=='C')|( S_1(i)=='A' & S_2(j)=='C');
488 Q1(i,j)=0.74e-19;
489 Q2(i,j)=0.74e-19;
490 else
491 if (S_1(i)=='C' & S_2(j)=='H')| (S_1(i)=='H' & S_2(j)=='C');
492 Q1(i,j)= 0.99e-19;
493 Q2(i,j)= 0.99e-19;
494 else
495 if (S_1(i)=='C' & S_2(j)=='K')|(S_1(i)=='K' & S_2(j)=='C');
496 Q1(i,j)= 1.34e-19;
497 Q2(i,j)= 1.34e-19;
498 else
499 if (S_1(i)=='C' & S_2(j)=='R')|(S_1(i)=='R' & S_2(j)=='C');
500 Q1(i,j)= 1.59e-19;
501 Q2(i,j)= 1.59e-19;
502 else
503 if (S_1(i)=='N' & S_2(j)=='N')|(S_1(i)=='N' & S_2(j)=='F')|...
504 (S_1(i)=='N' & S_2(j)=='Q')|(S_1(i)=='N' & S_2(j)=='Y')|...
505 (S_1(i)=='N' & S_2(j)=='S')|(S_1(i)=='N'& S_2(j)=='M')|...
506 (S_1(i)=='F' & S_2(j)=='N')|(S_1(i)=='Q' & S_2(j)=='N')|...
507 (S_1(i)=='Y' & S_2(j)=='N')|(S_1(i)=='S' & S_2(j)=='N')|...
508 (S_1(i)=='M'& S_2(j)=='N');
509 Q1(i,j)=0.74e-19;
510 Q2(i,j)=0.74e-19;
511 else
512 if (S_1(i)=='N' & S_2(j)=='H')|(S_1(i)=='H' & S_2(j)=='N')
513 Q1(i,j)= 0.99e-19;
514 Q2(i,j)= 0.99e-19;
515 else
516 if(S_1(i)=='N' & S_2(j)=='K')|(S_1(i)=='K' & S_2(j)=='N');
517 Q1(i,j)= 1.05e-19;
518 Q2(i,j)= 1.05e-19;
519 else
520 if (S_1(i)=='N' & S_2(j)=='R')|(S_1(i)=='R' & S_2(j)=='N');
```

```
521 Q1(i,j)= 1.1e-19;
522 Q2(i,j)= 1.1e-19;
523 else
524 if ((S_1(i)=='F' & S_2(j)=='F')|(S_1(i)=='F' & S_2(j)=='Q'));
525 Q1(i,j)=0.74e-19;
526 Q2(i,j)=0.74e-19;
527 else
528 if ((S_1(i)=='F' & S_2(j)=='Y')|(S_1(i)=='F' & S_2(j)=='S')|..
529 (S_1(i)=='F' & S_2(j)=='M')|(S_1(i)=='Q' & S_2(j)=='F')|...
530 (S_1(i)=='Y' & S_2(j)=='F'));
531 Q1(i,j)=0.74e-19;
532 Q2(i,j)=0.74e-19;
533 else
534 if (S_1(i)=='S' & S_2(j)=='F')|(S_1(i)=='M' & S_2(j)=='F');
535 Q1(i,j)=0.74e-19;
536 Q2(i,j)=0.74e-19;
537 else
538 if (S_1(i)=='F' & S_2(j)=='H')|(S_1(i)=='H' & S_2(j)=='F');
539 Q1(i,j)= 0.99e-19;
540 Q2(i,j)= 0.99e-19;
541 else
542 if (S_1(i)=='F' & S_2(j)=='K')|(S_1(i)=='K' & S_2(j)=='F');
543 Q1(i,j)= 1.05e-19;
544 Q2(i,j)= 1.05e-19;
545 else
546 if (S_1(i)=='F' & S_2(j)=='R')|(S_1(i)=='R' & S_2(j)=='F');
547 Q1(i,j)= 1.1e-19;
548 Q2(i,j)= 1.1e-19;
549 else
550 % Q
551 if (S_1(i)=='Q' & S_2(j)=='H')|(S_1(i)=='H' & S_2(j)=='Q');
552 Q1(i,j)= 0.99e-19;
553 Q2(i,j)= 0.99e-19;
554 else
555 if (S_1(i)=='Q' & S_2(j)=='K')|(S_1(i)=='K' & S_2(j)=='Q');
556 Q1(i,j)= 1.05e-19;
557 Q2(i,j)= 1.05e-19;
558 else
559 if (S_1(i)=='Q' & S_2(j)=='R')|(S_1(i)=='R' & S_2(j)=='Q');
560 Q1(i,j)= 1.1e-19;
561 Q2(i,j)= 1.1e-19;
562 else
563 % Y
564 if (S_1(i)=='Q' & S_2(j)=='H')|(S_1(i)=='H' & S_2(j)=='Q');
565 Q1(i,j)= 0.99e-19;
566 Q2(i,j)= 0.99e-19;
567 else
568 if (S_1(i)=='Y' & S_2(j)=='K')|(S_1(i)=='K' & S_2(j)=='Y')
569 Q1(i,j)= 1.05e-19;
570 Q2(i,j)= 1.05e-19;
571 else
572 if (S_1(i)=='Y' & S_2(j)=='R')|(S_1(i)=='R' & S_2(j)=='Y');
```

```
573 Q1(i,j)= 1.1e-19;
574 Q2(i,j)= 1.1e-19;
575 else
576 if (S_1(i)=='S' & S_2(j)=='H')|(S_1(i)=='H' & S_2(j)=='S');
577 Q1(i,j)= 0.99e-19;
578 Q2(i,j)= 0.99e-19;
579 else
580 if (S_1(i)=='S' & S_2(j)=='K')|(S_1(i)=='K' & S_2(j)=='S');
581 Q1(i,j)= 1e-19;
582 Q2(i,j)= 1e-19;
583 else
584 if (S_1(i)=='S' & S_2(j)=='R')|(S_1(i)=='R' & S_2(j)=='S');
585 Q1(i,j)= 1.1e-19;
586 Q2(i,j)= 1.1e-19;
587 else
588 if (S_1(i)=='M' & S_2(j)=='H')|(S_1(i)=='H' & S_2(j)=='M');
589 Q1(i,j)= 0.99e-19;
590 Q2(i,j)= 0.99e-19;
591 else
592 if (S_1(i)=='M' & S_2(j)=='K')|(S_1(i)=='K' & S_2(j)=='M');
593 Q1(i,j)= 1e-19;
594 Q2(i,j)= 1e-19;
595 else
596 if (S_1(i)=='M' & S_2(j)=='R')|(S_1(i)=='R' & S_2(j)=='M');
597 Q1(i,j)= 1.1e-19;
598 Q2(i,j)= 1.1e-19;
599 else
600 if (S_1(i)=='T' & S_2(j)=='H')|(S_1(i)=='H' & S_2(j)=='T');
601 Q1(i,j)= 0.99e-19;
602 Q2(i,j)= 0.99e-19;
603 else
604 if (S_1(i)=='T' & S_2(j)=='K')|(S_1(i)=='K' & S_2(j)=='T');
605 Q1(i,j)= 1e-19;
606 Q2(i,j)= 1e-19;
607 else
608 if (S_1(i)=='T' & S_2(j)=='R')|(S_1(i)=='R' & S_2(j)=='T');
609 Q1(i,j)= 1.05e-19;
610 Q2(i,j)= 1.05e-19;
611 else
612 if (S_1(i)=='I' & S_2(j)=='H')|(S_1(i)=='H' & S_2(j)=='I');
613 Q1(i,j)= 0.99e-19;
614 Q2(i,j)= 0.99e-19;
615 else
616 if (S_1(i)=='I' & S_2(j)=='K')|(S_1(i)=='K' & S_2(j)=='I');
617 Q1(i,j)= 1e-19;
618 Q2(i,j)= 1e-19;
619 else
620 if (S_1(i)=='I' & S_2(j)=='R')|(S_1(i)=='R' & S_2(j)=='I');
621 Q1(i,j)= 1.05e-19;
622 Q2(i,j)= 1.05e-19;
623 else
624 if (S_1(i)=='G' & S_2(j)=='H')|(S_1(i)=='H' & S_2(j)=='G');
```

```
625 Q1(i,j)= 0.99e-19;
626 Q2(i,j)= 0.99e-19;
627 else
628 if (S_1(i)=='G' & S_2(j)=='K')|(S_1(i)=='K' & S_2(j)=='G');
629 Q1(i,j)= 1e-19;
630 Q2(i,j)= 1e-19;
631 else
632 if (S_1(i)=='G' & S_2(j)=='R')|(S_1(i)=='R' & S_2(j)=='G');
633 Q1(i,j)= 1.05e-19;
634 Q2(i,j)= 1.05e-19;
635 else
636 if (S_1(i)=='V' & S_2(j)=='H')|(S_1(i)=='H' & S_2(j)=='V');
637 Q1(i,j)= 0.99e-19;
638 Q2(i,j)= 0.99e-19;
639 else
640 if (S_1(i)=='V' & S_2(j)=='K')|(S_1(i)=='K' & S_2(j)=='V');
641 Q1(i,j)= 1e-19;
642 Q2(i,j)= 1e-19;
643 else
644 if (S_1(i)=='V' & S_2(j)=='R')|(S_1(i)=='R' & S_2(j)=='V');
645 Q1(i,j)= 1.05e-19;
646 Q2(i,j)= 1.05e-19;
647 else
648 if (S_1(i)=='W' & S_2(j)=='H')|(S_1(i)=='H' & S_2(j)=='W');
649 Q1(i,j)= 0.99e-19;
650 Q2(i,j)= 0.99e-19;
651 else
652 if (S_1(i)=='W' & S_2(j)=='K')|(S_1(i)=='K' & S_2(j)=='W');
653 Q1(i,j)= 1e-19;
654 Q2(i,j)= 1e-19;
655 else
656 if (S_1(i)=='W' & S_2(j)=='R')|(S_1(i)=='R' & S_2(j)=='W');
657 Q1(i,j)= 1.05e-19;
658 Q2(i,j)= 1.05e-19;
659 else
660 if (S_1(i)=='L' & S_2(j)=='H')|(S_1(i)=='H' & S_2(j)=='L');
661 Q1(i,j)= 0.99e-19;
662 Q2(i,j)= 0.99e-19;
663 else
664 if (S_1(i)=='L' & S_2(j)=='K')|(S_1(i)=='K' & S_2(j)=='L');
665 Q1(i,j)= 1e-19;
666 Q2(i,j)= 1e-19;
667 else
668 if (S_1(i)=='L' & S_2(j)=='R')|(S_1(i)=='R' & S_2(j)=='L');
669 Q1(i,j)= 1.05e-19;
670 Q2(i,j)= 1.05e-19;
671 else
672 if (S_1(i)=='A' & S_2(j)=='H')|(S_1(i)=='H' & S_2(j)=='A');
673 Q1(i,j)= 0.99e-19;
674 Q2(i,j)= 0.99e-19;
675 else
676 if (S_1(i)=='A' & S_2(j)=='K')|(S_1(i)=='K' & S_2(j)=='A');
```

```
677 Q1(i,j)= 1e-19;
678 Q2(i,j)= 1e-19;
679 else
680 if (S_1(i)=='A' & S_2(j)=='R')|(S_1(i)=='R' & S_2(j)=='A');
681 Q1(i,j)= 1.05e-19;
682 Q2(i,j)= 1.05e-19;
683 else
684 if (S_1(i)=='P' & S_2(j)=='H')|(S_1(i)=='H' & S_2(j)=='P');
685 Q1(i,j)= 0.99e-19;
686 Q2(i,j)= 0.99e-19;
687 else
688 if (S_1(i)=='P' & S_2(j)=='K')|(S_1(i)=='K' & S_2(j)=='P');
689 Q1(i,j)= 0.82e-19;
690 Q2(i,j)= 0.82e-19;
691 else
692 if (S_1(i)=='P' & S_2(j)=='R')|(S_1(i)=='R' & S_2(j)=='P');
693 Q1(i,j)= 0.96e-19;
694 Q2(i,j)= 0.96e-19;
695 else
696 if (S_1(i)=='H' & S_2(j)=='H');
697 Q1(i,j)= 0.82e-19;
698 Q2(i,j)= 0.82e-19;
699 else
700 if (S_1(i)=='H' & S_2(j)=='K')|(S_1(i)=='K' & S_2(j)=='H');
701 Q1(i,j)= 0.82e-19;
702 Q2(i,j)= 0.82e-19;
703 else
704 if (S_1(i)=='H' & S_2(j)=='R')|(S_1(i)=='R' & S_2(j)=='H');
705 Q1(i,j)= 0.74e-19;
706 Q2(i,j)= 0.74e-19;
707 else
708 if (S_1(i)=='K' & S_2(j)=='K');
709 Q1(i,j)= 0.54e-19;
710 Q2(i,j)= 0.54e-19;
711 else
712 if (S_1(i)=='K' & S_2(j)=='R')|(S_1(i)=='R' & S_2(j)=='K');
713 Q1(i,j)= 0.41e-19;
714 Q2(i,j)= 0.41e-19;
715 else
716 if (S_1(i)=='R' & S_2(j)=='R');
717 Q1(i,j)= 0.16e-19;
718 Q2(i,j)= 0.16e-19;
719 else
720 Q1(i,j)= 0.824e-19;
721 Q2(i,j)= 0.824e-19;
722 end
723 end
724 end
725 end
726 end
727 end
728 end
```

```
729 end
730 end
731 end
732 end
733 end
734 end
735 end
736 end
737 end
738 end
739 end
740 end
741 end
742 end
743 end
744 end
745 end
746 end
747 end
748 end
749 end
750 end
751 end
752 end
753 end
754 end
755 end
756 end
757 end
758 end
759 end
760 end
761 end
762 end
763 end
764 end
765 end
766 end
767 end
768 end
769 end
770 end
771 end
772 end
773 end
774 end
775 end
776 end
777 end
778 end
779 end
780 end
```

```
781 end
782 end
783 end
784 end
785 end
786 end
787 end
788 end
789 end
790 end
791 end
792 end
793 end
794 end
795 end
796  Q3=[];
797  Q4=[];
798  R1=[];
799  R2=[];
800 for i=1:length(S_1);
801 if (S_1(i)=='A');
802 R1(i)=0.6e-9;
803 else
804 if (S_1(i)=='R');
805 R1(i)=0.809e-9;
806 else
807 if (S_1(i)=='N');
808 R1(i)=0.682e-9;
809 else
810 if (S_1(i)=='D');
811 R1(i)=0.665e-9;
812 else
813 if (S_1(i)=='C');
814 R1(i)=0.629e-9;
815 else
816 if (S_1(i)=='Q');
817 R1(i)=0.725e-9;
818 else
819 if (S_1(i)=='E');
820 R1(i)=0.714e-9;
821 else
822 if (S_1(i)=='G');
823 R1(i)=0.537e-9;
824 else
825 if (S_1(i)=='H');
826 R1(i)=0.732e-9;
827 else
828 if (S_1(i)=='I');
829 R1(i)=0.735e-9;
830 else
831 if (S_1(i)=='L');
832 R1(i)=0.734e-9;
```

```
833 else
834 if (S_1(i)=='K');
835 R1(i)=0.737e-9;
836 else
837 if (S_1(i)=='M');
838 R1(i)=0.741e-9;
839 else
840 if (S_1(i)=='F');
841 R1(i)=0.781e-9;
842 else
843 if (S_1(i)=='P');
844 R1(i)=0.672e-9;
845 else
846 if (S_1(i)=='S');
847 R1(i)=0.615e-9;
848 else
849 if (S_1(i)=='T');
850 R1(i)=0.659e-9;
851 else
852 if (S_1(i)=='W');
853 R1(i)=0.826e-9;
854 else
855 if (S_1(i)=='Y');
856 R1(i)=0.781e-9;
857 else
858 if (S_1(i)=='V');
859 R1(i)=0.694e-9;
860 end
861 end
862 end
863 end
864 end
865 end
866 end
867 end
868 end
869 end
870 end
871 end
872 end
873 end
874 end
875 end
876 end
877 end
878 end
879 end
880 for j=1:length(S_2);
881 if (S_2(j)=='A');
882 R2(j)=0.6e-9;
883 else
884 if (S_2(j)=='R');
```

```
937 else
938 if (S_2(j)=='V');
939 R2(j)=0.694e-9;
940 end
941 end
942 end
943 end
944 end
945 end
946 end
947 end
948 end
949 end
950 end
951 end
952 end
953 end
954 end
955 end
956 end
957 end
958 end
959 end
960 end
961 end
962  Ra=0.6e-9;
963  Rr=0.809e-9;
964  Rn=0.682e-9;
965  Rd=0.665e-9;
966  Rc=0.629e-9;
967  Rq=0.725e-9;
968  Re=0.714e-9;
969  Rg=0.725e-9;
970  Rh=0.732e-9;
971  Ri=0.735e-9;
972  Rl=0.734e-9;
973  Rk=0.737e-9;
974  Rm=0.741e-9;
975  Rf=0.781e-9;
976  Rp=0.672e-9;
977  Rs=0.615e-9;
978  Rt=0.659e-9;
979  Rw=0.826e-9;
980  Ry=0.781e-9;
981  Rv=0.694e-9;
982 for i=1:length(S_1);
983 for j=1:length(S_2);
984 if (S_1(i)=='R'& S_2(j)=='D');
985     h(i,j)=.15*10^(-9)+Rr+Rd;
986 else
987 if (S_1(i)=='R'& S_2(j)=='E');
988      h(i,j)=.15*10^(-9)+Rr+Re;
```

```
989  else
990  if (S_1(i)=='D'& S_2(j)=='R');
991      h(i,j)=.15*10^(-9)+Rd+Rr;
992  else
993  if (S_1(i)=='D'& S_2(j)=='H');
994     h(i,j)=.15*10^(-9)+Rd+Rh;
995  else
996  if (S_1(i)=='D'& S_2(j)=='R');
997      h(i,j)=.15*10^(-9)+Rd+Rr;
998  else
999  if (S_1(i)=='D'& S_2(j)=='H');
1000     h(i,j)=.15*10^(-9)+Rd+Rh;
1001 else
1002 if (S_1(i)=='D'& S_2(j)=='K');
1003     h(i,j)=.15*10^(-9)+Rd+Rk;
1004 else
1005 if (S_1(i)=='E')& (S_2(j)=='R');
1006     h(i,j)=.15*10^(-9)+Re+Rr;
1007 else
1008 if (S_1(i)=='E'& S_2(j)=='H');
1009    h(i,j)=.15*10^(-9)+Re+Rh;
1010 else
1011 if (S_1(i)=='E'& S_2(j)=='K');
1012    h(i,j)=.15*10^(-9)+Re+Rk;
1013 else
1014 if (S_1(i)=='H'& S_2(j)=='D')
1015     h(i,j)=.15*10^(-9)+Rh+Rd;
1016 else
1017 if (S_1(i)=='H'& S_2(j)=='E')
1018     h(i,j)=.15*10^(-9)+Rh+Re;
1019 else
1020 if (S_1(i)=='R'& S_2(j)=='R')
1021     h(i,j)=.4*10^(-9)+Rr+Rr;
1022 else
1023 if (S_1(i)=='R'& S_2(j)=='H')
1024    h(i,j)=.4*10^(-9)+Rr+Rh;
1025 else
1026 if (S_1(i)=='R'& S_2(j)=='H')
1027     h(i,j)=.4*10^(-9)+Rr+Rh;
1028 else
1029 if (S_1(i)=='R'& S_2(j)=='K')
1030     h(i,j)=.4*10^(-9)+Rr+Rk;
1031 else
1032 if (S_1(i)=='D'& S_2(j)=='E');
1033     h(i,j)=.4*10^(-9)+Rd+Re;
1034 else
1035 if (S_1(i)=='D'& S_2(j)=='D');
1036     h(i,j)=.4*10^(-9)+Rd+Rd;
1037 else
1038 if (S_1(i)=='H'& S_2(j)=='R')
1039     h(i,j)=.4*10^(-9)+Rh+Rr;
1040 else
```

```
1041 if (S_1(i)=='H'& S_2(j)=='H')
1042      h(i,j)=.4*10^(-9)+Rh+Rh;
1043 else
1044 if (S_1(i)=='H'& S_2(j)=='K')
1045       h(i,j)=.4*10^(-9)+Rh+Rk;
1046 else
1047 if (S_1(i)=='K'& S_2(j)=='R')
1048     h(i,j)=.4*10^(-9)+Rk+Rr;
1049 else
1050 if (S_1(i)=='K'& S_2(j)=='H')
1051      h(i,j)=.4*10^(-9)+Rk+Rh;
1052 else
1053 if (S_1(i)=='K'& S_2(j)=='K')
1054      h(i,j)=.4*10^(-9)+Rk+Rk;
1055 else
1056 if (S_1(i)=='N'& S_2(j)=='Q')
1057     h(i,j)=.25*10^(-9)+Rn+Rq;
1058 else
1059 if (S_1(i)=='N'& S_2(j)=='S')
1060     h(i,j)=.25*10^(-9)+Rn+Rs;
1061 else
1062 if (S_1(i)=='N'& S_2(j)=='Y')
1063      h(i,j)=.25*10^(-9)+Rn+Ry;
1064 else
1065 if(S_1(i)=='Q'& S_2(j)=='S')|(S_1(i)=='Q')& (S_2(j)=='Y');
1066      h(i,j)=.25*10^(-9)+Rq+Rs;
1067 else
1068 if (S_1(i)=='Q')& (S_2(j)=='Y');
1069       h(i,j)=.25*10^(-9)+Rq+Ry;
1070 else
1071 if (S_1(i)=='S'& S_2(j)=='Y');
1072      h(i,j)=.25*10^(-9)+Rs+Ry;
1073 else
1074      h(i,j)=1.76*10^(-9);
1075 end
1076 end
1077 end
1078 end
1079 end
1080 end
1081 end
1082 end
1083 end
1084 end
1085 end
1086 end
1087 end
1088 end
1089 end
1090 end
1091 end
1092 end
```

```
1093 end
1094 end
1095 end
1096 end
1097 end
1098 end
1099 end
1100 end
1101 end
1102 end
1103 end
1104 end
1105 end
1106 end
1107
1108 function[A]=electrostatic(Q1,Q2, R1,R2,h,M,N,N1,epsilon)
1109 for i=1:N
1110     for j=1:M
1111         if R1(i)>R2(j)
1112             gamma(i,j)=R1(i)/R2(j);
1113         else
1114             if  R1(i)<R2(j)
1115                 gamma(i,j)=R2(j)/R1(i);
1116               else if R1(i)==R2(j);
1117      gamma(i,j)=R2(j)/R1(i);
1118           end
1119             end
1120         end
1121         if h(i,j)>(R1(i)+R2(j))
1122             r(i,j)=h(i,j)/(R1(i)+R2(j));
1123         else if  h(i,j)<=(R1(i)+R2(j))
1124             r(i,j)=(R1(i)+R2(j))/h(i,j);
1125         end
1126         end
1127     y(i,j)=(((r(i,j)^2*(1+gamma(i,j))^2)-...
1128 (1+(gamma(i,j))^2))/(2*gamma(i,j)));
1129     beta(i,j)=acosh(y(i,j));
1130     z(i,j)=exp(-beta(i,j));
1131     S12=0;
1132     S22=0;
1133     S11=0;
1134     for k=1:N1
1135         gamma1(i,j)=R2(j)/R1(i);
1136         S_1(k)=(z(i,j)^k)/(((1-z(i,j)^(2*k)))*...
1137 ((gamma(i,j)+y(i,j))-(y(i,j)^2-1)^(1/2)*...
1138 (1+z(i,j)^(2*k))/(1-z(i,j)^(2*k))));
1139         S11=S11+S_1(k);
1140         S_2(k)=(z(i,j)^(2*k))/(1-(z(i,j)^(2*k)));
1141         S12=S12+S_2(k);
1142         S_3(k)=(z(i,j)^k)/(((1-z(i,j)^(2*k)))*...
1143 ((1-gamma(i,j)*y(i,j))-gamma(i,j)*(y(i,j)^2-1)^(1/2)*...
1144 (1+z(i,j)^(2*k))/(1-z(i,j)^(2*k))));
```

```
1145            S22=S22+S_3(k);
1146        end
1147        epsilon0=8.85418781762*10^(-12);
1148    c11(i,j)=(2*gamma(i,j)*((y(i,j)^2-1)^(1/2))).*S11;
1149    c22(i,j)=(2*gamma(i,j)*((y(i,j)^2-1)^(1/2))).*S22;
1150    c12(i,j)=-((2*gamma(i,j)*...
1151    ((y(i,j)^2-1))^(1/2))/(r(i,j)*(1+gamma(i,j)))).*S12;
1152    delta(i,j)=((c11(i,j)*c22(i,j)-c12(i,j)^2));
1153        k=1/(4*pi*epsilon0);
1154        k1=1/(4*pi*epsilon0*epsilon);
1155            alpha(i,j)=Q2(j)/Q1(i);
1156        if R1(i)>R2(j)
1157            gamma(i,j)=R1(i)/R2(j);
1158    W1(i,j)=((1/k1)*R2(j)*gamma(i,j))*...
1159    ((1+gamma(i,j))/(2*alpha(i,j)))*...
1160    ((alpha(i,j)^2*c11(i,j)-2*alpha(i,j)*...
1161    c12(i,j)+c22(i,j))/delta(i,j));
1162            else if (R1(i)<R2(j))
1163                gamma(i,j)=R2(j)/R1(i);
1164    W1(i,j)=((1/k1)*R1(i)*gamma(i,j))*...
1165    ((1+gamma(i,j))/(2*alpha(i,j)))*...
1166    ((alpha(i,j)^2*c11(i,j)-2*alpha(i,j)*...
1167    c12(i,j)+c22(i,j))/delta(i,j));
1168         else if R1(i)==R2(j);
1169    W1(i,j)=((1/k1)*R1(i)*gamma(i,j))*...
1170    ((1+gamma(i,j))/(2*alpha(i,j)))*...
1171    ((alpha(i,j)^2*c11(i,j)-2*alpha(i,j)*...
1172    c12(i,j)+c22(i,j))/delta(i,j));
1173                end
1174                end
1175        end
1176        W2(i,j)=(k*(Q1(i)*Q2(j)))/(R1(i)+R2(j));
1177        A1(i,j)=W1(i,j);
1178        A2(i,j)=W2(i,j);
1179         A(i,j)=A1(i,j)/A2(i,j);
1180
1181        end
1182    end
1183    return
1184
1185    function[cond2]=condmy(A)
1186    [U,S,V]=SVD_2(A);
1187    lambda_max=max(diag(S));
1188    lambda_min=min(diag(S));
1189    cond_1=((lambda_max)/(lambda_min));
1190    cond2=(log(cond_1))/(log(10));
1191    return
1192    function [Uout,Sout,Vout] = SVD_2(A)
1193          m = size(A,1);
1194          n = size(A,2);
1195          U = eye(m);
1196          V = eye(n);
```

```
1197        e = eps*fro(A);
1198        while (sum(abs(A(~eye(m,n)))) > e)
1199        for i = 1:n
1200        for j = i+1:n
1201            [J1,J2] = jacobi(A,m,n,i,j);
1202            A = mtimes(J1,mtimes(A,J2));
1203            U = mtimes(U,J1');
1204            V = mtimes(J2',V);
1205        end
1206        for j = n+1:m
1207            J1 = jacobi2(A,m,n,i,j);
1208            A = mtimes(J1,A);
1209            U = mtimes(U,J1');
1210        end
1211        end
1212        end
1213        S = A;
1214        if (nargout < 3)
1215           Uout = diag(S);
1216        else
1217            Uout = U; Sout = times(S,eye(m,n)); Vout = V;
1218        end
1219        end
1220      function [J1,J2] = jacobi(A,m,n,i,j)
1221         B = [A(i,i), A(i,j); A(j,i), A(j,j)];
1222         [U,S,V] = tinySVD(B); %
1223         J1 = eye(m);
1224         J1(i,i) = U(1,1);
1225         J1(j,j) = U(2,2);
1226         J1(i,j) = U(2,1);
1227         J1(j,i) = U(1,2);
1228         J2 = eye(n);
1229         J2(i,i) = V(1,1);
1230         J2(j,j) = V(2,2);
1231         J2(i,j) = V(2,1);
1232         J2(j,i) = V(1,2);
1233      end
1234      function J1 = jacobi2(A,m,n,i,j)
1235         B = [A(i,i), 0; A(j,i), 0];
1236         [U,S,V] = tinySVD(B);
1237         J1 = eye(m);
1238         J1(i,i) = U(1,1);
1239         J1(j,j) = U(2,2);
1240         J1(i,j) = U(2,1);
1241         J1(j,i) = U(1,2);
1242      end
1243
1244      function [Uout,Sout,Vout] = tinySVD(A)
1245 t=rdivide((minus(A(1,2),A(2,1))),(plus(A(1,1),A(2,2))));
1246        c = rdivide(1,sqrt(1+t^2));
1247        s = times(t,c);
1248        R = [c,-s;s,c];
```

```
1249         M = mtimes(R,A);
1250         [U,S,V] = tinySymmetricSVD(M);
1251         U = mtimes(R',U);
1252         if (nargout < 3)
1253             Uout = diag(S);
1254         else
1255             Uout = U; Sout = S; Vout = V;
1256         end
1257         end
1258
1259 function [Uout,Sout,Vout] = tinySymmetricSVD(A)
1260         if (A(2,1) == 0)
1261             S = A;
1262             U = eye(2);
1263             V = U;
1264         else
1265             w = A(1,1);
1266             y = A(2,1);
1267             z = A(2,2);
1268             ro = rdivide(minus(z,w),times(2,y));
1269 t2=rdivide(sign(ro),plus(abs(ro),sqrt(plus(times(ro,ro),1))));
1270             t = t2;
1271             c = rdivide(1,sqrt(plus(1,times(t,t))));
1272             s = times(t,c);
1273             U = [c, -s; s, c];
1274             V = [c,  s;-s, c];
1275             S = mtimes(U,mtimes(A,V));
1276             U = U';
1277             V = V';
1278         end
1279         [U,S,V] = fixSVD(U,S,V);
1280         if (nargout < 3)
1281             Uout = diag(S);
1282         else
1283             Uout = U; Sout = S; Vout = V;
1284         end
1285         end
1286
1287     function [U,S,V] = fixSVD(U,S,V)
1288         Z = [sign(S(1,1)),0; 0,sign(S(2,2))]; %
1289         U = mtimes(U,Z);
1290         S = mtimes(Z,S);
1291         if (S(1,1) < S(2,2))
1292             P = [0,1;1,0];
1293             U = mtimes(U,P);
1294             S = mtimes(P,mtimes(S,P));
1295             V = mtimes(P,V);
1296         end
1297         end
```

```
1298
1299        function f = fro(M)
1300          f = sqrt(sum(sum(times(M,M))));
1301        end
1302        function s = sign(x)
1303            if (x > 0)
1304                s = 1;
1305            else
1306                s = -1;
1307            end
1308        end
```

8.6 Matlab Script Algorithm 4 for Finding the Optimal Composition of the Amino Acid Composition of Peptides Used as a Therapy

Input parameters:

1. S_1, S_{20} are amino acid sequences of biological complexes ($S_1 \geq S_{20}$)
2. epsilon is the dielectric constant of the medium
3. sh1 is step shift
4. sh2 is is the length of the frame

Output parameters:

lg(cond(W) is the common logarithm of the condition number of the matrix W, where its elements are composed of the electrostatic potential energy which is created based on the interaction between pair of amino acid residues of biological complexes.

Compute:

lg(cond(W) is the common logarithm of the condition number of the matrix W, which will allow a prediction the reactivity of the studied biological complexes.

```
1  clc
2  clear all
3  format long e
4  %BCL-2 WT 10-233
5  S_100=[ 'N'  'R'  'E'  'I' 'V'  'M'   'K'  'Y' ...
6  'I'  'H'  'Y'  'K'  'L'  'S'  'Q'  'R'  'G' ...
7  'Y'  'E'  'W' 'D'  'A'  'G'  'D'  'V'  ]
8  %Bax 59-74
9  S_20=[     'L'  'S'  'E'  'C'  'L'  'K'  'R' 'I' ...
10  'G'  'D'  'E'  'L'  'D'  'S'  'N'  'M'   ]
11 %-----------------------------------------
12 sh1=10;
13 sh2=1;
14 n_el=4;
15 epsilon=80;
16 len_S20=length(S_20);
17 len_S100=length(S_100);
18 N1=5*len_S100;
19 del_len=len_S100-len_S20;
20 X=[];
21 Out=[];
22 V=[];
23 F=[];
24 br=ceil(del_len/sh2)-1;
25 ost=len_S100-br*sh2-len_S20;
26 if ost~=0
27 OSTATOK=[S_100(len_S100-ost+1:len_S100)];
28 end
29 for i=1:br+1
30     if i~=br+1
31 X=[S_100(1:sh1) S_100(sh1+i*sh2+1:sh1+i*sh2+1+len_S20-sh1-1)];
32     else
33 X=[S_100(1:sh1) S_100(len_S100-(len_S20-sh1)+1:len_S100)];
34     end
35     S_1=X;
36     num=i;
37 N=length(S_1);
38 M=length(S_20);
39 S_2=S_20;
40 Q1=[];
41 Q2=[];
42 R1=[];
43 R2=[];
44 [S_1,S_2,Q1,Q2,R1,R2,h]=potential(S_1,S_2,N1,N,M);
45 [A]=electrostatic(Q1,Q2, R1,R2,h,M,N,N1,epsilon);
46 [cond2]=condmy(A)
47 Out=[Out; X];
48 F=[F {num, S_1,S_2,(real(cond2))}'];
49  end
50 len_X=length(X);
51 len_Out=length(Out);
52 F;
```

```
53  barX=cell2mat(F(1,:));
54  barY=cell2mat(F(4,:));
55  SortF = sortrows(F',4);
56  barX_sort=cell2mat(SortF(:,1));
57  barY_sort=cell2mat(SortF(:,4));
58  minelem=[SortF(1:n_el,1) SortF(1:n_el,2)...
59  SortF(1:n_el,3) SortF(1:n_el,4)]
60  figure();
61  bar(barX,barY)
62  hold on
63  for i=1:n_el
64  bar(cell2mat(SortF(i,1)),cell2mat(SortF(i,4)),'red')
65  end
66  set(0,'DefaultTextInterpreter', 'latex');
67  set(0,'DefaultTextFontSize',14,'DefaultTextFontName',...
68  'Arial Cyr');
69  xlabel('\bf Numer aminoacid residual');
70  set(0,'DefaultTextFontSize',14,'DefaultTextFontName',...
71  'Arial Cyr');
72  ylabel('lg(cond(W))');
73  figure();
74  plot(barX,barY,'ok')
75  hold on
76  for i=1:n_el
77  plot(cell2mat(SortF(i,1)),cell2mat(SortF(i,4)),'*r')
78  end
79  set(0,'DefaultTextInterpreter', 'latex');
80  set(0,'DefaultTextFontSize',14,'DefaultTextFontName',...
81  'Arial Cyr');
82  xlabel('\bf Numer aminoacid residual');
83  set(0,'DefaultTextFontSize',14,'DefaultTextFontName',...
84  'Arial Cyr');
85  ylabel('lg(cond(W))');
86  %-----------------------------------------------
87  function [S_1,S_2,Q1,Q2,R1,R2,h]=potential(S_1,S_2,N1,N,M);
88  N=length(S_1);
89  M=length(S_2);
90  Q1=[];
91  Q2=[];
92  R1=[];
93  R2=[];
94  for i=1:length(S_1);
95  for j=1:length(S_2);
96  if (S_1(i)=='D' & S_2(j)=='E')| (S_1(i)=='E' & S_2(j)=='D');
97  Q1(i,j)= 0.16e-19;
98  Q2(i,j)= 0.16e-19;
99  else
100 if (S_1(i)=='D' & S_2(j)=='D');
101 Q1(i,j)= 0.07e-19;
102 Q2(i,j)= 0.07e-19;
103 else
104 if (S_1(i)=='D' & S_2(j) =='C')|(S_1(i)=='C' & S_2(j) =='D');
```

```
105 Q1(i,j)= 0.05e-19;
106 Q2(i,j)= 0.05e-19;
107 else
108 if (S_1(i)=='D' &S_2(j)=='N')|(S_1(i)=='N' &S_2(j)=='D')|...
109 (S_1(i)=='D' &S_2(j)=='F')|(S_1(i)=='D' &S_2(j)=='Y')|...
110 (S_1(i)=='D' &S_2(j)=='Q')|(S_1(i)=='D' &S_2(j)=='S')|...
111 (S_1(i)=='F' &S_2(j)=='D')|(S_1(i)=='Y' &S_2(j)=='D')|...
112 (S_1(i)=='Q' &S_2(j)=='D')|(S_1(i)=='S' &S_2(j)=='D');
113 Q1(i,j)= 0.57e-19;
114 Q2(i,j)= 0.57e-19;
115 else
116 if ((S_1(i)=='D' & S_2(j)=='M')|(S_1(i)=='D' & S_2(j)=='T')|...
117 (S_1(i)=='D' & S_2(j)=='I')|(S_1(i)=='D' & S_2(j)=='G')|...
118 (S_1(i)=='D' & S_2(j)=='V')|(S_1(i)=='D' & S_2(j)=='W')|...
119 (S_1(i)=='D' & S_2(j)=='L')|(S_1(i)=='D' & S_2(j)=='A')|...
120 (S_1(i)=='M' & S_2(j)=='D')|(S_1(i)=='T' & S_2(j)=='D')|...
121 (S_1(i)=='I' & S_2(j)=='D')|(S_1(i)=='G' & S_2(j)=='D')|...
122 (S_1(i)=='V' & S_2(j)=='D')|(S_1(i)=='W' & S_2(j)=='D')|...
123 (S_1(i)=='L' & S_2(j)=='D')|(S_1(i)=='A' & S_2(j)=='D'));
124 Q1(i,j)= 0.64e-19;
125 Q2(i,j)= 0.64e-19;
126 else
127 if ((S_1(i)=='D'& S_2(j)=='P')|(S_1(i)=='P'& S_2(j)=='D'));
128 Q1(i,j)= 0.78e-19;
129 Q2(i,j)= 0.78e-19;
130 else
131 if ((S_1(i)=='D' & S_2(j)=='H')|(S_1(i)=='H'& S_2(j)=='D'));
132 Q1(i,j)= 0.99e-19;
133 Q2(i,j)= 0.99e-19;
134 else
135 if ((S_1(i)=='D'& S_2(j)=='K')|(S_1(i)=='K'& S_2(j)=='D'));
136 Q1(i,j)= 1.4e-19;
137 Q2(i,j)= 1.4e-19;
138 else
139 if ((S_1(i)=='D' & S_2(j)=='R')|(S_1(i)=='R'& S_2(j)=='D'));
140 Q1(i,j)= 1.59e-19;
141 Q2(i,j)= 1.59e-19;
142 else
143 if ((S_1(i)=='E'&S_2(j)=='E'));
144 Q1(i,j)= 0.16e-19;
145 Q2(i,j)= 0.16e-19;
146 else
147 if ((S_1(i)=='E'  & S_2(j)=='C')|(S_1(i)=='E' & S_2(j)=='F')|...
148 (S_1(i)=='E' & S_2(j)=='N')|(S_1(i)=='C'  & S_2(j)=='E')|...
149 (S_1(i)=='F' & S_2(j)=='E')|(S_1(i)=='N' & S_2(j)=='E'));
150 Q1(i,j)= 0.55e-19;
151 Q2(i,j)= 0.55e-19;
152 else
153 if  ((S_1(i)=='E' & S_2(j)=='Q')|(S_1(i)=='E' & S_2(j)=='Y')|...
154 (S_1(i)=='E' & S_2(j)=='S')|(S_1(i)=='E' & S_2(j)=='M')|...
155 (S_1(i)=='E' & S_2(j)=='T')|(S_1(i)=='E' & S_2(j)=='I')|...
156 (S_1(i)=='E' & S_2(j)=='G')|(S_1(i)=='E' & S_2(j)=='V')|...
```

```
157 (S_1(i)=='E' & S_2(j)=='W')|(S_1(i)=='E' & S_2(j)=='L')|...
158 (S_1(i)=='E' & S_2(j)=='A')|(S_1(i)=='Q' & S_2(j)=='E')|...
159 (S_1(i)=='Y' & S_2(j)=='E')| (S_1(i)=='S' & S_2(j)=='E')|...
160 (S_1(i)=='M' & S_2(j)=='E')|(S_1(i)=='T' & S_2(j)=='E')|...
161 (S_1(i)=='I' & S_2(j)=='E')| (S_1(i)=='G' & S_2(j)=='E')|...
162 (S_1(i)=='V' & S_2(j)=='E')|(S_1(i)=='W' & S_2(j)=='E')|...
163 (S_1(i)=='L' & S_2(j)=='E')|(S_1(i)=='A' & S_2(j)=='E'));
164 Q1(i,j)= 0.64e-19;
165 Q2(i,j)= 0.64e-19;
166 else
167 if ((S_1(i)=='E' & S_2(j)=='P' )|(S_1(i)=='P' & S_2(j)=='E'));
168 Q1(i,j)= 0.78e-19;
169 Q2(i,j)= 0.78e-19;
170 else
171 if ((S_1(i)=='E' & S_2(j)=='H')|(S_1(i)=='H' &S_2(j)=='E'));
172 Q1(i,j)= 0.99e-19;
173 Q2(i,j)= 0.99e-19;
174 else
175 if (S_1(i)=='E'& S_2(j)=='K')| (S_1(i)=='K'& S_2(j)=='E');
176 Q1(i,j)= 1.34e-19;
177 Q2(i,j)= 1.34e-19;
178 else
179 if (S_1(i)=='E' & S_2(j)=='R')|(S_1(i)=='R' & S_2(j)=='E');
180 Q1(i,j)= 1.58e-19;
181 Q2(i,j)= 1.58e-19;
182 else
183 if (S_1(i)=='C' & S_2(j)=='C')|(S_1(i)=='C' & S_2(j)=='F')|...
184 (S_1(i)=='C' & S_2(j)=='Q')|(S_1(i)=='C'& S_2(j)=='Y')|...
185 (S_1(i)=='C' & S_2(j)=='S')|(S_1(i)=='C' & S_2(j)=='M')|...
186 (S_1(i)=='C' & S_2(j)=='T')|(S_1(i)=='C' & S_2(j)=='I')|...
187 (S_1(i)=='C' & S_2(j)=='G')|(S_1(i)=='C' & S_2(j)=='V')|...
188 (S_1(i)=='C' & S_2(j)=='W')|(S_1(i)=='C' & S_2(j)=='L')|...
189 (S_1(i)=='C' & S_2(j)=='L')|(S_1(i)=='C' & S_2(j)=='A')|...
190 (S_1(i)=='F' & S_2(j)=='C')|(S_1(i)=='Q' & S_2(j)=='C')|...
191 (S_1(i)=='Y'& S_2(j)=='C')|(S_1(i)=='S' & S_2(j)=='C')|...
192 (S_1(i)=='M' & S_2(j)=='C')|(S_1(i)=='T' & S_2(j)=='C')|...
193 (S_1(i)=='I' & S_2(j)=='C')|(S_1(i)=='G' & S_2(j)=='C')|...
194 (S_1(i)=='V' & S_2(j)=='C')|(S_1(i)=='W' & S_2(j)=='C')|...
195 (S_1(i)=='L' & S_2(j)=='C')|( S_1(i)=='A' & S_2(j)=='C');
196 Q1(i,j)=0.74e-19;
197 Q2(i,j)=0.74e-19;
198 else
199 if (S_1(i)=='C' & S_2(j)=='H')| (S_1(i)=='H' & S_2(j)=='C');
200 Q1(i,j)= 0.99e-19;
201 Q2(i,j)= 0.99e-19;
202 else
203 if (S_1(i)=='C' & S_2(j)=='K')|(S_1(i)=='K' & S_2(j)=='C');
204 Q1(i,j)= 1.34e-19;
205 Q2(i,j)= 1.34e-19;
206 else
207 if (S_1(i)=='C' & S_2(j)=='R')|(S_1(i)=='R' & S_2(j)=='C');
208 Q1(i,j)= 1.59e-19;
```

```
209 Q2(i,j)= 1.59e-19;
210 else
211 if (S_1(i)=='N' & S_2(j)=='N')|(S_1(i)=='N' & S_2(j)=='F')|...
212 (S_1(i)=='N' & S_2(j)=='Q')|(S_1(i)=='N' & S_2(j)=='Y')|...
213 (S_1(i)=='N' & S_2(j)=='S')|(S_1(i)=='N'& S_2(j)=='M')|...
214 (S_1(i)=='F' & S_2(j)=='N')|(S_1(i)=='Q' & S_2(j)=='N')|...
215 (S_1(i)=='Y' & S_2(j)=='N')|(S_1(i)=='S' & S_2(j)=='N')|...
216 (S_1(i)=='M'& S_2(j)=='N');
217 Q1(i,j)=0.74e-19;
218 Q2(i,j)=0.74e-19;
219 else
220 if (S_1(i)=='N' & S_2(j)=='H')|(S_1(i)=='H' & S_2(j)=='N')
221 Q1(i,j)= 0.99e-19;
222 Q2(i,j)= 0.99e-19;
223 else
224 if(S_1(i)=='N' & S_2(j)=='K')|(S_1(i)=='K' & S_2(j)=='N');
225 Q1(i,j)= 1.05e-19;
226 Q2(i,j)= 1.05e-19;
227 else
228 if (S_1(i)=='N' & S_2(j)=='R')|(S_1(i)=='R' & S_2(j)=='N');
229 Q1(i,j)= 1.1e-19;
230 Q2(i,j)= 1.1e-19;
231 else
232 if ((S_1(i)=='F' & S_2(j)=='F')|(S_1(i)=='F' & S_2(j)=='Q'));
233 Q1(i,j)=0.74e-19;
234 Q2(i,j)=0.74e-19;
235 else
236 if ((S_1(i)=='F' & S_2(j)=='Y')|(S_1(i)=='F' & S_2(j)=='S')|..
237 (S_1(i)=='F' & S_2(j)=='M')|(S_1(i)=='Q' & S_2(j)=='F')|...
238 (S_1(i)=='Y' & S_2(j)=='F'));
239 Q1(i,j)=0.74e-19;
240 Q2(i,j)=0.74e-19;
241 else
242 if (S_1(i)=='S' & S_2(j)=='F')|(S_1(i)=='M' & S_2(j)=='F');
243 Q1(i,j)=0.74e-19;
244 Q2(i,j)=0.74e-19;
245 else
246 if (S_1(i)=='F' & S_2(j)=='H')|(S_1(i)=='H' & S_2(j)=='F');
247 Q1(i,j)= 0.99e-19;
248 Q2(i,j)= 0.99e-19;
249 else
250 if (S_1(i)=='F' & S_2(j)=='K')|(S_1(i)=='K' & S_2(j)=='F');
251 Q1(i,j)= 1.05e-19;
252 Q2(i,j)= 1.05e-19;
253 else
254 if (S_1(i)=='F' & S_2(j)=='R')|(S_1(i)=='R' & S_2(j)=='F');
255 Q1(i,j)= 1.1e-19;
256 Q2(i,j)= 1.1e-19;
257 else
258 % Q
259 if (S_1(i)=='Q' & S_2(j)=='H')|(S_1(i)=='H' & S_2(j)=='Q');
260 Q1(i,j)= 0.99e-19;
```

```
261 Q2(i,j)= 0.99e-19;
262 else
263 if (S_1(i)=='Q' & S_2(j)=='K')|(S_1(i)=='K' & S_2(j)=='Q');
264 Q1(i,j)= 1.05e-19;
265 Q2(i,j)= 1.05e-19;
266 else
267 if (S_1(i)=='Q' & S_2(j)=='R')|(S_1(i)=='R' & S_2(j)=='Q');
268 Q1(i,j)= 1.1e-19;
269 Q2(i,j)= 1.1e-19;
270 else
271 % Y
272 if (S_1(i)=='Q' & S_2(j)=='H')|(S_1(i)=='H' & S_2(j)=='Q');
273 Q1(i,j)= 0.99e-19;
274 Q2(i,j)= 0.99e-19;
275 else
276 if (S_1(i)=='Y' & S_2(j)=='K')|(S_1(i)=='K' & S_2(j)=='Y')
277 Q1(i,j)= 1.05e-19;
278 Q2(i,j)= 1.05e-19;
279 else
280 if (S_1(i)=='Y' & S_2(j)=='R')|(S_1(i)=='R' & S_2(j)=='Y');
281 Q1(i,j)= 1.1e-19;
282 Q2(i,j)= 1.1e-19;
283 else
284 if (S_1(i)=='S' & S_2(j)=='H')|(S_1(i)=='H' & S_2(j)=='S');
285 Q1(i,j)= 0.99e-19;
286 Q2(i,j)= 0.99e-19;
287 else
288 if (S_1(i)=='S' & S_2(j)=='K')|(S_1(i)=='K' & S_2(j)=='S');
289 Q1(i,j)= 1e-19;
290 Q2(i,j)= 1e-19;
291 else
292 if (S_1(i)=='S' & S_2(j)=='R')|(S_1(i)=='R' & S_2(j)=='S');
293 Q1(i,j)= 1.1e-19;
294 Q2(i,j)= 1.1e-19;
295 else
296 if (S_1(i)=='M' & S_2(j)=='H')|(S_1(i)=='H' & S_2(j)=='M');
297 Q1(i,j)= 0.99e-19;
298 Q2(i,j)= 0.99e-19;
299 else
300 if (S_1(i)=='M' & S_2(j)=='K')|(S_1(i)=='K' & S_2(j)=='M');
301 Q1(i,j)= 1e-19;
302 Q2(i,j)= 1e-19;
303 else
304 if (S_1(i)=='M' & S_2(j)=='R')|(S_1(i)=='R' & S_2(j)=='M');
305 Q1(i,j)= 1.1e-19;
306 Q2(i,j)= 1.1e-19;
307 else
308 if (S_1(i)=='T' & S_2(j)=='H')|(S_1(i)=='H' & S_2(j)=='T');
309 Q1(i,j)= 0.99e-19;
310 Q2(i,j)= 0.99e-19;
311 else
312 if (S_1(i)=='T' & S_2(j)=='K')|(S_1(i)=='K' & S_2(j)=='T');
```

```
313 Q1(i,j)= 1e-19;
314 Q2(i,j)= 1e-19;
315 else
316 if (S_1(i)=='T' & S_2(j)=='R')|(S_1(i)=='R' & S_2(j)=='T');
317 Q1(i,j)= 1.05e-19;
318 Q2(i,j)= 1.05e-19;
319 else
320 if (S_1(i)=='I' & S_2(j)=='H')|(S_1(i)=='H' & S_2(j)=='I');
321 Q1(i,j)= 0.99e-19;
322 Q2(i,j)= 0.99e-19;
323 else
324 if (S_1(i)=='I' & S_2(j)=='K')|(S_1(i)=='K' & S_2(j)=='I');
325 Q1(i,j)= 1e-19;
326 Q2(i,j)= 1e-19;
327 else
328 if (S_1(i)=='I' & S_2(j)=='R')|(S_1(i)=='R' & S_2(j)=='I');
329 Q1(i,j)= 1.05e-19;
330 Q2(i,j)= 1.05e-19;
331 else
332 if (S_1(i)=='G' & S_2(j)=='H')|(S_1(i)=='H' & S_2(j)=='G');
333 Q1(i,j)= 0.99e-19;
334 Q2(i,j)= 0.99e-19;
335 else
336 if (S_1(i)=='G' & S_2(j)=='K')|(S_1(i)=='K' & S_2(j)=='G');
337 Q1(i,j)= 1e-19;
338 Q2(i,j)= 1e-19;
339 else
340 if (S_1(i)=='G' & S_2(j)=='R')|(S_1(i)=='R' & S_2(j)=='G');
341 Q1(i,j)= 1.05e-19;
342 Q2(i,j)= 1.05e-19;
343 else
344 if (S_1(i)=='V' & S_2(j)=='H')|(S_1(i)=='H' & S_2(j)=='V');
345 Q1(i,j)= 0.99e-19;
346 Q2(i,j)= 0.99e-19;
347 else
348 if (S_1(i)=='V' & S_2(j)=='K')|(S_1(i)=='K' & S_2(j)=='V');
349 Q1(i,j)= 1e-19;
350 Q2(i,j)= 1e-19;
351 else
352 if (S_1(i)=='V' & S_2(j)=='R')|(S_1(i)=='R' & S_2(j)=='V');
353 Q1(i,j)= 1.05e-19;
354 Q2(i,j)= 1.05e-19;
355 else
356 if (S_1(i)=='W' & S_2(j)=='H')|(S_1(i)=='H' & S_2(j)=='W');
357 Q1(i,j)= 0.99e-19;
358 Q2(i,j)= 0.99e-19;
359 else
360 if (S_1(i)=='W' & S_2(j)=='K')|(S_1(i)=='K' & S_2(j)=='W');
361 Q1(i,j)= 1e-19;
362 Q2(i,j)= 1e-19;
363 else
364 if (S_1(i)=='W' & S_2(j)=='R')|(S_1(i)=='R' & S_2(j)=='W');
```

```
365 Q1(i,j)= 1.05e-19;
366 Q2(i,j)= 1.05e-19;
367 else
368 if (S_1(i)=='L' & S_2(j)=='H')|(S_1(i)=='H' & S_2(j)=='L');
369 Q1(i,j)= 0.99e-19;
370 Q2(i,j)= 0.99e-19;
371 else
372 if (S_1(i)=='L' & S_2(j)=='K')|(S_1(i)=='K' & S_2(j)=='L');
373 Q1(i,j)= 1e-19;
374 Q2(i,j)= 1e-19;
375 else
376 if (S_1(i)=='L' & S_2(j)=='R')|(S_1(i)=='R' & S_2(j)=='L');
377 Q1(i,j)= 1.05e-19;
378 Q2(i,j)= 1.05e-19;
379 else
380 if (S_1(i)=='A' & S_2(j)=='H')|(S_1(i)=='H' & S_2(j)=='A');
381 Q1(i,j)= 0.99e-19;
382 Q2(i,j)= 0.99e-19;
383 else
384 if (S_1(i)=='A' & S_2(j)=='K')|(S_1(i)=='K' & S_2(j)=='A');
385 Q1(i,j)= 1e-19;
386 Q2(i,j)= 1e-19;
387 else
388 if (S_1(i)=='A' & S_2(j)=='R')|(S_1(i)=='R' & S_2(j)=='A');
389 Q1(i,j)= 1.05e-19;
390 Q2(i,j)= 1.05e-19;
391 else
392 if (S_1(i)=='P' & S_2(j)=='H')|(S_1(i)=='H' & S_2(j)=='P');
393 Q1(i,j)= 0.99e-19;
394 Q2(i,j)= 0.99e-19;
395 else
396 if (S_1(i)=='P' & S_2(j)=='K')|(S_1(i)=='K' & S_2(j)=='P');
397 Q1(i,j)= 0.82e-19;
398 Q2(i,j)= 0.82e-19;
399 else
400 if (S_1(i)=='P' & S_2(j)=='R')|(S_1(i)=='R' & S_2(j)=='P');
401 Q1(i,j)= 0.96e-19;
402 Q2(i,j)= 0.96e-19;
403 else
404 if (S_1(i)=='H' & S_2(j)=='H');
405 Q1(i,j)= 0.82e-19;
406 Q2(i,j)= 0.82e-19;
407 else
408 if (S_1(i)=='H' & S_2(j)=='K')|(S_1(i)=='K' & S_2(j)=='H');
409 Q1(i,j)= 0.82e-19;
410 Q2(i,j)= 0.82e-19;
411 else
412 if (S_1(i)=='H' & S_2(j)=='R')|(S_1(i)=='R' & S_2(j)=='H');
413 Q1(i,j)= 0.74e-19;
414 Q2(i,j)= 0.74e-19;
415 else
416 if (S_1(i)=='K' & S_2(j)=='K');
```

```
417 Q1(i,j)= 0.54e-19;
418 Q2(i,j)= 0.54e-19;
419 else
420 if (S_1(i)=='K' & S_2(j)=='R')|(S_1(i)=='R' & S_2(j)=='K');
421 Q1(i,j)= 0.41e-19;
422 Q2(i,j)= 0.41e-19;
423 else
424 if (S_1(i)=='R' & S_2(j)=='R');
425 Q1(i,j)= 0.16e-19;
426 Q2(i,j)= 0.16e-19;
427 else
428 Q1(i,j)= 0.824e-19;
429 Q2(i,j)= 0.824e-19;
430 end
431 end
432 end
433 end
434 end
435 end
436 end
437 end
438 end
439 end
440 end
441 end
442 end
443 end
444 end
445 end
446 end
447 end
448 end
449 end
450 end
451 end
452 end
453 end
454 end
455 end
456 end
457 end
458 end
459 end
460 end
461 end
462 end
463 end
464 end
465 end
466 end
467 end
468 end
```

```
469 end
470 end
471 end
472 end
473 end
474 end
475 end
476 end
477 end
478 end
479 end
480 end
481 end
482 end
483 end
484 end
485 end
486 end
487 end
488 end
489 end
490 end
491 end
492 end
493 end
494 end
495 end
496 end
497 end
498 end
499 end
500 end
501 end
502 end
503 end
504  Q3=[];
505  Q4=[];
506  R1=[];
507  R2=[];
508 for i=1:length(S_1);
509 if (S_1(i)=='A');
510 R1(i)=0.6e-9;
511 else
512 if (S_1(i)=='R');
513 R1(i)=0.809e-9;
514 else
515 if (S_1(i)=='N');
516 R1(i)=0.682e-9;
517 else
518 if (S_1(i)=='D');
519 R1(i)=0.665e-9;
520 else
```

```
521 if (S_1(i)=='C');
522 R1(i)=0.629e-9;
523 else
524 if (S_1(i)=='Q');
525 R1(i)=0.725e-9;
526 else
527 if (S_1(i)=='E');
528 R1(i)=0.714e-9;
529 else
530 if (S_1(i)=='G');
531 R1(i)=0.537e-9;
532 else
533 if (S_1(i)=='H');
534 R1(i)=0.732e-9;
535 else
536 if (S_1(i)=='I');
537 R1(i)=0.735e-9;
538 else
539 if (S_1(i)=='L');
540 R1(i)=0.734e-9;
541 else
542 if (S_1(i)=='K');
543 R1(i)=0.737e-9;
544 else
545 if (S_1(i)=='M');
546 R1(i)=0.741e-9;
547 else
548 if (S_1(i)=='F');
549 R1(i)=0.781e-9;
550 else
551 if (S_1(i)=='P');
552 R1(i)=0.672e-9;
553 else
554 if (S_1(i)=='S');
555 R1(i)=0.615e-9;
556 else
557 if (S_1(i)=='T');
558 R1(i)=0.659e-9;
559 else
560 if (S_1(i)=='W');
561 R1(i)=0.826e-9;
562 else
563 if (S_1(i)=='Y');
564 R1(i)=0.781e-9;
565 else
566 if (S_1(i)=='V');
567 R1(i)=0.694e-9;
568 end
569 end
570 end
571 end
572 end
```

```
573 end
574 end
575 end
576 end
577 end
578 end
579 end
580 end
581 end
582 end
583 end
584 end
585 end
586 end
587 end
588 for j=1:length(S_2);
589 if (S_2(j)=='A');
590 R2(j)=0.6e-9;
591 else
592 if (S_2(j)=='R');
593 R2(j)= 0.809e-9;
594 else
595 if (S_2(j)=='N');
596 R2(j)=0.682e-9;
597 else
598 if (S_2(j)=='D');
599 R2(j)=0.665e-9;
600 else
601 if (S_2(j)=='C');
602 R2(j)=0.629e-9;
603 else
604 if (S_2(j)=='Q');
605 R2(j)=0.725e-9;
606 else
607 if (S_2(j)=='E');
608 R2(j)=0.714e-9;
609 else
610 if (S_2(j)=='G');
611 R2(j)=0.537e-9;
612 else
613 if (S_2(j)=='H');
614 R2(j)=0.732e-9;
615 else
616 if (S_2(j)=='I');
617 R2(j)=0.735e-9;
618 else
619 if(S_2(j)=='L');
620 R2(j)=0.734e-9;
621 else
622 if (S_2(j)=='K')
623 R2(j)=0.737e-9;
624 else
```

```
625 if (S_2(j)=='M')
626 R2(j)=0.741e-9;
627 else
628 if (S_2(j)=='F')
629 R2(j)=0.781e-9;
630 else
631 if (S_2(j)=='P');
632 R2(j)=0.672e-9;
633 else
634 if (S_2(j)=='S');
635 R2(j)=0.615e-9;
636 else
637 if (S_2(j)=='T');
638 R2(j)=0.659e-9;
639 else
640 if (S_2(j)=='W');
641 R2(j)=0.826e-9;
642 else
643 if (S_2(j)=='Y');
644 R2(j)=0.781e-9;
645 else
646 if (S_2(j)=='V');
647 R2(j)=0.694e-9;
648 end
649 end
650 end
651 end
652 end
653 end
654 end
655 end
656 end
657 end
658 end
659 end
660 end
661 end
662 end
663 end
664 end
665 end
666 end
667 end
668 end
669 end
670  Ra=0.6e-9;
671  Rr=0.809e-9;
672  Rn=0.682e-9;
673  Rd=0.665e-9;
674  Rc=0.629e-9;
675  Rq=0.725e-9;
676  Re=0.714e-9;
```

```
677  Rg=0.725e-9;
678  Rh=0.732e-9;
679  Ri=0.735e-9;
680  Rl=0.734e-9;
681  Rk=0.737e-9;
682  Rm=0.741e-9;
683  Rf=0.781e-9;
684  Rp=0.672e-9;
685  Rs=0.615e-9;
686  Rt=0.659e-9;
687  Rw=0.826e-9;
688  Ry=0.781e-9;
689  Rv=0.694e-9;
690 for i=1:length(S_1);
691 for j=1:length(S_2);
692 if (S_1(i)=='R'& S_2(j)=='D');
693     h(i,j)=.15*10^(-9)+Rr+Rd;
694 else
695 if (S_1(i)=='R'& S_2(j)=='E');
696       h(i,j)=.15*10^(-9)+Rr+Re;
697 else
698 if (S_1(i)=='D'& S_2(j)=='R');
699       h(i,j)=.15*10^(-9)+Rd+Rr;
700 else
701 if (S_1(i)=='D'& S_2(j)=='H');
702     h(i,j)=.15*10^(-9)+Rd+Rh;
703 else
704 if (S_1(i)=='D'& S_2(j)=='R');
705       h(i,j)=.15*10^(-9)+Rd+Rr;
706 else
707 if (S_1(i)=='D'& S_2(j)=='H');
708       h(i,j)=.15*10^(-9)+Rd+Rh;
709 else
710 if (S_1(i)=='D'& S_2(j)=='K');
711       h(i,j)=.15*10^(-9)+Rd+Rk;
712 else
713 if (S_1(i)=='E')& (S_2(j)=='R');
714       h(i,j)=.15*10^(-9)+Re+Rr;
715 else
716 if (S_1(i)=='E'& S_2(j)=='H');
717     h(i,j)=.15*10^(-9)+Re+Rh;
718 else
719 if (S_1(i)=='E'& S_2(j)=='K');
720     h(i,j)=.15*10^(-9)+Re+Rk;
721 else
722 if (S_1(i)=='H'& S_2(j)=='D')
723       h(i,j)=.15*10^(-9)+Rh+Rd;
724 else
725 if (S_1(i)=='H'& S_2(j)=='E')
726       h(i,j)=.15*10^(-9)+Rh+Re;
727 else
728 if (S_1(i)=='R'& S_2(j)=='R')
```

```
729        h(i,j)=.4*10^(-9)+Rr+Rr;
730   else
731   if (S_1(i)=='R'& S_2(j)=='H')
732       h(i,j)=.4*10^(-9)+Rr+Rh;
733   else
734   if (S_1(i)=='R'& S_2(j)=='H')
735        h(i,j)=.4*10^(-9)+Rr+Rh;
736   else
737   if (S_1(i)=='R'& S_2(j)=='K')
738        h(i,j)=.4*10^(-9)+Rr+Rk;
739   else
740   if (S_1(i)=='D'& S_2(j)=='E');
741        h(i,j)=.4*10^(-9)+Rd+Re;
742   else
743   if (S_1(i)=='D'& S_2(j)=='D');
744        h(i,j)=.4*10^(-9)+Rd+Rd;
745   else
746   if (S_1(i)=='H'& S_2(j)=='R')
747        h(i,j)=.4*10^(-9)+Rh+Rr;
748   else
749   if (S_1(i)=='H'& S_2(j)=='H')
750        h(i,j)=.4*10^(-9)+Rh+Rh;
751   else
752   if (S_1(i)=='H'& S_2(j)=='K')
753         h(i,j)=.4*10^(-9)+Rh+Rk;
754   else
755   if (S_1(i)=='K'& S_2(j)=='R')
756       h(i,j)=.4*10^(-9)+Rk+Rr;
757   else
758   if (S_1(i)=='K'& S_2(j)=='H')
759        h(i,j)=.4*10^(-9)+Rk+Rh;
760   else
761   if (S_1(i)=='K'& S_2(j)=='K')
762        h(i,j)=.4*10^(-9)+Rk+Rk;
763   else
764   if (S_1(i)=='N'& S_2(j)=='Q')
765       h(i,j)=.25*10^(-9)+Rn+Rq;
766   else
767   if (S_1(i)=='N'& S_2(j)=='S')
768       h(i,j)=.25*10^(-9)+Rn+Rs;
769   else
770   if (S_1(i)=='N'& S_2(j)=='Y')
771        h(i,j)=.25*10^(-9)+Rn+Ry;
772   else
773   if(S_1(i)=='Q'& S_2(j)=='S')|...
774   (S_1(i)=='Q')& (S_2(j)=='Y');
775        h(i,j)=.25*10^(-9)+Rq+Rs;
776   else
777   if (S_1(i)=='Q')& (S_2(j)=='Y');
778         h(i,j)=.25*10^(-9)+Rq+Ry;
779   else
780   if (S_1(i)=='S'& S_2(j)=='Y');
```

```
781         h(i,j)=.25*10^(-9)+Rs+Ry;
782     else
783         h(i,j)=1.76*10^(-9);
784     end
785     end
786     end
787     end
788     end
789     end
790     end
791     end
792     end
793     end
794     end
795     end
796     end
797     end
798     end
799     end
800     end
801     end
802     end
803     end
804     end
805     end
806     end
807     end
808     end
809     end
810     end
811     end
812     end
813     end
814     end
815     end
816
817     function[A]=electrostatic(Q1,Q2, R1,R2,h,M,N,N1,epsilon)
818     for i=1:N
819         for j=1:M
820             if R1(i)>R2(j)
821                 gamma(i,j)=R1(i)/R2(j);
822             else
823                 if  R1(i)<R2(j)
824                     gamma(i,j)=R2(j)/R1(i);
825                   else if R1(i)==R2(j);
826          gamma(i,j)=R2(j)/R1(i);
827               end
828                 end
829             end
830             if h(i,j)>(R1(i)+R2(j))
831                 r(i,j)=h(i,j)/(R1(i)+R2(j));
832             else if  h(i,j)<=(R1(i)+R2(j))
```

```
833                r(i,j)=(R1(i)+R2(j))/h(i,j);
834            end
835            end
836       y(i,j)=(((r(i,j)^2*(1+gamma(i,j))^2)-...
837   (1+(gamma(i,j))^2))/(2*gamma(i,j)));
838       beta(i,j)=acosh(y(i,j));
839       z(i,j)=exp(-beta(i,j));
840       S12=0;
841       S22=0;
842       S11=0;
843       for k=1:N1
844            gamma1(i,j)=R2(j)/R1(i);
845            S_1(k)=(z(i,j)^k)/(((1-z(i,j)^(2*k)))*...
846   ((gamma(i,j)+y(i,j))-(y(i,j)^2-1)^(1/2)*...
847   (1+z(i,j)^(2*k))/(1-z(i,j)^(2*k))));
848            S11=S11+S_1(k);
849            S_2(k)=(z(i,j)^(2*k))/(1-(z(i,j)^(2*k)));
850            S12=S12+S_2(k);
851            S_3(k)=(z(i,j)^k)/(((1-z(i,j)^(2*k)))*...
852   ((1-gamma(i,j)*y(i,j))-gamma(i,j)*(y(i,j)^2-1)^(1/2)*...
853   (1+z(i,j)^(2*k))/(1-z(i,j)^(2*k))));
854            S22=S22+S_3(k);
855       end
856       epsilon0=8.85418781762*10^(-12);
857  c11(i,j)=(2*gamma(i,j)*((y(i,j)^2-1)^(1/2))).*S11;
858  c22(i,j)=(2*gamma(i,j)*((y(i,j)^2-1)^(1/2))).*S22;
859  c12(i,j)=-((2*gamma(i,j)*...
860  ((y(i,j)^2-1))^(1/2))/(r(i,j)*(1+gamma(i,j)))).*S12;
861  delta(i,j)=((c11(i,j)*c22(i,j)-c12(i,j)^2));
862       k=1/(4*pi*epsilon0);
863       k1=1/(4*pi*epsilon0*epsilon);
864            alpha(i,j)=Q2(j)/Q1(i);
865       if R1(i)>R2(j)
866            gamma(i,j)=R1(i)/R2(j);
867  W1(i,j)=((1/k1)*R2(j)*gamma(i,j))*...
868  ((1+gamma(i,j))/(2*alpha(i,j)))*...
869  ((alpha(i,j)^2*c11(i,j)-2*alpha(i,j)*...
870  c12(i,j)+c22(i,j))/delta(i,j));
871            else if (R1(i)<R2(j))
872                 gamma(i,j)=R2(j)/R1(i);
873  W1(i,j)=((1/k1)*R1(i)*gamma(i,j))*...
874  ((1+gamma(i,j))/(2*alpha(i,j)))*...
875  ((alpha(i,j)^2*c11(i,j)-2*alpha(i,j)*...
876  c12(i,j)+c22(i,j))/delta(i,j));
877        else if R1(i)==R2(j);
878  W1(i,j)=((1/k1)*R1(i)*gamma(i,j))*...
879  ((1+gamma(i,j))/(2*alpha(i,j)))*...
880  ((alpha(i,j)^2*c11(i,j)-2*alpha(i,j)*...
881  c12(i,j)+c22(i,j))/delta(i,j));
882                 end
883                 end
884       end
```

```
885      W2(i,j)=(k*(Q1(i)*Q2(j)))/(R1(i)+R2(j));
886      A1(i,j)=W1(i,j);
887      A2(i,j)=W2(i,j);
888       A(i,j)=A1(i,j)/A2(i,j);
889
890      end
891  end
892  return
893
894
895  function[cond2]=condmy(A)
896  [U,S,V]=SVD_2(A);
897  lambda_max=max(diag(S));
898  lambda_min=min(diag(S));
899  cond_1=((lambda_max)/(lambda_min));
900  cond2=(log(cond_1))/(log(10));
901  return
902  function [Uout,Sout,Vout] = SVD_2(A)
903          m = size(A,1);
904          n = size(A,2);
905          U = eye(m);
906          V = eye(n);
907          e = eps*fro(A);
908          while (sum(abs(A(~eye(m,n)))) > e)
909          for i = 1:n
910          for j = i+1:n
911              [J1,J2] = jacobi(A,m,n,i,j);
912              A = mtimes(J1,mtimes(A,J2));
913              U = mtimes(U,J1');
914              V = mtimes(J2',V);
915          end
916          for j = n+1:m
917              J1 = jacobi2(A,m,n,i,j);
918              A = mtimes(J1,A);
919              U = mtimes(U,J1');
920          end
921          end
922          end
923          S = A;
924          if (nargout < 3)
925             Uout = diag(S);
926          else
927              Uout = U; Sout = times(S,eye(m,n)); Vout = V;
928          end
929          end
930        function [J1,J2] = jacobi(A,m,n,i,j)
931           B = [A(i,i), A(i,j); A(j,i), A(j,j)];
932           [U,S,V] = tinySVD(B); %
933           J1 = eye(m);
934           J1(i,i) = U(1,1);
935           J1(j,j) = U(2,2);
936           J1(i,j) = U(2,1);
```

```
937         J1(j,i) = U(1,2);
938         J2 = eye(n);
939         J2(i,i) = V(1,1);
940         J2(j,j) = V(2,2);
941         J2(i,j) = V(2,1);
942         J2(j,i) = V(1,2);
943     end
944     function J1 = jacobi2(A,m,n,i,j)
945         B = [A(i,i), 0; A(j,i), 0];
946         [U,S,V] = tinySVD(B);
947         J1 = eye(m);
948         J1(i,i) = U(1,1);
949         J1(j,j) = U(2,2);
950         J1(i,j) = U(2,1);
951         J1(j,i) = U(1,2);
952     end
953     function [Uout,Sout,Vout] = tinySVD(A)
954 t=rdivide((minus(A(1,2),A(2,1))),(plus(A(1,1),A(2,2))));
955       c = rdivide(1,sqrt(1+t^2));
956       s = times(t,c);
957       R = [c,-s;s,c];
958       M = mtimes(R,A);
959       [U,S,V] = tinySymmetricSVD(M);
960       U = mtimes(R',U);
961       if (nargout < 3)
962           Uout = diag(S);
963       else
964           Uout = U; Sout = S; Vout = V;
965       end
966       end
967 function [Uout,Sout,Vout] = tinySymmetricSVD(A)
968       if (A(2,1) == 0)
969          S = A;
970          U = eye(2);
971          V = U;
972       else
973          w = A(1,1);
974          y = A(2,1);
975          z = A(2,2);
976          ro = rdivide(minus(z,w),times(2,y));
977 t2=rdivide(sign(ro),plus(abs(ro),sqrt(plus(times(ro,ro),1))));
978          t = t2;
979          c = rdivide(1,sqrt(plus(1,times(t,t))));
980          s = times(t,c);
981          U = [c, -s; s, c];
982          V = [c,  s;-s, c];
983          S = mtimes(U,mtimes(A,V));
984          U = U';
985          V = V';
986       end
987       [U,S,V] = fixSVD(U,S,V);
988       if (nargout < 3)
```

```
989             Uout = diag(S);
990         else
991             Uout = U; Sout = S; Vout = V;
992         end
993         end
994     function [U,S,V] = fixSVD(U,S,V)
995       Z = [sign(S(1,1)),0; 0,sign(S(2,2))]; %
996       U = mtimes(U,Z);
997       S = mtimes(Z,S);
998       if (S(1,1) < S(2,2))
999           P = [0,1;1,0];
1000          U = mtimes(U,P);
1001          S = mtimes(P,mtimes(S,P));
1002          V = mtimes(P,V);
1003      end
1004      end
1005    function f = fro(M)
1006      f = sqrt(sum(sum(times(M,M))));
1007    end
1008    function s = sign(x)
1009       if (x > 0)
1010           s = 1;
1011       else
1012           s = -1;
1013       end
1014       end
```

References

1. J. Ferlay, H.R. Shin, F. Bray, D. Forman, C. Mathers, D.M. Parkin, Estimates of worldwide burden of cancer in 2008: GLOBOCAN 2008. Int. J. Cancer **127**(12), 2893–2917 (2010)
2. J. Ferlay, E. Steliarova-Foucher, J. Lortet-Tieulent, S. Rosso, J.W. Coebergh, H. Comber, D. Forman, F. Bray, Cancer incidence and mortality patterns in Europe: estimates for 40 countries in 2012. Eur. J. Cancer **49**(6), 1374–1403 (2013)
3. M. Arnold, M.S. Sierra, M. Laversanne, I. Soerjomataram, A. Jemal, F. Bray, Global patterns and trends in colorectal cancer incidence and mortality. Gut **66**(4), 683–691 (2017)
4. J. Thundimadathil, Cancer treatment using peptides: current therapies and future prospects. J. Amino Acids (2012)
5. R. Domalaon, B. Findlay, M. Ogunsina, G. Arthur, F. Schweizer, Ultrashort cationic lipopeptides and lipopeptoids: evaluation and mechanistic insights against epithelial cancer cells. Peptides **84**, 58–67 (2016)
6. S.R. Dennison, F. Harris, D.A. Phoenix, The interactions of aurein 1.2 with cancer cell membranes. Biophys. Chem. **127**(1–2), 78–83 (2007)
7. M.R. Felicio, O.N. Silva, S. Goncalves, N.C. Santos, O.L. Franco, Peptides with dual antimicrobial and anticancer activities. Front. Chem. **5** (2017)
8. A. Jemal, F. Bray, M.M. Center, J. Ferlay, E. Ward, D. Forman, Global cancer statistics. CA Cancer J. Clin. **61**(2), 69–90 (2011)
9. Baguley B. C. Multiple drug resistance mechanisms in Cancer// Mol. Biotechnol. 2010. V.46. N.3. Pp.308–316
10. L.A. Mandell, P. Ball, G. Tillotson, Antimicrobial safety and tolerability: differences and dilemmas. Clin. Infect. Dis. **32**(1), S72–9 (2001)

11. C.R. Lincke, A.M. van der Bliek, G.J. Schuurhuis, T. van der Velde-Koerts, J.J. Smit, P. Borst, Multidrug resistance phenotype of human BRO melanoma cells transfected with a wild-type human mdr1 complementary DNA. Cancer Res. **50**(6), 1779–1785 (1990)
12. C.A. Arias, B.E. Murray, Antibiotic-resistant bugs in the 21st century - a clinical super-challenge. N. Engl. J. Med. **360**(5), 439–443 (2009)
13. D. Kakde, D. Jain, V. Shrivastava, R. Kakde, A.T. Patil, Cancer therapeutics-opportunities, challenges and advances in drug delivery. J. Appl. Pharm. Sci. **1**, 1–10 (2011)
14. S. Mocellin, P. Pilati, D. Nitti, Peptide-based anticancer vaccines: recent advances and future perspectives. Curr. Med. Chem. **16**(36), 4779–4796 (2009)
15. J.N. Francis, M. Larch, Peptide-based vaccination: where do we stand? Curr. Opin. Allergy Clin. Immunol. **5**(6), 537–543 (2005)
16. Y.-F. Xiao, M.-M. Jie, B.-S. Li, C.-J. Hu, R. Xie, B. Tang, S.-M. Yang, Peptide-based treatment: a promising cancer therapy. J. Immunol. Res. **2015**. Article ID 761820 (2015) 13 pp
17. Y. Yoshitake, D. Fukuma, A. Yuno, M. Hirayama, H. Nakayama, T. Tanaka, M. Nagata, Y. Takamune, K. Kawahara, Y. Nakagawa, R. Yoshida, A. Hirosue, H. Ogi, A. Hiraki, H. Jono, A. Hamada, K. Yoshida, Y. Nishimura, Y. Nakamura, M. Shinohara, Phase II clinical trial of multiple peptide vaccination for advanced head and neck cancer patients revealed induction of immune responses and improved OS. Clin. Cancer. Res. **21**(2), 312–321 (2015)

Index

T. Koshlan and K. Kulikov, *Mathematical Modeling of Protein Complexes*,
Biological and Medical Physics, Biomedical Engineering,
https://doi.org/10.1007/978-3-319-98304-2

Printed in the United States
By Bookmasters